214
CF-TFN
NORTH STAR
AIR LINES

The Canadair North Star

£3-00

MANUFACTURED BY
Canadair Ltd.
MONTREAL
NORTH STAR

The Canadair North Star

Larry Milberry

CANAV Books

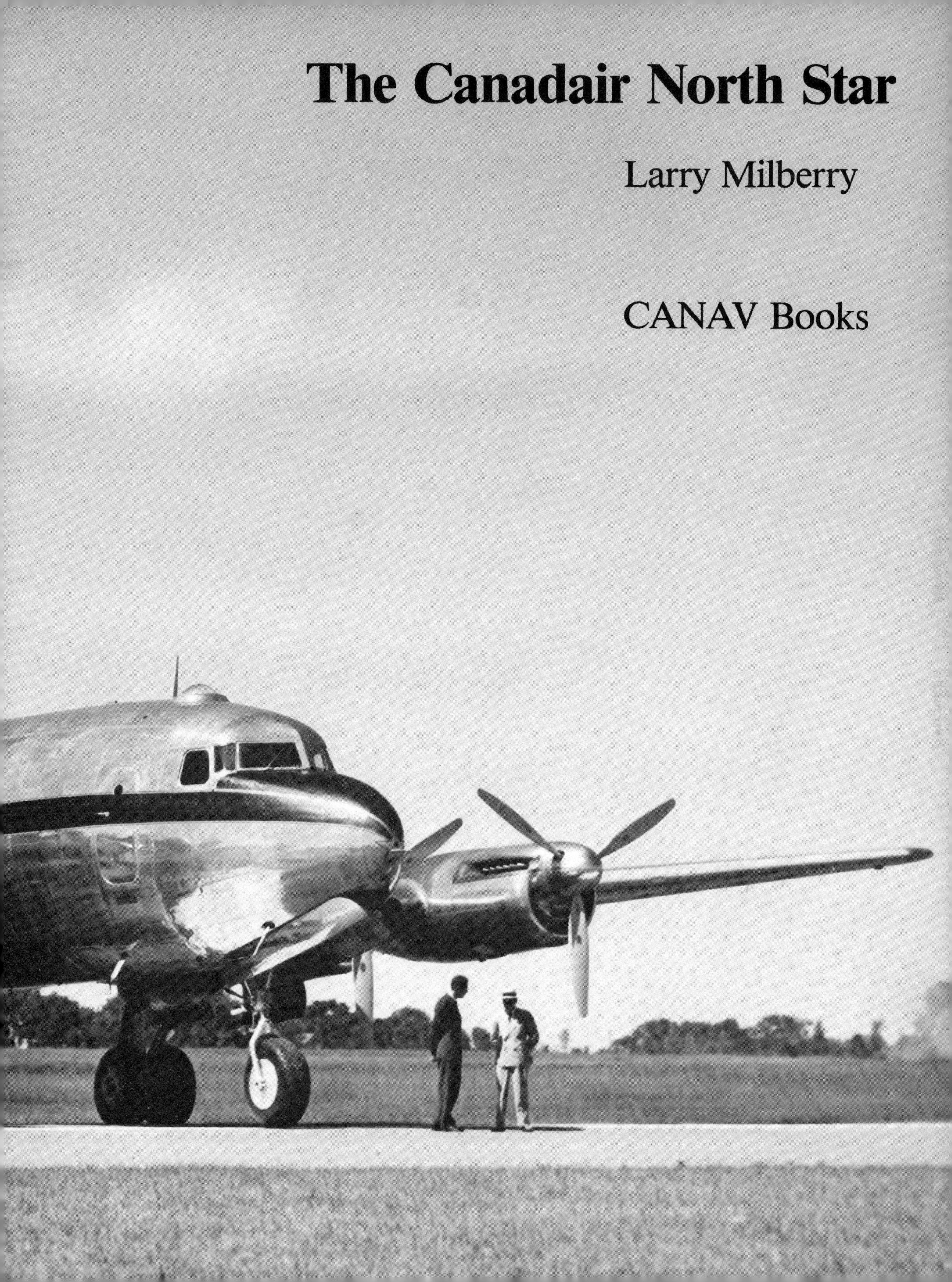

1 2 3 4 5 7 6 5 4 3 2

Printed and bound in Canada by the Bryant Press Limited, Toronto

Published by
CANAV Books
51 Balsam Avenue
Toronto M4E 3B6
Canada

Canadian Cataloguing in Publication Data
Milberry, Larry, 1943-
The Canadair North Star

Includes index.
Bibliography: p.
ISBN 0-9690703-1-4

1. Canadair North Star (Airplane) - History.
I. Title.
HE9815.A35M54 387.7'3343 C82-095033-5

Designed and edited by Robin Brass, Toronto
Typesetting by Arlene Weber, Second Story Graphics, Toronto
Photo retouching by Stephen Ng, SNG Retouching Studio, Toronto; and Colin Clark, Toronto
Repro assembly by Rob Devine and Betty Mason, Printart, Toronto
Additional proofreading, Ralph Clint, Toronto

Front endpaper
CF-TFN of TCA cruises above clouds in this fine portrait. (Air Canada Archives)

Title page
The prototype North Star poses on the ramp as Ben Franklin and Bob Brush chat below. (Canadair Ltd.)

Below
Work begins on the prototype North Star at Canadair. (Canadair Ltd.)

Opposite page
Preparing to mount a Merlin engine on the North Star prototype. (Canadair Ltd.)

Overleaf
Always photogenic, CF-TEN-X gleams on the tarmac. (Canadair Ltd. via J.T. Bain)

Back endpaper
Detailed photograph showing an early North Star flight deck configuration. (Canadair Ltd.)

Contents

Foreword

The Canadair North Star was one of the most significant undertakings of Canada's postwar aircraft industry, but writing a foreword for the book about it was a challenging assignment. It is almost ancient history now. Some of those who made the project possible have gone, and I cannot hope to do justice to the many who worked so hard and collectively to make the airplane a success.

The North Star was a major step forward in Canadian aircraft manufacturing and operations. Sure, there were trials and tribulations, but at the time all new airliners were having their problems.

When the North Star first appeared, the politicians had a field day. Opponents tried hard to destroy the project; but that didn't stop major airlines from all over the world trying to buy the airplane. It was unfortunate that the contract with the Douglas Company for airframe engineering limited sales of Canadian-produced aircraft to Canada and Britain.

I am being careful to avoid names in these few words. I'm leaving that to Larry Milberry. There are so many in government, at Canadair and elsewhere in the industry, with the airlines and air force, without whom the North Star could never have been built or become the success it eventually proved to be.

As this book will show, the North Star turned Trans-Canada Air Lines into a major world operator. It gave Canadian Pacific Airlines its opening onto the international scene. For British Overseas Airways Corporation, where it was known as the Argonaut, it was a Godsend. And it gave the Royal Canadian Air Force a much-needed heavy transport for use on such big jobs as the Korean airlift. For all the details, read on.

James T. Bain

Preface

The Canadair North Star, like my book about the CF-100, tells the story of a great Canadian achievement. It's another two-sided story—about an airplane, but also about people. Canada's aviation past is rich in such stories waiting to be researched, written and published in a way to do honour both to the machine and to those who conceived, promoted, designed, built and flew it. Hence the appearance of this book. Whatever your interest in the North Star, there will be plenty here for you: the North Star's conception with Trans-Canada Air Lines, the politics, the engineering, the new airplane's teething problems, then its service history with airlines, the Royal Canadian Air Force and many other operators.

The Canadair North Star is the result of detailed research into every aspect of the airplane, and the co-operation of hundreds of people anxious to contribute to a worthy aeronautical subject. There are so many "thank you's" due, and as with *Aviation in Canada* and *The Avro CF-100*, it is in the co-operation of my numerous cohorts that the success of this book lies. To start with, I must thank Beth Buchanan who established Air Canada's fine Archives in Montreal. Beth has been in aviation since Canadian Airways days and joined TCA in 1944. For many years she was about as close to the top of the airline as anyone could be—she was Gordon R. McGregor's private secretary from 1945 to 1968.

From the beginning of this project, Beth opened all essential doors for me, including arranging for access to the McGregor and Bain files in the Public Archives of Canada. Dozens of other valuable papers came my way from Beth, as did access to the superb collection of photographs in the Air Canada Archives, and introduction to many key TCA/Air Canada people in the North Star story.

James T. Bain, "father" of the North Star, was another real blessing to the project, providing raw material, proof reading, criticisms and photos. What can a writer say, having the brains of the operation on his side!

Others involved over the long term with the project include Ralph Clint who spent 25 years with TCA/Air Canada and is an aviation history buff with a keen sense for accuracy. He checked out many a lead and provided great assistance with proof reading. Captain B.G. Cramp of Crawley, UK, provided some original insights into operation of the Argonaut by independent airlines like British Midland Airways. N.K. Found of Found Brothers Aviation was a constant source of encouragement, information and constructive criticism. Ronald J. Fox of Toronto was another keen participant and never ceased coming up with exotic tidbits in his role as North Star history sleuth. Peter R. Keating of Ickenham, UK, also played an ongoing role, especially as a source of photos. As longtime aircrew with BOAC/BA, Peter photographed North Stars and Argonauts at airports around the world, and generously shared these with me, adding a special richness to the photographic coverage. He also stirred up a lot of interest among former BOAC personnel, many of whom made contributions to the book. Alexander J. Lilly of Moncton, copilot on the first North Star flight, and Canadair's chief test pilot, opened the doors on many vital aspects of the North Star story. Captain Don Lowry of Air Canada was there whenever called upon with answers to all sorts of questions. Some of the results of Tony Merton Jones' extensive North Star/Argonaut research have also gone into rounding off this book. Tony has been digging into propliner history from his London base for years and has produced detailed coverage of UK independent airlines. Jack McNulty of Hamilton has gone all out in support of my research efforts. I'm happy to say that this is my third book to be brightened with Jack's work. Stephen Piercey of Addlestone, UK, and publisher of the stellar quarterly *Propliner*, was another major participant in getting the North Star/Argonaut story together. A fan of this great airplane since he was a boy, Stephen has travelled thousands of miles in search of North Star lore, to places like Frobisher Bay, Sarasota and El Salvador. The results of his digging are included in my book, with Stephen's kind approval. To his consternation, I must also give special note to Merlin "Moe" Reddy who started this whole thing by introducing me to airports, airshows and photography when I was ten years old. Our regular outings to Malton airport in Moe's '54 Ford and things like our day trips to Buffalo by Convair and DC-6 set the stage for what you see in the following pages. As usual, Peter Robertson of the Public Archives bent over backwards from the photographic end of things. And as usual, LCol Bev Totman, DND Directorate of Information Services, has been in the background getting doors opened whenever called upon. Special thanks are also due to Air Canada, CP Air and the Canadian Forces for providing transportation on several occasions this year.

Besides these helpers are many other individuals, associations and firms. I have done my best to include them all in the following list: David R. Adamson (412 Sqdn.), D.D. Allyn, William G. Anglin, D.E. Anderson (Air Canada), AVCO International, R.J. "Ron" Baker (Air Canada), Robert Bausch, Eric W. Bendall, R.N. Bennett (Air Canada), Pierre

CF-TEN-X, one propeller feathered, prepares to alight at Cartierville. (Canadair Ltd. via J.T. Bain)

Berton, John Bosma (Air Canada), J.R. Bowker (Air Canada), Molly Brass, R.W. "Bob" Bradford, British Aerospace, British Airways, Jim Bruce, Robert P. "Bob" Brush, Bob Bullough, Jeff Burch, A.M. "Al" Burns (Air Canada), W.J. "Robbie" Burns (BOAC), Canadair Ltd., Peter Cane (BOAC), J.B. Carter (BOAC), W. Glen Cawker (Air Canada), J.C. Charleson, R.J. Childerhose, Harry Clarke (Air Canada), Don Connolly (426 Sqdn.), Philip Cooksey, A.G. "Gerry" Couillard, Hugh Crawford, R.W. "Bob" Dale (BOAC), Paul Dane (BOAC), W/C Peter Delaney, Ron Dennis (Air Canada), DND Directorate of Flight Safety, DND Directorate of History, DND Directorate of Information Services, John Dix, Harold G. Dondenaz (Air Canada), Douglas Aircraft (Long Beach), Eric Drinkwater, Jim Duncan (Air Canada), J.T. "Jack" Dyment (Air Canada), A.W. Eagle (BOAC), R.M. "Bob" Edwards (426 Sqdn.), Ernest A. Evans (BOAC), W/C T.J. Evans, Flight International, Neil Gaunt, Ian Geddes (Canadair), R. Earl Gerow (Air Canada), Clayton H. Glenn (Air Canada), Peter W. Gooch, E.L. Gosling (BOAC), J.M.G. Gradidge, Robert S. Grant, Ian Gray (CP Air), Robert G. Halford, Ian M. Hamer, Fred Hatch (Directorate of History), Colin Harman, E.G. "Gene" Harris, Mike Harris (Air Canada), Paul Hartman, Stan Hegstrom (Air Canada), V.A. Hodgkinson (BOAC), Capt. Joe Houlden (DFS), George Howard (426 Sqdn.), Allan Hunt (Air Canada), W.G. "Bill" Hunt (BOAC), L.W. "Les" Hussey (412 Sqdn.), A.J. Jackson, George Jackson (426 Sqdn.), George Jenks, Virgil Jones, Walter Kent (Air Canada), Jack Kern, LAAS International, Maurice Labine (Air Canada), Russell K. Lamm, D.W. Lamont (Air Canada), Bill Lavers (Air Canada), Mel Lee (CP Air), Ralph Leek (Air Canada), G/C Z.L. Leigh, Donald Levy (CADC), Hank Lillybeck, Roger Lindsay, George Lothian (Air Canada), Nick MacDonald-Wolochatiuk, Hugh B. Main (CP Air), Norm Malayney, Carson Malcolm, Laurie Manchee (Air Canada), Paul Marshall (Directorate of History), Garth Martin (426 Sqdn.), David Menard, R.T. "Dick" Merrick (412 Sqdn.), Carl Millard, Jas. Miller (BOAC), J.R. Moir (CP Air), K.M. Molson, LCol B.T. Montgomery, G.S. "Monty" Montgomery (Air Canada), Hayward Moores (426 Sqdn.), Peter Mossman, Larry Motiuk (426 Sqdn.), Ian Macdonald, Ian MacFarlane, Bruce J. McColl, D.J. "Des" McGill, G.W. McLaren (Air Canada), J.G. McKeachie (CP Air), Ted McLaughlin (426 Sqdn.), Bud McNair (426 Sqdn.), Don McVicar, NRC Flight Research Lab., George R. Nicol, Vic Parker (BOAC), J.S. "Stew" Parmelee (426 Sqdn.), Col W.J. Penfold (426 Sqdn.), R.C. "Bob" Penrose (Air Canada), C. Gordon Peters, Fred Phillips, D.N. Platt (Air Canada), Mrs. C.A. Proudfoot, LCol Dave Purich (DFS), Wendell Reid (Air Canada), R.G.K. "Bob" Rice (BOAC), Rolls-Royce Ltd., J.L. "Lindy" Rood (Air Canada), Ted P. Schneider, H.W. "Herb" Seagrim (Air Canada), Ken F. Smy, Glay Sperling, Dick Spratt (426 Sqdn.), L.C. Stevenson (CP Air), Ralph G. Stopps, Jim Storms, Jan Stroomenbergh, Lorne Tapp (426 Sqdn.), the late Harry Tate, Frank A. Taylor (BOAC), David J. Thompson, G.A. "Gerry" Turner, Fernando Vachon (CP Air), Archie Vanhee (CPA), J.E. "Jerry" Vernon, Carl Vincent, Col G.G. Waddell (426 Sqdn.), John Waldie, Leslie Waller, John Wegg, Clif Wenzel (426 Sqdn.), W.J. "Bill" Wheeler, R.H. "Bud" Whelan (412 Sqdn.), Wilf White, Harry H. Whiteman, John A. Whittle, D.J. Wiley (Air Canada), Les Wilkinson, R.A.R. Wilson, Graham Wragg, Glenn T. Wright (PAC), J.R. Young (BOAC), Zoë Brass, W.H. Meaden, Matthew, Katie, Simon and Stephanie Milberry (envelope stuffing and stamp licking), Bob Mitchell (Air Canada).

Thanks to everyone!

Larry Milberry

North Star Beginnings

Commercial Aviation Spreads Its Wings

The decade of spectacular pioneer flights after World War 1, including the 1919 transatlantic crossings by the R-34 airship and the Vickers Vimy of Alcock and Brown, the 1924 round-the-world journey of three US Army aircraft and the 1927 New York-Paris solo flight by Charles Lindbergh, encouraged commercial aviation in the 1930s to look to international operations.

Before long, airships and flying boats were developed for transoceanic service. A succession of disasters soon ended the career of the airships, but the flying boats were more successful. Imperial Airways' Short flying boats were introduced throughout the British Empire, while Pan Am's Martin, Boeing and Sikorsky "boats" were soon at work on Atlantic, Caribbean and Pacific services. At the same time, transcontinental aviation was making great headway as modern designs such as the Boeing 247, Lockheed 10 and Douglas DC-3 appeared and airlines rushed to adopt them — including Canada's newly formed Trans-Canada Air Lines.

TCA was established in 1937 to provide a coast-to-coast air service, replacing various existing local services. Principal among these was Canadian Airways, which operated on the west coast, the Prairies, central Canada and the Maritimes. Each was a regional operation, the regions not being connected. James Richardson, founder and owner, was anxious to turn his airline into the national carrier, but his ambitions were thwarted by the government in Ottawa, which went ahead with TCA on a state-owned, monopoly basis. Like nearly everything else that happened in Canadian aviation for about twenty years, TCA owed its existence to C.D. Howe, then Minister of Transport.

C.D. Howe, in many respects the father of modern aviation in Canada, wielded great power in Ottawa and at various times headed such key ministries as Transport, Munitions and Supply, and Reconstruction and Supply. (via R. Halford)

TCA built up a fleet of Lockheed 10, 14 and 18 airliners to provide service to all major centres between Halifax and Victoria and quickly developed into a first-class operation. It focused on domestic and US service, but also kept an eye on the future. Larger aircraft would soon be needed domestically, and intercontinental service was inevitable.

TCA began developing specifications for these future needs. A company document of February 1945 notes: "A specification for a postwar aeroplane for TCA was first prepared in 1940, using as a basis the best features from the specifications for three of the most recently designed commercial aeroplanes at that time. The TCA specification was altered frequently during the subsequent years as experience dictated, or as new components were designed, or as new features were developed. The three specifications first studied were for the original Douglas DC-4, the Boeing Stratoliner and the Curtiss CW-20. The original DC-4 specification was prepared by the combined efforts of the five leading United States airlines and the Douglas Company with the objective of providing in an aeroplane all the features believed at that time to be desirable in view of the economic depression that prevailed . . . The Boeing Stratoliner, a smaller aeroplane with most of the features of the DC-4, was then produced to better suit the conditions that prevailed at the time. The Curtiss CW-20 was designed to provide competition for the Douglas DC-3 in the field of two-engined aeroplanes."

In 1943 TCA sent a study team to Boeing, Consolidated, Douglas and Lockheed, scouting for a postwar design. The team had guidelines based on such priorities as safety, speed, reliability, comfort, operating costs, passenger capacity and general suitability for TCA's needs.

Heading the team was Edinburgh-born J.T. "Jim" Bain. He had joined the RAF in 1920 and trained at the RAF Engineering College, Cranwell. He later worked for several aviation firms before he was brought over from British Airways in the spring of 1938 to join TCA. He had been hired by H. Oliver West of TCA, and when West returned to the United States in October 1939, Bain became superintendent of maintenance and overhaul. He duly progressed to superintendent and finally to director of engineering and maintenance, an office he held until his retirement from the airline in 1965.

Whereas Bain's training and experience were of the practical kind, the company also took on an aeronautical engineer, J.T. "Jack" Dyment. Dyment had graduated in 1929 as a civil engineer from the University of

A Pan American Airways Boeing Clipper flying boat at Shediac, New Brunswick. Such aircraft pioneered commercial aviation on the Atlantic in the late 1930s. (New Brunswick Museum)

Canadian Airways' Junkers W-33 CF-AQW. Aircraft such as these were the backbone of regional and bush flying operations in Canada before the founding of TCA in 1937. (CP Air)

TCA built its original fleet on the Lockheed family of twin-engine airliners. These served Canadian cities from 1937 onward. This Lockheed 18 Lodestar is seen over Halifax. (Air Canada Archives)

Lancaster CF-CMX on the tarmac at Dorval. Together with the BOAC Liberators these aircraft established the first year-round air transport service on the North Atlantic. (Air Canada Archives)

Toronto, and under the RCAF's Provisional Pilot Officers program had won his wings at Camp Borden in 1927. In 1928 he wrote a thesis in England on problems of low-speed control of aircraft. In 1929 he went to work at Dearborn on the Ford Tri-motor project; under him the Ford was first fitted with the Townend Ring around its uncowled engines, thus improving its performance. In 1931 Jack Dyment joined the Aeronautical Engineering Division of the Department of National Defence in Ottawa where he was involved largely in stress analysis with such aircraft as the Puss Moth, Norseman, Sekani and Northrop Delta. He too joined TCA in 1938, becoming chief engineer.

Transatlantic Pioneering

Only a few years after it became feasible to fly commercially across this continent, World War 2 made it imperative to fly aircraft regularly and in large numbers across the Atlantic Ocean. An aircraft construction industry sprang up to build warplanes for Britain, and the airlines, with their facilities, personnel and expertise, were conscripted into the war effort. While TCA was searching for the right airplane for its postwar needs, it was gaining valuable experience in transatlantic operations providing hundreds of personnel to operate the Canadian Government Trans-Atlantic Air Service. This service was announced in Parliament on June 16, 1943, by C.D. Howe:

"Mr. Speaker, I desire to inform the house regarding the establishment of a transatlantic air service. Arrangements have been completed for the immediate establishment of a Canadian government wartime transatlantic air service to carry mail to and from the Canadian armed forces in the British Isles and to transport members of the Canadian armed forces and of Canadian government departments and technicians engaged in the production of war materials. The new war service is not a commercial or a permanent one. It will carry no fare-paying passengers. Space on the aircraft from Canada will be allocated by the deputy minister of transport in Ottawa; space on aircraft westward by the Canadian high commissioner in London.

"The growing strength of the Canadian armed forces overseas and their increased activities have made it necessary to establish this quick and effective means of communication with the United Kingdom for men and materials. The need for a speedy and regular troops air mail service is particularly pressing. Over a quarter of a million members of the Canadian armed forces are now in the United Kingdom. During the past year, due to lack of space on aircraft, they have unfortunately not been receiving mail regularly and quickly. It is expected that the initial flight of the new service will occur in July."

TCA was chosen to operate the CGTAS for the government and one British-built Lancaster, CF-CMS, modified for civil use, was initially allotted for operation from Dorval. The airline was already involved in a similar operation, running a maintenance base and providing some flight crews for BOAC's transatlantic Liberator service. TCA's publication, *Between Ourselves*, made this comment about the company's involvement with the BOAC service: "Besides making a direct and essential contribution to the war effort, the work thus performed affords an opportunity for the officers and maintenance personnel of the Company to obtain a first hand working knowledge of four-engined equipment. This is a knowledge which it is hoped may be put to use in the Company's own service after the war, when four-engined aircraft become standard equipment for long distance flights."

Transatlantic flight was still in the pioneering stage. Before the war, a mere four years earlier, only Pan Am and Imperial Airways had any great experience with transatlantic flying, but not on a year-round basis. In the urgency of ferrying bombers to Britain a great deal was learned in a short time. The uncertainty and adventure of such flying may be felt in this excerpt written by veteran TCA Captain George Lothian. Here he describes a return ferry flight in a BOAC Liberator after he had delivered a bomber to England. Such was the sophistication of transatlantic flying in the early 1940s:

There was a well-justified saying in Ferry Command: "You don't get paid for the trip over, you get paid for the ride back." This was no reflection on the crews responsible for the trip home but on the means of transportation.

It could mean boarding a ship in a slow convoy travelling through sub-infested waters for two or three weeks; or on the other hand, making the return in the bowels of a converted Liberator bomber on a two-stage journey encompassing as many as 20 hours, in cabin temperatures of -25 -30°F.

To add zest to one's anticipation of the latter means of transportation, three Liberators had been lost during the past two months, each carrying its quota of 22 people, the majority of them ferry crews. One had hit Goat Fell on the top of the island of Arran; another had run off the runway on take-off at Ayr, after which it struggled along in the mud until it hit the railway embankment. The third had been coming in from Gibraltar when it was shot down by an enthusiastic Polish fighter pilot who mistook it for German.

Somehow, the evening for departure always seemed to come up with all the ingredients that make a good setting for ghastly murder.

There the Liberator would sit, dimly visible through the falling rain and mist which shrouded both aircraft and the darkly glinting puddles on the blacked-out tarmac. It was easy to tell yourself that other people in other places were going through much more harrowing experiences, but somehow this one became very personal.

The reluctant passengers — or bodies, as they became in that distasteful term used by the RAF — boarded the aircraft through a hatch in the bomb bay area. The floor of the bomb bay, lined with plywood, was strewn with sleeping bags, but most people preferred to take their place on one of the long wooden benches running along each side during the early stages of proceedings.

As there were no windows, the 22 "bodies" tended to sit and stare morosely at the far wall. All talking ceased the instant the engines were started and the Lib began to rock as it taxied to the runway. Twenty-two pairs of ears pricked up noticeably and the listening was so intense one could almost feel it as the engines were run up for the magneto checks. When they were opened up for the endless take-off it was not uncommon to see people crossing themselves; and there was always some sadist aboard who would start to intone a countdown.

There was a theory that if the Lib was not off the ground when he got to a certain number you had had it. I never learned for certain what the final number was as we were always airborne "and counting."

Once launched into the sluggish climb the bodies began sorting themselves out on the floor for the endless 12 or 13 hours to the next stop at Gander. After what seemed a hundred years the aircraft finally arrived at Dorval.

George Lothian

CGTAS Lancaster CF-CMV modified with elongated nose and tail. It entered service July 22, 1944, and was sold in 1947 to Flight Refuelling Ltd. in the UK. Before being scrapped, it served on the Berlin Airlift. (Air Canada Archives)

The Lancaster assigned to the Canadian Government Trans-Atlantic Air Service was soon joined by others built at Malton and the scale of the operation increased. This story in *Between Ourselves* in 1945 describes activity in the CGTAS and the "luxury" offered to travellers of the day:

Swelling the Trans-Atlantic Fleet

On the runways of Dorval Airport these days are four strange Lancaster aircraft, sleek and trim and very new. They are additions to the fleet of the Trans-Atlantic Air Service.

Having emerged from the Victory Aircraft factory at Toronto only a few weeks ago, they have just completed the thorough test-flying and "shaking down" necessary to qualify them for Trans-Atlantic Service.

But these Lancasters are "new" in more than the sense of being recently built. From point of view of interior arrangement and passenger accommodation they differ from any other Lancaster, military or commercial version, ever to take the skies.

Interior view of a CGTAS Lancaster at Dorval in July 1945. The aircraft accommodated 10 passengers. For the times, it appears to have been nicely appointed. (Air Canada Archives)

The "Lancs" used for the past two years on the Trans-Atlantic Service have been the product of an elaborate conversion process designed to fit them for the carriage of very heavy mail and freight loads. They are anything but passenger aircraft, although accommodation has been available for a limited number of trans-oceanic travellers.

Chronology of TCA's Transatlantic Involvement 1938-47

1938-1939	Handled the Canadian arrangements for the flying boat service of Imperial Airways.
1940-1941	Handled maintenance and modifications at St. Hubert for Canadian Pacific Air Service's transatlantic ferry operations.
Mar.-July 1941	Several TCA pilots resigned from the airline to fly aircraft overseas for ATFERO (Atlantic Ferry Organization).
May 1941	TCA Engineering, Dorval, undertook conversion, servicing and maintenance of LB-30 Liberators used by BOAC to transport ferry crews back to Canada from the U.K.
Sept.-Dec. 1941	J.L. Rood and G.B. Lothian loaned to Ferry Command.
Sept. 1942	Ferry Command operations moved from St. Hubert to Dorval.
Oct. 1942	After loss of three Liberators on return ferry duties, three TCA crews loaned to Return Ferry Service. These included Captains M.B. Barclay, J.L. Rood and George Lothian; First Officers W.R. Bell, H.I. Hayes and A.H. Palmer; Radio Officers G. Nettleton, A.J. Blackwood and J.A. Webster; Flight Engineers F.D. Burgess, J. Wilson and R.R. Spafford.
Mar.-May 1943	Lancaster R5727, pattern aircraft for Victory Aircraft, flown by TCA crews on Moncton-Goose Bay freight operation and returned to Dorval May 5 for maintenance/modification.
May-July 1943	J.L. Rood and George Lothian loaned to RCAF to instruct on long-range flying on Liberators at Dartmouth and Gander.
May 15-16, 1943	TCA crew with RCAF navigator ferried R5727 to Manchester for mods.
July 1-2, 1943	R5727 returned to Dorval as CF-CMS to serve in newly formed CGTAS. First eastbound flight on the 3500-mile route July 22 with crew of R.F. George, Art Rankin, G. Nettleton and S/L J.R. Gilmore. Carried 2600 lb. mail and 8 passengers, on 12 hr. 26 min. non-stop crossing.
Oct. 8, 1943	Record 11 hr. 56 min. crossing Dorval-Prestwick with 4700 lb. mail and 3 passengers: C.D. Howe, H.J. Symington and J.R. Baldwin.
Aug. 1944	Three round trips weekly being operated, five Lancasters in service.
Dec. 28, 1944	Disappearance over Atlantic of CF-CMU in first CGTAS serious loss. Over 500 CGTAS crossings to date.
Summer 1945	Commencement of ticket sales to civilian passengers wishing to fly CGTAS.
Dec. 27, 1945	Dorval-Havana flight by CF-CMY with VIPs.
Jan. 9, 1946	CF-CMW Prestwick-Dorval-Vancouver 28 hrs., 23 min. CF-CMW to mark Vancouver's Diamond Jubilee.
Sept. 16, 1946	CGTAS service extended to London.
May 1, 1947	CGTAS became TCA (Atlantic) Ltd.

CGTAS Lancasters CF-CMX and CF-CMZ on the ramp at Santa Maria in the Azores. (G.W. McLaren)

Crew of the first TCA transatlantic charter flight (via Lancaster) seen in Paris in May 1946. Left to right are Radio Operator Frank Brusatore, Purser Doug V. Miller, Captain Gil McLaren, F/O J.R. Shaw and Navigating Officer Pat Roy. (via G.W. McLaren)

TCA President H.J. Symington presents Captain George Lothian with a gift commemorating his 100th transatlantic crossing with the CGTAS. Two other 100-crossing crew seen in the photo are Navigator Harold Thomae and, on the right, Al Blackwood, both of TCA at this time. (Air Canada Archives)

Now, however, with the trans-Atlantic carriage of military mail declining, and with passenger traffic bulking ever-larger in importance, there is need for aircraft fitted-out specifically to meet the new conditions. Hence the four unique Lancasters at Dorval.

Victory Aircraft did a fine job on them. Through the elimination of a portion of the bomb-bay of the original military type, accommodation has been provided for ten passengers in a roomy, "stand-up" cabin.

The entire aircraft has been insulated and sound-proofed with fibreglass of the latest type.

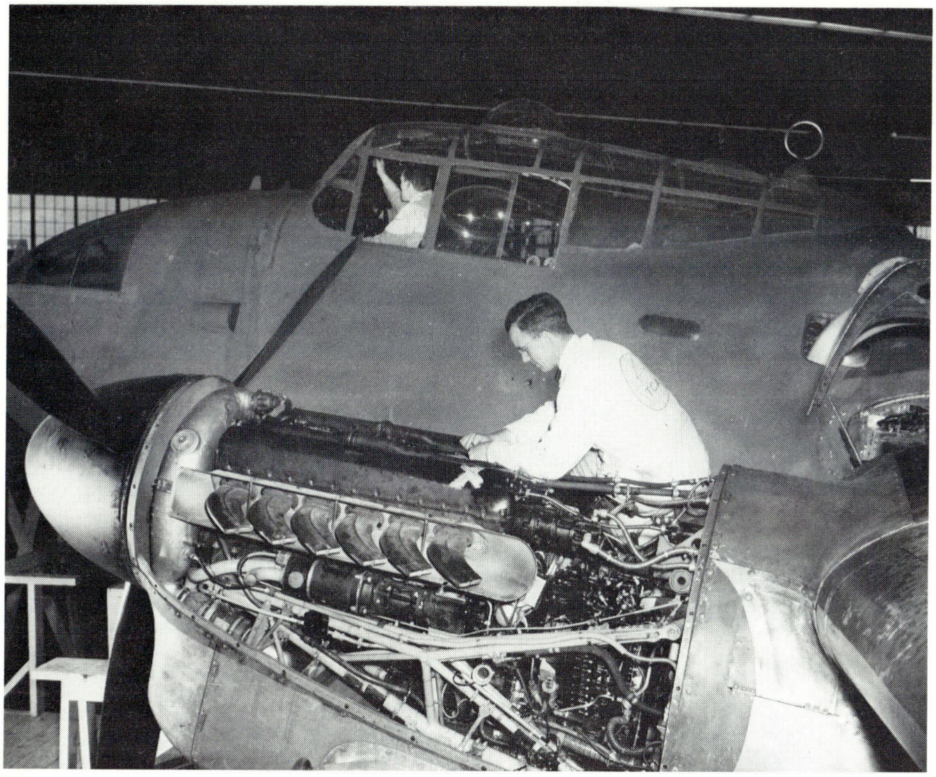

Maintenance by TCA staff on CF-CMS, the first CGTAS Lancaster at Dorval, October 1943. Note the open stacks of the Merlin engine. (Air Canada Archives)

Cabin heating is of a high standard, a gasoline-operated combustion heater having replaced the earlier method of extracting heat from the liquid coolant in the engines.

Individual reading lights, call buttons, ash trays, oxygen outlets and all the other incidentals familiar to the airline passenger have been provided.

There is a galley equipped for the serving of hot meals. Close behind it is the cabin attendant's desk and chair.

The upholstery and interior color design have been very effectively worked out by a TCA interior decorator. Ceilings and walls are yellow with the lower side-wall panels rust. A mahogany coloured carpet covers the cabin floor, while the standard TCA seats are finished in a light green.

The improvements engineered into these new aircraft embrace even more than passenger facilities. The flight crew compartment has been enlarged to provide additional room for the navigating and radio officers. Use has been made of many of the very latest developments in aircraft instruments and accessories.

While cargo space has been reduced to some extent to permit increased passenger accommodation, it remains ample for immediate Trans-Atlantic mail and baggage needs.

These deluxe Lancasters are, of course, still interim equipment, basically of military design, and will serve only until the time when we secure four-engined aircraft built for purely commercial purposes. But, of their kind, they are distinctive, by no means taking second place to the similar British Lancastrians. They are tributes to Canadian workmanship and to the excellence of Canadian aircraft manufacture.

Between Ourselves, No.32, 1945

Choosing the New Airliner

The TCA study team that toured the major US aircraft builders in 1943 found that there was no suitable large aircraft that would be available in the near future. They had covered the whole aviation scene and had looked into a wide range of proposals that included the Douglas DC-4, Lockheed Constellation, civilianized Boeing B-29, Consolidated B-32, and even the proposed civilianized C-74 Globemaster then designated as the DC-7. As well, Britain's Avro York was on the TCA list. Of all the types considered, only the DC-4 and the Constellation were serious contenders.

The Douglas DC-4 had its beginnings when design commenced in February 1936 on a four-engined airliner to have double the passenger capacity of the DC-3. A triple-finned unpressurized 52-seat prototype resulted and first flew on June 7, 1938. It was soon doing proving flights coast-to-coast, but these showed it to be too large and uneconomical for existing market needs. The prototype was sold to Japan and work began on a revised edition. The DC-4A thus appeared and made its first flight on February 14, 1942. By then America was in the war and production of the DC-4 was diverted to the United States Army Air Force. Aircraft were modified for military needs with such changes as a much-increased fuel capacity. As such, the DC-4 was redesignated the C-54 and the first one joined the Air Force in June 1942. By war's end, a series of C-54s had been developed, with 1163 aircraft in all being turned out, and the aircraft had established a reputation as one of the world's finest modern transports.

The Lockheed Constellation had been designed to meet the needs of Transcontinental and Western Air, later Trans World Airlines. This airline was run by Howard Hughes, a man noted for his interest in aviation development. Design had begun in June 1939 with the first aircraft flying January 9, 1943. Heavier than the DC-4, this new transport was also diverted to the military and suitably modified. It was redesignated C-69 and the first one was turned over to the Air Force in April 1944. It received civilian certification from the Civil Aeronautics Board in December 1945.

To TCA in 1943, the C-54 and C-69 were strictly military transports with uncertain peacetime futures. Of the

TCA studies for a postwar airliner favoured the Douglas DC-4. Derived from the military C-54, it seemed to TCA to hold the greatest promise as an airliner to serve both domestic and overseas routes. Forty years later, the DC-4 is still part of the aviation scene in Canada. This Millardair DC-4 is loading freight at Quebec City. (Larry Milberry)

two, TCA viewed the C-54 more favourably. Though faster, the C-69 would not be economical on TCA's domestic routes, especially on short interurban services. Its Wright R-3350 Cyclone engines were viewed as overly complex and, at 2200 hp, hard on fuel.

Meanwhile it was known that Douglas was continuing to develop a civilian DC-4. This appeared immediately at war's end as the DC-4-1009, of which 79 were to be built. The larger and pressurized DC-6 was also in the works, being developed at the behest of American and United Airlines as their reply to TWA's Constellations. But in 1943 TCA had no knowledge of this project, which was being kept under wraps in the US.

TCA opted for the DC-4 as the basis for its new airliner. The results of TCA's intensive search were summarized in a February 1945 internal report: "After a study of the detailed design of the structure, the factors of safety, the performance, the accommodations etc., it was decided that the Douglas DC-4 would be a more economical aeroplane to operate than the Constellation and would still have adequate performance, though it would not fly as fast. . . . The robust structural features, and the expected low maintenance and operating costs of the DC-4 aeroplane, however, favoured it very much and weighed heavily in the final selection of the aeroplane . . . The airframe of the Douglas DC-4, as a result of its world-wide use as the C-54 during the past two years, has been through its teething troubles and is now in a state where little maintenance trouble can be expected. The experience of TCA with previous Lockheed aeroplanes . . indicates that a considerable period will elapse before all the minor deficiencies are eliminated from the maintenance standpoint in the Constellation."

In 1944 the Canadian government obtained a licence from Douglas to build an aircraft incorporating both DC-4 and DC-6 components. By then the DC-6 was known, and in some TCA documents of the time the airline's new aircraft was even referred to as the DC-6. It was to be powered by the engine ultimately preferred by TCA and the resulting aircraft was to suit domestic, trans-border and trans-atlantic routes.

Which Engine?

The choice of an engine was to result in one of the minor political uproars of modern Canadian history and feelings run high to this day. A TCA document of July 13, 1944, by Jack Dyment entitled "Engine Selection for the Douglas DC-4" lists the engines considered: the Pratt and Whitney R-2000-9 and R-2800-C; the Rolls-Royce RM-14-SM and RM-20-SM (Merlins) and Griffin; the Bristol HE-10 Hercules; and the Wright R-2600-22. In due course all but the Pratt and Whitney R-2800-C and the Rolls-Royce RM-14-SM Merlin were eliminated.

The document describes the performance criteria the engine had to satisfy and explains the airline's choice:

It is known from the personal experiences of TCA pilots that it is necessary to be able to fly at any desired altitude up to at least 25,000 feet in order to maintain the degree of regularity desired over the North Atlantic. The weather conditions, therefore, dictate the cruising altitude that must be attainable. The weather affects the choice of alternate airports, so it also dictates the distance that the airplane must be capable of flying.

Cruising power is the most important criterion in the selection of the engine. Sufficient cruising power must be available to permit the fully loaded airplane to fly in an attitude of maximum lift/drag at the maximum desired altitude. The maintenance costs of the engine will increase rapidly as the continuous power drawn approaches the rated power of the engines. An ample margin is therefore required above the power actually used to keep the costs down . . .

A weight analysis, engine performance analysis, ultimate performance analysis and economic analysis were made from which it was decided that the Rolls-Royce Merlin 620 would be the most suitable for the TCA airplane.

TCA was also impressed with the Lockheed Constellation for possible postwar use. This KLM "Connie" is seen arriving at Toronto on November 15, 1959. (Larry Milberry)

Early artist's conception of TCA's new transport to be based on the DC-4. (via R.G. Stopps)

The characteristics of this engine were such that it was considered suitable for domestic service too, thus simplifying the interchange of equipment between the services. . . .

Only the Rolls-Royce RM-14-SM and the Pratt and Whitney R-2800-C engines have sufficient power at altitude to provide the performance required by Trans-Canada Airlines for the trans-Atlantic post war operation.

The rated power of the Rolls-Royce RM-14-SM is higher than the other engines under consideration at altitudes above 20,000 feet where the maximum possible power is essential for realistic trans-Atlantic service.

The cruising power used to obtain equal performance will therefore be at a lower percentage of rated power with the RM-14-SM than with any other engine.

The cruising power required at 25,000 feet, if the aircraft takes off at 73,000 pounds and climbs directly to altitude, assuming 150 bhp will be required for pressurization, will be 88% of rated power for the RM-14-SM and 91% for the R-2800-C.

The power available for takeoff, climb and low altitude operations is greater with the Pratt and Whitney R-2800-C but adequate power is also available from the RM-14-SM for these conditions.

The weight of all four RM-14-SM power plants is 13,200 pounds compared with 14,955 pounds for the R-2800-C power plants . . . The fuel consumption is greater with the RM-14-SM than with the R-2800-C. The miles per pound at 25,000 feet altitude with a 50 mph headwind over a typical trans-Atlantic trip is approximately 0.150 with the RM-14-SM compared with 0.159 miles per pound with the R-2800-C, assuming the drag of the aeroplane is the same with both types of power plants. Assuming an increase in maximum speed of 15-20 mph due to the expected reduction in drag with the RM-14-SM nacelles, this engine should be able to fly 0.156 miles per pound.

The difference in fuel consumption for the two engines would be greatly reduced at the lower altitudes if Rolls-Royce developed the third speed for their supercharger as they expect to do. . .

The payload ton-miles per Imperial gallon that can be carried on a typical trans-Atlantic flight is 3.32 with the RM-14-SM installed and 2.53 with the R-2800-C installed. This value has a major place in the selection of engines.

The direct operating cost per ton-mile for a typical trans-Atlantic flight is $0.295 with the RM-14-SM as compared with $0.392 for the R-2800-C . . . The maximum speed with normal rated powers at critical engine altitudes is expected to be 351 mph with the Pratt and Whitney R-2800-C and 349 mph with the RM-14-SM, assuming

Two views of Rolls-Royce's civil Merlin engine. J.T. Bain writes, "The Merlin was built as a self-contained, standardized unit. As needed, it could be simply and quickly exchanged. This was an aviation first, envied and later copied by many manufacturers." (Air Canada Archives)

equal aircraft drag for both engines. If the expected drag decrease with the RM-14-SM installed is considered, speed with this engine would be increased to 370 mph. All three of these speeds are greater than the normal rated max. speed of the Constellation at critical engine altitude. . . .

The cost of maintenance of the two types of power plants can not be forecast with any degree of accuracy. On one hand we have an engine designed for operation at high military powers but with no commercial operating experience to indicate the time that can be permitted between overhauls. Sufficient is to say that as a result of Trans-Canada Air Lines present operating experience with the Merlin 38 engines and after viewing many service engines being overhauled at the Glasgow plant, we are quite confident that we should be able to obtain a service life of between 800 and 1000 hours with the RM-14-SM engines. Again, no experience is available to ascertain the probable overall life of the RM-14-SM. There is no reason to believe that its life would be shorter than that of the R-2800-C because of its long period of development and appearance after a number of existing overhaul periods.

On the other hand, we have the R-2800-C which is essentially designed for commercial use and which should operate for 800 hours between overhauls but which may be considerably less in view of the operating at a higher percentage of rated power. It is fairly certain from past experience that the cost of cowl flap maintenance will be very high.

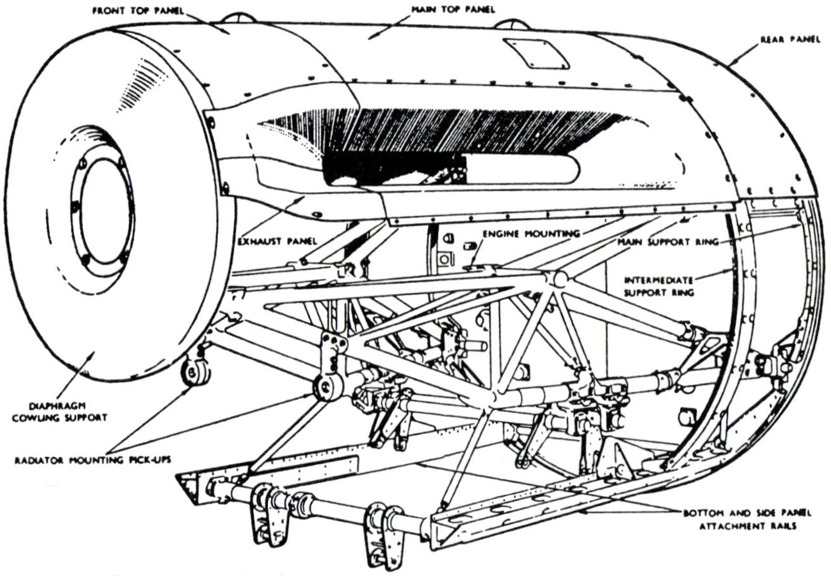

North Star M2 cowling support structure.

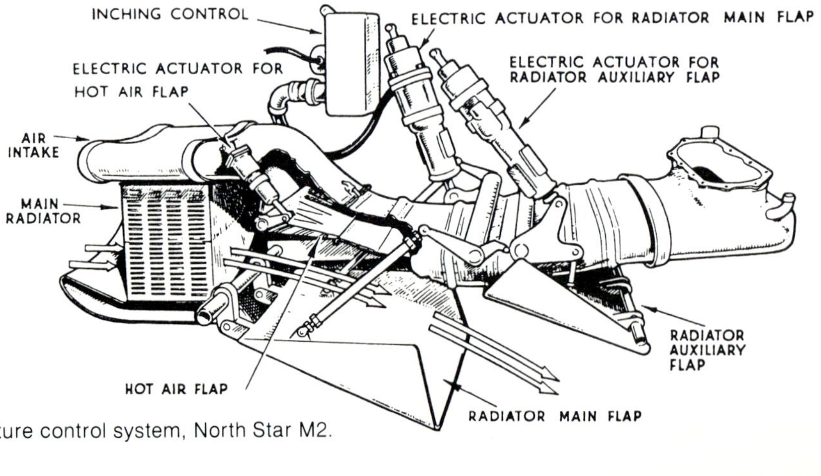

Temperature control system, North Star M2.

Features for interchangeability of components are better for the RM-14-SM than on the R-2800-C, that is, spark plugs can be changed quicker, a whole block can be removed, the piston rings examined and the block replaced within two hours.

One of the most important items in the selection of engines is that the Rolls-Royce company will unconditionally guarantee a reasonable service life of their engines and direct operation costs at least comparable with a good design of air cooled installation.

Another very important item in the selection of engines is that the Rolls-Royce company have a very extensive Development Section at Hucknall where the entire power plant is designed and tested to give the optimum performance of the engines. As yet the American engine manufacturers are unable to assume the responsibility for the design of the entire power plant. It is expected that with the facilities that Pratt and Whitney are now adding to their establishment, that in a period of a few years they too will be able to control the entire design of the power plant. This point of being able to fix the responsibility for the elimination of troubles cannot be too strongly stressed. Our experience to date with cowl flaps, exhaust systems, air intakes, carburetor icing systems, ignition systems etc., has clearly demonstrated the inadequacies of the American practice of permitting the aeroplane manufacturer to design part of the power plant, the engine manufacturer other parts, and the accessory manufacturer still others. The evasion of responsibility by the various manufacturers for inadequacies in design has been a continuous source of trouble and expense for Trans-Canada Air Lines. . . .

It is our opinion based upon this investigation that the most suitable engine for use by Trans-Canada Air Lines in the DC-4 aeroplane will be the Rolls-Royce RM-14-SM engine.

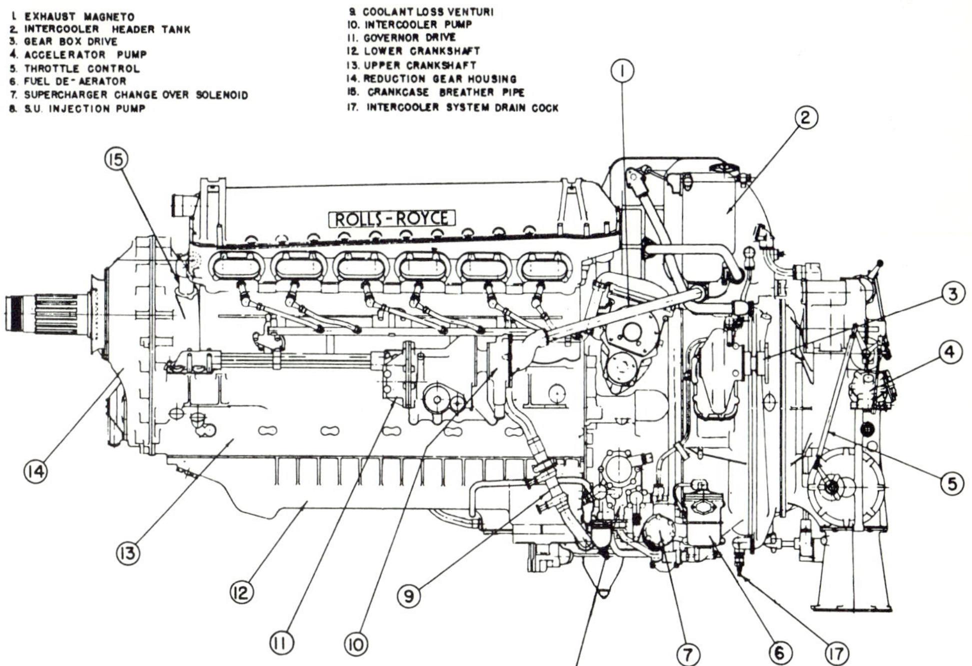

Merlin engine details, left side.

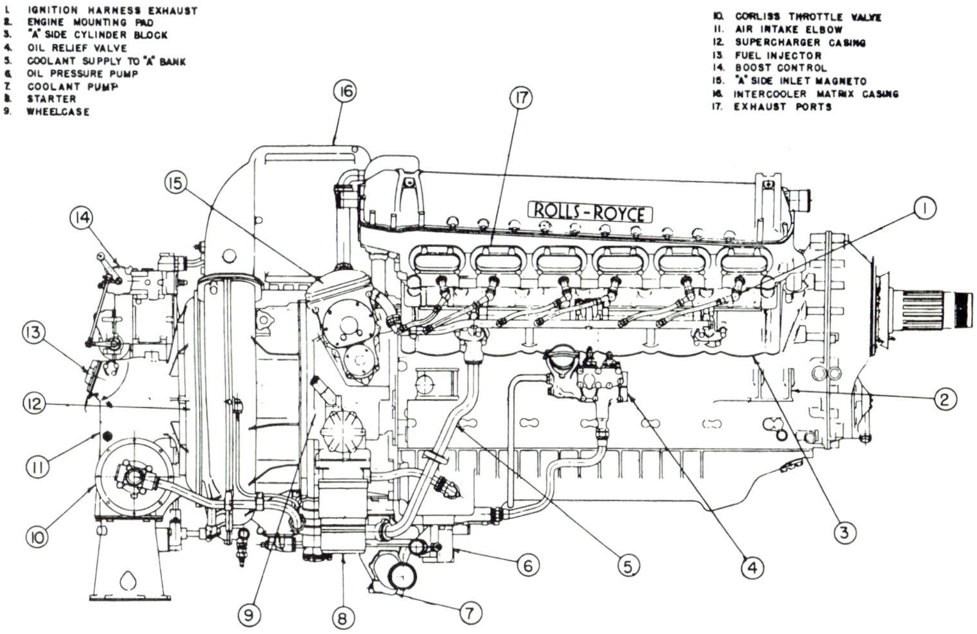

Merlin engine details, right side.

The Rolls-Royce Merlin

The Rolls-Royce Merlin family of engines was born in the 1930s. They powered the RAF to victory in the Battle of Britain as the engine of the Hurricane and Spitfire, and later carried Bomber Command's Lancasters and other types on raids over enemy territory that would eventually cripple Nazi Germany. The Merlin was produced under licence by Packard in the US and powered the legendary P-51 Mustang. All in all, it carried a formidable reputation. Late in the war Rolls-Royce began development of its civilian Merlin series and these engines were chosen for the Avro Tudor and the North Star.

Unlike most aero engines of the day, which were radial and air-cooled, the 27-litre Merlin had its 12 cylinders in a V arrangement and was liquid-cooled. The following description of the Merlin is taken from a Rolls-Royce document:

The Merlin is a 12-cylinder pressure liquid-cooled "V" engine, with its two cylinder blocks inclined at 60°. Early engines had single-stage single-speed superchargers, later types, two-speed single-stage superchargers, and for power at altitude, the two-speed two-stage supercharged engine with liquid-cooled intercooler was developed. All superchargers are centrifugal and a pressure-injection fuel pump delivers fuel into the eye of the supercharger.

Each cylinder block comprises a light alloy cylinder skirt and cylinder head, steel liners, flanged at their upper end, being inserted into the cylinder skirt to form the cylinder bores. Coolant circulates around the effective area of the liner. The liners float at their lower ends in the crankcase, the block being held in compression by the cylinder holding-down studs.

Each cylinder has four valves, two inlet and two sodium-cooled exhaust. All valves in each cylinder block are operated from a single centrally-disposed overhead camshaft through spherical-ended tappets. The crankshaft runs in seven lead-bronze lined main bearings. . .

The simple single spur reduction gear at the front of the engine is kept independent of crankshaft malalignment by a floating splined coupling. At the rear of the engine the wheelcase houses the inclined drives to each camshaft, the magnetos and the main coolant pump. The engine sump houses the pressure oil pump and scavenge oil pumps for engine lubrication.

The two-speed, two-stage supercharger unit is attached to the rear of the wheelcase. The two centrifugal rotors are arranged in tandem and are driven from the rear end of the crankshaft through three centrifugally-loaded clutch wheels which absorb the high inertia loading resulting from rapid acceleration and deceleration. . . When the supercharger is in M.S. (moderate supercharge) gear the drive is taken through one clutch wheel only and when F.S. (full supercharge) gear is engaged the drive is through the remaining two wheels. . .

The engine-driven pump circulates coolant round a closed circuit via the cylinder blocks, header tank, radiator and back to the pump. The coolant is a water-glycol mixture, non-freezing, and pressurized to prevent boiling. . . The main radiator occupies the port half of the radiator assembly. . . Control of main coolant temperature is achieved by use of radiator air exit flaps.

The liquid-cooled intercooler reduces the charge temperature at high power condition, and thus prevents detonation. It also ensures good volumetric efficiency. The heat exchanger between supercharger delivery and induction pipe extracts heat from the charge, and circulation of coolant through the intercooler radiator maintains a good ratio of heat transfer. . . The intercooler radiator is positioned between the main radiator and oil cooler.

Production power plants are fully proof run on a hanger, and controls set, prior to delivery. This ensures that, after installation, the unit is cleared for flight without intensive ground and air testing. In addition the four-point slinging is designed so that an operator may have available a spare power plant with propellor also fitted and proof run, thus reducing the time for an unscheduled power plant change to less than half an hour. . .

The complete stripping of a power plant for overhaul has been superseded by a system of "lifting" the various components and sub-assemblies, thus assisting the operator to obtain maximum utilization from a given number of power plants.

The wartime experience of some TCA pilots flying the CGTAS Lancasters left them mistrustful of the Merlin engine. They regarded it as a military power plant — fine to run full-bore on quick trips over enemy territory then to scrap after a few hundred hours, but not suitable for airline service. Others regarded the choice as a political one, a friendly gesture to assist Britain in the lean postwar years. A statement in Parliament on June 16, 1948, seems to support this view: "The Canadian plane has a British power plant which was put in as a measure of cooperation with the British government. . ." The speaker was C.D. Howe, but the depth of technical studies would seem to contradict him. Some, though, even went so far as to suggest the airline was cooking the figures in favour of the Merlin and/or against the competing engines. Political opponents of the government were quick to cash in on the controversy. Whatever the reasons for the choice of the Merlin engine, it proved to be an advantage when the opportunity arose to sell the aircraft to the British.

Canadair at War's End

During the war Minister of Munitions and Supply C.D. Howe moved his own people into key positions in the Canadian aircraft industry. Among them was the dynamic, if somewhat unorthodox, Benjamin Franklin, whom he hired to help manage and spur along war production at Canadian Vickers. Like many of Howe's hand-picked men, Franklin was hired not because of his expertise in aircraft production (he had none) but for his abilities in organization, management and administration. Franklin had previously sold farm machinery and insurance! Franklin's chief assistant was Ralph G. Stopps, who had been with Canadian Vickers' aircraft division since 1936. By 1942 Stopps was production manager at its plant in Montreal where Consolidated PBY patrol bombers were being built.

In June 1942 Franklin sent Stopps to supervise construction of a new plant at Cartierville and to be plant engineer. In a short three months the new plant began production of PBY components and in time was building complete aircraft. PBY production was phased out at the old Canadian Vickers plant on the Montreal waterfront and PBY flight testing moved to Cartierville from St. Hubert. Soon the Cartierville plant was running so efficiently that the Consolidated operations in San Diego and New Orleans placed large orders for hull wing sections and other components. These were quickly produced and shipped to the US, the company realizing handsome profits from the deal.

Towards the end of the war the directors of Canadian Vickers decided to get out of the aircraft business and sold their Cartierville operation to the government. At this point Ben Franklin, T.R. McLagan of Canada Steamship Lines and others formed Canadair Ltd. to manage the Cartierville plant on a management fee basis. At the same time they became aware of the government's plans to build a large commercial and military transport plane for the postwar market. A little-known aspect of this story is that C.D. Howe had apparently already designated Boeing of Canada at Vancouver as prime contractor for the new transport. Boeing was a major producer of PBYs and B-29 components. Franklin claimed that his operation had been doing a much more efficient job building PBYs and suggested to C.D. Howe that his firm could better handle the transport. Howe replied that the Cartierville plant was still needed for wartime production. Franklin, the story goes, made a wager with Howe that he could obtain approval in the UK to release Cartierville's production capacity from wartime requirements. He flew

These Consolidated RY-3s were ferried from Dorval to Cartierville at war's end, where they were cannibalized and scrapped by Canadair. The photo is dated May 23, 1946. (Canadair Ltd. via W.G. Anglin)

An example of the 369 Cansos built by Canadian Vickers/Canadair during World War 2. The capability to mass produce the Canso enabled Canadair to move smoothly into its DC-3 conversion program and then the North Star. This Canso was seen at Trenton on July 1, 1961. (Larry Milberry)

straight to the UK on a Liberator and within a few days was back with the agreement in writing. Howe stuck to his end of the bargain and the contract went to Canadair.

With the transport contract secured, Canadair next pursued a major venture in the US. Ralph Stopps was sent to Santa Monica to study the DC-4 production line. While there he had a discussion with Donald Douglas Sr. about the future of the DC-3. He learned that Douglas was to dispose of most of its C-47/DC-3 assets including a plant at Oklahoma City. The C-54 plant at Parkridge, Illinois, was also up for disposal as were other operations such as the C-47 wing production facility at Pullman's in Chicago.

Ralph Stopps returned to Montreal and met with Howe and Franklin, explaining that over $120 million worth of Douglas assets could be bought very cheaply. Stopps was quickly designated Canadair's representative and sent back to the States, where he went straight to work, bidding on large lots on the basis of $40.00 a ton for tools and $200.00 a ton for aircraft parts. Later the lots were broken down for the benefit of smaller bidders, but Canadair cornered the bulk of the lots, including virtually complete C-47s and C-54s. At one point Stopps allowed a number of wing sets to slip by. These were picked up by Republic Aircraft, which was converting DC-3s for airline use. But the wings were incomplete and Canadair had already acquired the jigs and tools needed to finish the wings. These it gladly leased to Republic for a tidy profit.

In all, the Parkridge and Pullman deals netted Canadair some 600 carloads of material, including 60 to 70 C-54 fuselages. The latter arrived at Cartierville by flatcar. The material purchased at Oklahoma City was trucked to Montreal.

In this way Canadair gained its postwar beachhead. And while Douglas was being cleaned out in the US (including all original C-47/DC-3 engineering drawings), other deals were made in the US and UK for over 200 surplus C-47s. These were ferried to Canada, some in very rickety condition, and a big program was begun to convert them for commercial service. For its own corporate use Cana-

CF-DXU, one of over 200 C-47s converted for commercial use by Canadair, served as the company's general workhorse for over 30 years. More recently it was flying in the Arctic for Kenn Borek Air, but crashed on takeoff from Inuvik June 22, 1981. (Larry Milberry)

Two DC-3s turned out by Canadair, one belonging to USCAN Engineering of Yellowknife and the other to the Argentine operator ZONDA. (Canadair Ltd. via W.G. Anglin)

dair acquired a US surplus C-47 at Dorval for $15,000. The airplane had just 16 hours total time! Registered CF-DXU, it was to serve Canadair for over 30 years. In all, the DC-3 conversion program was very good business for Canadair, and for over 20 years it was to profit from the production and sale of DC-3 parts for customers around the world.

Jim Bain has a few reflections on this era at Canadair:

At the end of the war the US government stopped all contracted production and offered everything that had been produced for sale at scrap value. Benny Franklin decided that he should buy all that "scrap" and build some aircraft for the airlines. He was financially backed by Electric Boat of East Hartford, Connecticut.

Franklin went to see H.J. Symington to interest him in purchasing some aircraft. Symington sent for me. I insisted (with the benefit of all our research) that TCA must have pressurized aircraft and that the scrap parts would by no means be suitable.

Next, Franklin went to C.D. Howe and pleaded his case to keep the Canadair plant working. He wanted Howe to tell TCA that they could get along with an unpressurized version of the DC-4, made from C-54 "scrap."

Howe sent for Symington who took me with him. We successfully defined and defended our requirements for a pressurized design.

Over a period of weeks, there followed some ins and outs with Franklin, the RCAF and Mr. Howe in which TCA did not participate, but one day Franklin came to my Dorval office with a contract that would keep Canadians working by building some unpressurized aircraft for the airline which would be turned over to the RCAF when the pressurized version could be delivered. Apparently Mr. Symington had been obliged to accept this deal.

Building the pressurized aircraft meant alteration to existing jigs and the manufacturing of the necessary new ones to the Douglas engineering specifications for a pressurized DC-4. As a result of all these preliminaries, Howe asked Symington to put me into Canadair during the initial construction of the first aircraft. This was done and I became Benny Franklin's executive assistant for 15 months.

Engineering supervisor Ralph Stopps and pilots Fred King and Bill Anglin of Canadair just before the North Star era at Cartierville. (Canadair Ltd. via W.G. Anglin)

Pilot Bill Anglin had a close call on this "Canadair" DC-3. On takeoff, the landing gear prematurely retracted and the port propeller came through the cockpit. Anglin pulled his legs back onto the seat just as the propeller blades swished by where they had been a moment earlier. (Canadair Ltd. via W.G. Anglin)

Purchaser	*Nature of Job*	*Quantity*
A.B. Aerotransport, Sweden	C-47 to DC-3	3
Aviquipo, Mozambique	C-47 to DC-3	3
Aviquipo, Portugal	C-53 to DC-3	4
Aviquipo, Spain	C-53 Special	1
Colonial Airlines Inc., USA	C-47 to DC-3	3
Eldorado Mining & Smelting (1944) Ltd.	C-47 Special	1
Doris, Yellowknife	C-47 Special	1
Eastern Airlines, Inc., USA	C-53 to DC-3	15
French Air Mission	C-53 to DC-3	5
Linea Aeropostal Venezolana	C-47 to DC-3	3
Netherlands Purchasing Commission	C-47 Overhaul	12
Norseman Air Transport, USA	C-47 Special	1
Northeast Airlines Inc., USA	C-47 Overhaul	1
Royal Norwegian Purchasing Commission	C-53 and C-47 to DC-3	5
SABENA Belgium	C-49 and C-47 to DC-3	7
TCA	C-47 to DC-3	20
USCAN Engineering Ltd., Canada	C-47 Special	1
ZONDA, Argentina	C-47 to DC-3	1
	Total	87

Summary of Canadair's C-47/DC-3 conversion program from commencement in June 1945 to September 14, 1946, as noted in the House of Commons Debates, March 20, 1947.

The DC-4M Takes Shape

Long-time TCA engineering test pilot Ron J. Baker recently summarized the project as he viewed it during the pre-production stages in 1944-45:

We were very aware of weather problems and wanted an aircraft that could top most icing and other adverse conditions. This meant that we had to have an engine that could take us up to at least 20,000 feet. We were tired of using oxygen over the North Atlantic and the mountains so this meant that we would need a pressurized aircraft. Nothing suitable was on the market so we set about drawing up specifications and examining possibilities.

The DC-4 was a proven success in the USAF but it was slow and could not top the weather. There were plans at Douglas for a new aircraft which eventually became the DC-6; but in the meantime Rolls-Royce were developing an engine with more power than was used in the Lancaster. We felt that marrying this engine with a pressurized fuselage the size of a DC-4 would give us the aircraft we wanted.

The basic arrangements were made with Douglas and Rolls-Royce; then TCA got down to the details for such features as the cabin and cockpit. This was a major undertaking for a small company, but work went ahead in our Engineering Department under Jack Dyment.

We built a wooden mockup at Winnipeg and used it to develop ideas for the production airplane. Technical, flight and sales people all contributed. My own biggest interest was the flight deck. At the time there was no standardization of flight deck layout as there is now. Ideas formulated were transferred to Canadair so we eventually had an aircraft built to our own specifications. As to the mockup, I still admire the skilled work of the men from our carpentry shop. They built a real masterpiece.

As the project developed we spent a lot of time at Canadair. We all worked together very well and it was relatively easy to come to agreement and make modifications. We also spent time at Rolls-Royce at Hucknall and Derby where I had already gotten to know the people as I had taken Lancasters down there in wartime for engine changes.

Ron J. Baker

A TCA document from late in the war makes an interesting comment about the state of development of the new TCA airliner to that point: "With the original estimate of the war not ending before 1946, sufficient time was available to make all the modifications to the aeroplane desired by TCA to make it more suitable to Canadian service. It now appears that sufficient time will not be available. The entire situation and plans have, therefore, had to be revised and compromises made between accepting certain features as presently designed or modifying them as originally planned by TCA. . . It had been TCA's hope that subsequent modifications to the DC-4 would for the most part be unnecessary, by having them incorporated by the Canadian manufacturer of the aeroplane. As stated, this hope will only be partially fulfilled now due to the urgency of obtaining delivery of aeroplanes as quickly as possible. . . The aeroplane still remains the most

	049 Constellation	DC-4M	DC-6
Takeoff weight	86,250 lb	77,000 lb	80,500 lb
Landing weight	75,000 lb	58,500 lb	63,500 lb
Engine	Wright R-3350 Cyclone. 18 cyl.	Rolls-Royce RM-14-SM Merlin. 12 cyl.	Pratt & Whitney R-2800. 18 cyl.
Takeoff power	2200 bhp	1660 bhp	2100 bhp
Max. cruise power	1300 bhp	1112 bhp	1100 bhp
Altitude for max. cruise power	20,000 ft	20,100 ft	21,500 ft
Passengers	64	44	50
Cargo volume	450 ft^3	420 ft^3	483 ft^3
Max. speed	340 mph	353 mph	339 mph
Altitude for max. speed	18,000 ft	23,900 ft	18,500 ft
Max. cruise speed	303 mph	324 mph	315 mph
Altitude for max. cruise speed	20,000 ft	22,600 ft	23,600 ft

TCA comparison chart from 1944 for aircraft considered for postwar re-equipment. These figures changed as the aircraft developed.

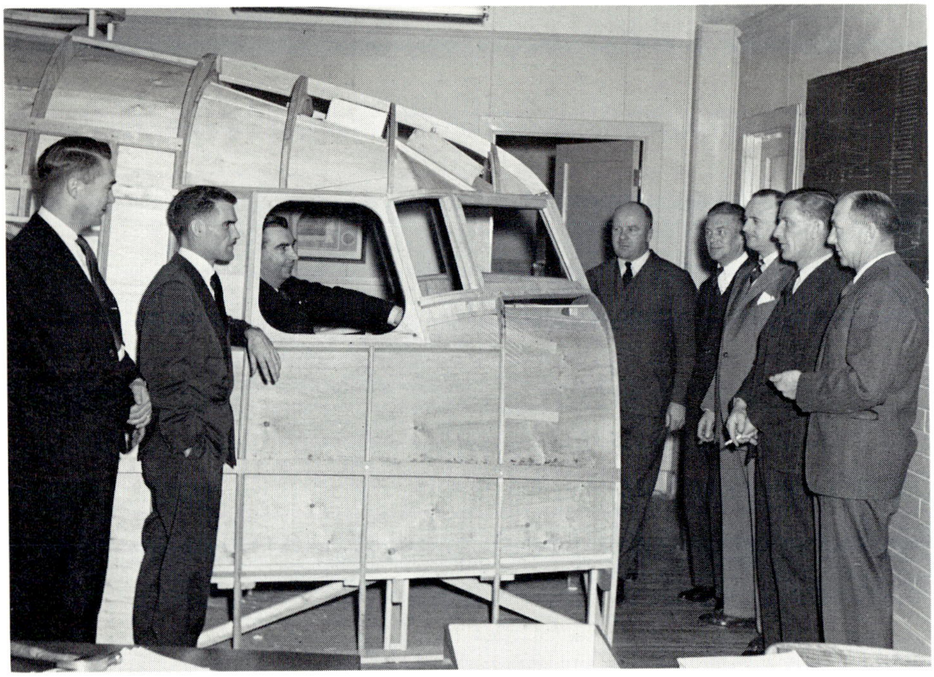

Wooden mockup of TCA's new airliner at Winnipeg in 1944. Seen are W.W. Fowler, Frank Young, J.T. Dyment, E.W. Stull, J.T. Bain, F.M. McGregor, M.B. Barclay and J.H. Tudhope, all of TCA. (Air Canada Archives)

Canadair DC-4M	$ 670,621
Curtiss C-46	$ 442,800
Lockheed Constellation	$ 987,182
Boeing Stratocruiser	$1,427,902

Aircraft costs noted by TCA, Nov. 14, 1945.

suitable that can be obtained for Canadian use in the immediate postwar period.''

From 1944 onward design work on the DC-4M, as the new airliner was designated, gathered speed, with most activity taking place at the TCA engineering facilities at Winnipeg and Dorval, and at Canadair in suburban Montreal. Through 1945 Canadair concentrated on gearing up for production.

Redesign work on the C-54/DC-4 was a major job for TCA, Douglas, Canadair and Rolls-Royce. Douglas provided 30,000 engineering drawings, many of which had to be modified, and Canadair had to produce thousands of new drawings. Canadair's chief engineer at the time, Peter W. Gooch, wrote of this stage of development: ''Very close liaison with Douglas was necessary, particularly in the early stages of the contract. A number of Canadair engineers spent periods of up to a year at the Douglas plant becoming familiar with the basic airplanes and with Douglas' methods of engineering. In this way TCA has been able to get an airplane designed to the well known Douglas standards yet engineered specifically to its own requirements. Similarly, Canadair engineers spent considerable time in England with Rolls-Royce.''

There was also close liaison with the RCAF. It too required a heavy transport for postwar use and was studying various designs, among them the Avro Tudor, a big Merlin-powered transport using the Lincoln bomber's wing. Five Malton-built Tudor IIFs were ordered in 1946 but by that October the order was cancelled. The Tudor was encountering development setbacks while Canadair was offering a more attractive design and guaranteeing quick delivery.

Re-engineering the C-54/DC-4

When Peter Gooch graduated with his Master's degree in aeronautical engineering from the University of Toronto in 1937, he was immediately hired by de Havilland in Toronto to work on the Tiger Moth program. In 1939 he joined the team working at Hadfield Hall, London Colney, in the UK, on a very secret project that resulted in the Mosquito bomber.

In 1941 Gooch was back in Canada working with Canadian Vickers in Montreal, having been hired by Benny Franklin and Roger McLagan. He worked on various aircraft before going out to Cartierville on the PBY program. In 1944, just as the new Canadair transport was beginning to take form, he was made chief engineer. He was soon completely involved with it and in making trips to Santa Monica where he and other Canadair people familiarized themselves with Douglas' technology. He recalls that Douglas' chief engineer, Ed Burton, was greatly impressed with the concept of a Merlin-powered DC-4 and thought that the airplane would have a great market potential. Peter Gooch was not to stay long in aviation engineering and in 1947 left Canadair to join Canadian Vickers Industrial Ltd. as president.

The challenge of converting the C-54/DC-4 to TCA's requirements is described here by Peter Gooch:

Installation of the Merlin power plant presented a number of design problems in the nacelle areas. It was considered desirable to maintain the original Douglas engine mount pickup location, firewall and nacelle.

The rigid specification on power plant interchangeability, replaceability and fire protection, however, required a new conception of nacelle installation. In order to meet the specification, the following factors were important: a) Removal from the power plant of all engine-driven accessories normally contributing to differences in power plants; b) Reducing to a minimum the number of connections between the power plant and the aircraft; c) Maximum accessibility to all power plant firewall connections; d) Location aft of the firewall of accessories whose failure might result in inflammable liquids being freed in the power plant.

Adaption of an auxiliary gearbox mounted on the firewall and driven from the engine by a flexible quick-disconnect shaft provided a means whereby these requirements could be satisfied. The gearbox mounts all the aircraft accessories and provides for the location of the hydraulic pump, a serious source of inflammable liquid supply, behind the firewall.

It provided a 25% reduction in the number of power plant connections to be broken for power plant removal. A complete nacelle mock-up was built to ascertain the optimum location of units and power plant firewall connections for best accessibility. Completion of this work permitted the

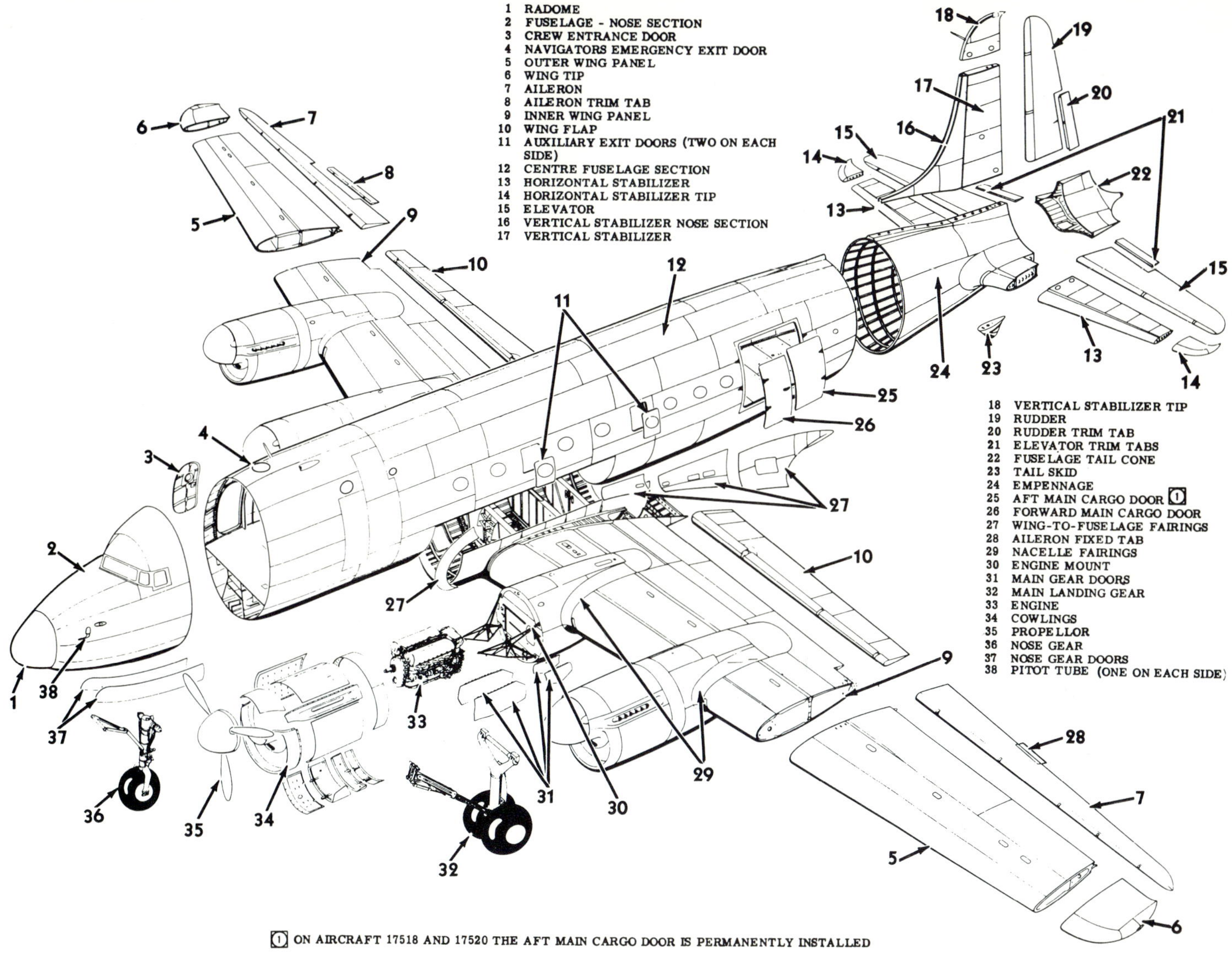

North Star M1 exploded view.

detail design of all nacelle installations to be carried out.

Special consideration was given to the prevention and extinction of fire in the power plant and nacelle areas. The recent rigid CAA requirements on fire protection were, in most cases, anticipated and maximum fire protection was designed into the critical areas . . .

A new nacelle oil system installation was designed to meet requirements of increased oil capacity and cold weather operation. Space limitations within the nacelle provided many difficulties in the design of an oil tank of sufficient capacity. In addition, the special features required in the tank for cold weather operation considerably increased the development work involved.

The fuel system in the nacelle area was completely redesigned to provide for the requirements of the power plant. A high pressure priming system was designed to meet cold weather starting conditions.

The use of the Merlin 620 engine necessitated several structural design changes, as the positions of the centre of gravity and the thrust line of the new engine were considerably different from the radial engines for which the structure had been designed. In addition, the increased power of the Merlin 620 resulted in higher flying speeds which in turn also produced higher loads on the structure.

The nacelles were redesigned to the new loads, at the same time maintaining the original loft lines of the Douglas nacelles. As an additional fire precaution, the main longerons, and all nacelle skins for 20 inches aft of the firewall were constructed from stainless steel. The main engine mount pickup fittings were redesigned as the Rolls-Royce engine mount used a sheer type connection instead of the tension type connection used by Douglas.

To handle the increased loads, the wing spar caps were made from 75ST material in place of 24ST, and heavier skin gauges were required inboard of the landing gear attachment. The integral fuel tanks which are located between front and centre spars in the centre section, and between centre spar and nose skin in the outer panels, have been designed to the latest Douglas DC-6 requirements.

They include the spar caps with the deep webs, the use of sleeve rivets, and many other features which resulted from the extensive service experience gained from the C-54 series aircraft.

The wing and fuselage were altered to suit an entirely new electrical and instrument arrangement. This includes the new main junction box and re-

North Star No. 104 on the production line at Canadair in 1946. It emerged as TCA M1 CF-TEM and later served for years as RCAF 17520 with 412 Squadron, before being written off at Hall Beach in the Arctic. The aircraft seen on the line are being assembled largely from components purchased from Douglas in the US. (Air Canada Archives)

arrangement of the crews' quarters and flight compartment. Similarly, the conversion from military requirements to passenger requirements necessitated changes in the window location, alteration to the floor for seat fittings, etc.

The increased flying speeds increased the loads on the empennage. As a result, the spars for both the horizontal and vertical stabilizer had to be strengthened.

The landing gear, flaps and nose wheel steering system were adapted from the DC-6 aircraft with minor service changes. However, the brake system was redesigned by Canadair incorporating four Bendix three-rotor brake assemblies and a Bendix power valve . . . .

Peter W. Gooch

As well, many other changes were incorporated in creating the modified Canadair design. It is important to note that the engineering for pressurization was done by Douglas. By this time it was clear that Canadair had in the works one of the finest modern airliner concepts. So good was it that Douglas was alarmed, and a TCA engineer involved at the time noted, "From our engineering studies, Douglas realized that we were going to have a better aircraft than they, and they stopped sending us the engineering necessary until the contract had been negotiated giving them the right to build the same kind of aircraft. They never did, of course, but proceeded to the DC-6 instead."

Production Begins

By early 1946 the first of the new Canadair airliners was completed and wearing the civil registration CF-TEN-X. The first aircraft on the production line were slated for delivery to the RCAF, these being unpressurized C-54GMs, or DC-4M1s. Due to pressing needs, six of these were to be turned over on loan to TCA for use on the North Atlantic. The usual pre-flight tests began at Cartierville, and by early summer taxi trials were under way. First flight was delayed by serious propeller and engine snags. On the first engine run-up, for example, one propeller inadvertently went into reverse.

TCA engineer Clayton H. Glenn later recalled these problems:"Originally it had been the intention to use reverse-pitch propellers for aerodynamic braking on the DC-4Ms, although the throttle system was left as in the standard DC-4. Reverse pitch, or reverse thrust as we knew it in later years, really required the use of separate sets of throttles so the aircraft could react to a yawing movement of the throttles, when in reverse, in the same way as when the throttles were in the forward position. In the case of the DC-4M the intent was to put on the reverse thrust by pushing the throttles forward again after first having pulled them to idle and having moved the propellers into reverse position. The system, had we kept it connected, would have undoubtedly led to some catastrophic situations on landing due to pilot confusion."

TCA decided not to have reverse pitch capability in its aircraft, nor did the RCAF aircraft have it (except the Rockcliffe Ice Wagon). But BOAC aircraft ordered later were equipped with this feature.

Propeller governor problems were solved by Glenn, who had been familiarizing himself with component manufacturers in the US. He had reached the point of having memorized many intricate relays, particularly in the "black box" that was to synchronize the Merlin engines. This unit was made by Woodward in Rockford, Illinois, and had been causing Canadair great difficulty. A few days before first flight, Glenn realized what was wrong. He explained to Jim Bain, who immediately shipped the troublesome unit to Rockford aboard the company's DC-3. There the changes recommended by Glenn were made, and the unit was returned to Canadair, installed and found to be working perfectly.

"North Star" It Is

By mid-July 1946 the new airliner was ready to fly, and during this period it was named. As Ron Baker of TCA remembers it: "One summer's day we were waiting around the flight test hangar. As I remember, some of those present were Jim Bain, who had been temporarily assigned to Canadair and was running the DC-4M project; Joan Paul, Jim's secretary; Frank Rousseau of TCA; and Clayton Glenn. People were suggesting various names for the new plane. I remember suggesting Polaris as I was quite interested in celestial navigation. Other names followed until someone said, 'Let's just call it the North Star.' That, quite simply, is how the North Star got its name."

In fact the "someone" mentioned by Baker was Benny Franklin, who had earlier coined the name Canadair.

North Star Test Pilots

To conduct the initial test flying on the North Star, Canadair borrowed the chief test pilot from Douglas in Santa Monica, Robert P. Brush. Bob Brush began flying in southern California in 1931. In 1935 he was flying in the US Army, operating such types as the P-12 and B-10; he then flew for the airlines for a brief period on DC-2s and Lockheed 10s. In the spring of 1939 he was hired by Douglas as a test pilot, a job he continued in for 16 years before going into sales for Douglas. During his test flying years he flew such aircraft as the Navy's SBD, the A-26 Invader, F3D Skyknight all-weather jet interceptor, the XB-43 (world's first jet bomber), the Globemaster heavy transport and the whole range of Douglas transports from DC-3 to DC-7. In a letter to aviator and writer Ernie Gann, Brush described his test flying years: "During 16 years as test pilot, chief experimental test pilot and director of flight operations, I had my share of experience with military and commercial aircraft and Douglas personnel at all levels. To many at the lower levels who worked as mechanics and inspectors on my aircraft, I owe a great deal because I'm alive today. On the other end of the personnel spectrum I often found young 'executives' as complicated and as unpredictable as the birds I was testing! They also, particularly the engineers and designers, often found me very critical of their

The first North Star at Cartierville before its completion. (Canadair via B. McColl)

Douglas test pilot Bob Brush and Canadair president Benny Franklin each looking quite pleased about things, mostly the gorgeous airplane behind them. (Canadair Ltd. via R. Brush)

Three men closely involved in North Star test flying: flight engineer Reg Thatcher, and pilots Bill Anglin and Al Lilly. (via W.G. Anglin)

airplane and stubborn as related to how it was going to be tested."

Such was the man of experience handed the job of testing the mettle of the North Star.

Canadair's chief test pilot was Alexander J. Lilly. He started flying at the Moose Jaw Flying Club in 1927, though it was several years before he won his wings. When the Depression came along, Al Lilly, who had been attending university, applied for a post with the RCMP. Considering that there were some 10,000 applicants for the 100 or so jobs available, he was rather amazed when he was taken on. He trained at Regina and was posted to northern Saskatchewan. There he operated mainly by canoe and dog team. A Canadian Airways Fox Moth was based nearby and on occasion he would charter it at $10.00 an hour and in a few hours accomplish what would take weeks to do paddling or mushing his dogs. In his spare time Lilly trained one of his dogs to take certain commands. An inspector noticed the dog's abilities and before long the RCMP had enlisted him to assist in establishing its Police Service Dog Section.

Just before the war Al Lilly replied to an advertisement inviting pilots to apply to Imperial Airways. Again to his surprise, he was hired. He left the RCMP and travelled to England, where he was assigned flying duties on the de Havilland Ensign. But war soon erupted and he was called back to Canada to aid in the Commonwealth air training scheme, initially as chief flying instructor at Moncton. Within a year he moved to Ferry Command HQ at St. Hubert as chief test pilot. He ferried numerous aircraft overseas, including six Hudsons and a B-24 during the summer of 1941, a feat which brought him a decoration. He also acted as a VIP pilot and on one trip westbound, flying a Boeing Clipper, he had as a passenger Winston Churchill — at least as far as everyone knew. Part way across the ocean, the crew realized their passenger was one of Churchill's doubles; the real Churchill was on his way to America aboard a battleship.

Near the end of the war Al Lilly, Wally Siple, King Parker and others purchased a number of surplus Stranraer flying boats and set up a freighting service from Tampa serving the Caribbean as far down as Colombia. Within a year, this venture folded and he took a job flying for Maritime Central Airways. This was short-lived, as an offer came from Canadair. With his experience as a wartime test pilot, his knowledge of the Merlin engine (he had many hours ferrying Lancasters and checking out pilots in them), and interestingly, his knowledge of aircraft certification gained from the Stranraer episode in Florida, Al Lilly was a valuable asset to Canadair. He was soon at work there, primarily with the DC-3 conversion scheme. But the North Star was just around the corner.

Bill Anglin was a pilot with 168 Squadron flying B-17s from Rockcliffe when he first got involved with the North Star. He was weathered in in Iceland, and it happened that a plane carrying Benny Franklin was also stuck there. Franklin and C.G. "Chubby" Power of Prime Minister King's War Committee were carrying vital papers back to Montreal from Rolls-Royce at Derby. Franklin asked Anglin for a ride but, as he was a civilian, couldn't be taken. However, Anglin did take the papers and took off in his supercharged B-17, which could get above the miserable weather.

The B-17 flew non-stop to Rockcliffe, from where Anglin headed to Cartierville on the train. He delivered the plans to Franklin's secretary, Marion Gaul, had a tour of the plant, then left. He continued his RCAF flying until severely injured in a disastrous crash of a Dakota in the UK. After a long convalescence he returned to flying and one day found himself

Crew on the North Star for its first flight: Clayton Glenn of TCA, Bob Brush of Douglas and Al Lilly and "Smokey" Harris of Canadair. (Air Canada Archives)

again at Cartierville. Marion Gaul remembered Anglin when she saw him and called out Franklin. The two went for lunch and Franklin offered Anglin a job flying the company's DC-3 and working in sales. Anglin was still in the air force, so Franklin called Chubby Power and Anglin was released from the RCAF overnight. He was soon flying at Canadair, making many trips to the States in connection with the big deals at Oklahoma City and Chicago. He was astounded at the massive amounts of material the company acquired from the Douglas plants and remembers the huge Oklahoma plant where the production line seemed to stretch for a mile.

During this period he and most others at Canadair learned what an irascible personality Franklin was. In one case a wind storm struck Cartierville, damaging some DC-3s. Franklin immediately fired the flight operations director. Another pilot, responsible for damaging a DC-3 on takeoff, was likewise fired on the spot. Had Bill Anglin not intervened, a top company flight engineer would also have been fired by Franklin. The engineer had damaged a DC-3 on run-up. Eventually Franklin was himself urged out of Canadair on account of his unorthodox dealings at Oklahoma City.

First Flight

In the late afternoon of July 15, 1946, the prototype North Star was on the runway at Cartierville doing more high-speed taxi trials. These had been going on for some days and the crew, especially Canadair's Al Lilly, were anxious to get the new plane into the air. As TEN-X sat at the end of the runway, Lilly suggested to Bob Brush that this was as good a time as any to fly the airplane. Also aboard were engineers W.L. "Smokey" Harris and Clayton Glenn, who was asked to go along because of his detailed knowledge of the propeller control unit. The decision was made and away went the North Star on its maiden flight. As noted in Al Lilly's log, the flight lasted for 25 minutes, with a routine landing being made back at Cartierville. Clayton Glenn recalls the event:

We had done a fair amount of taxi tests up and down the runway at high speed to check the controls. We burned out a number of sets of brakes in the process. On the evening of July 15 we were scheduled for more taxi tests. I was wearing a new gabardine suit as I was going out to dinner with a good friend, Jim Johnson, who was a General Electric representative. After we were aboard, Bob Brush let us know that there were parachutes stowed at the rear of the cabin and asked us to put them on.

So we flew the North Star. The gear was raised right after takeoff and the cockpit filled with smoke and the smell of burning rubber. This didn't faze Bob, who explained, "Don't panic. It's just the nose wheel spinning against the up brake."

Fortunately, Jim Johnson had his camera handy and took the only pictures of the first takeoff, then of us as we shot up the field. Ben Franklin was quite put off with the episode as he had been waiting so long for the flight and nobody had told him we were going to fly. The only people who had been alerted were those operating the fire engine.

The flight was short but gave us a chance to check the systems, controls, etc. and the wings for flutter and vibration. We returned to Cartierville just before dark as those were the days when it still had no runway lighting.

Needless to say Jim Bain was delighted and took us to dinner at Ruby Foo's where we were served the largest steaks I've ever seen, as thick, it seemed, as the Montreal telephone book!

Clayton H. Glenn

Tests and Demonstrations

The North Star was back in the air July 17, then on Saturday, July 20, was publicly unveiled at a ceremony at Canadair. For the occasion there was a short demonstration flight and an interesting manoeuvre on the ground described here by Bob Brush: "To please Ben Franklin and the dignitaries present I taxied toward the reviewing platform, directly at it and with sufficient speed so I could apply the braking needed to compress the nose-wheel strut several inches. This lowered the beautiful North Star's nose and the four Merlins enough to provide a graceful bow to the audience. It thrilled Franklin tremendously."

Days of intensive test flights followed. As noted by Bob Brush, some of these included: stall tests for characteristics and for accurate speed determination; speed vs power for evaluating cruise speed and range; determination of takeoff and landing distances, and of climb performance. Flight notes provided by Brush for this book include: "Stall tests indicated that the stall characteristics, particularly power on clean, were not as good as for the DC-4 or DC-6. This may have been

A gallery of CF-TEN-X including superb in flight poses and one of roll-out day at Cartierville, July 20, 1946. (Canadair Ltd.)

TEN-X
NORTH STAR
CF-TEN-X
Canadair Ltd.
CF-

TEN-X
CF-TEN-X
Canadair Ltd.
CF-

TCA President H.J. Symington and Canadair President Benjamin Franklin at the official christening of the North Star in July 1946. (Air Canada Archives)

Initial test flying phase of the North Star as conducted in 1946 by Bob Brush of Douglas Aircraft and Al Lilly of Canadair.

July 15	:25 hours	July 24	3:00 hours
17	1:15	25	1:25
	1:00		2:00
19	:30	26	2:00
20	:10	27	:50
22	2:00	28	1:20
	:20		:25

The North Star's first cross-country demonstration flight was flown in September 1946. Those aboard were Captain Al Lilly, Ron Baker, Jack Hood (flight engineer), Smokey Harris (flight engineer), D. Black, R.M. Crooks, R. Agar, H.J. Symington, C.D. Howe, Eric Warlow-Davies (Rolls-Royce), J. Hall (Rolls-Royce) and Jim Bain (TCA/Canadair).

Sept. 14	Dorval-Toronto	1:30 hours
	Toronto-Ft. William	2:10
	Ft. William-Winnipeg	1:40
	Winnipeg-Lethbridge	3:00
	Lethbridge-Vancouver	2:10
15	Vancouver-Santa Monica	4:50
17	Santa Monica-Vancouver	4:30
19	Vancouver-Dorval	8:45

due to the very clean nacelles compared to those of radial engines. Maximum cruise and long-range cruise were slightly below estimates although the aerodynamic design of the nacelles was superb. Fuel consumption was slightly over book. One or two in-flight engine shut-downs were encountered and I also seem to recall minor propeller or governor problems. Cabin noise from the two beautiful inboard Merlins was identified as a problem, and exhaust stack redesign was started. Stability and control were excellent after only modest changes." After just a month's involvement with the North Star, Bob Brush returned to Douglas and all further test flying was now directed by Al Lilly.

Soon others were checking out on the North Star, including new Canadair pilots hired for the program, Department of Transport inspectors and TCA pilots. Included were Hugh Noseworthy, Russ Crooks, Joe McElrea, Bill Anglin and Des Murphy.

Eric Bendall joined Canadair early in 1946. He had flown in the bush before the war, then was a Liberator captain for the RAF flying from Dorval during the war. He remembers one unusual aspect of the North Star program: "By the time the DC-3 conversion program was well established, efforts switched to the North Star. This entailed weekly flights to Douglas at Santa Monica in the DC-3, which kept us very busy. At that time Canadair didn't have a milling machine so we had to take the raw North Star wing spars, which were made in Kingston, out to Douglas for milling, then fly the finished product back to Montreal. These spars were very long, reaching from inside the cockpit of the DC-3 (just missing the undercarriage levers) to the toilet compartment past the last bulkhead. The spars were difficult to tie down and were quite a worry in turbulent conditions. I flew some 15 round trips to Santa Monica, along with many other trips in the US."

Early in the North Star test program it was decided to fly coast-to-coast on a publicity tour, and to show it off to Douglas at Santa Monica. C.D. Howe, ever the master of public relations, wanted every opportunity to get coverage for this special project of his. To all aboard, the noise problem was obvious, and on the way back to Dorval President Symington of TCA told Jim Bain to get to work on it immediately. By this time Ben Franklin had been replaced as Canadair's president by H. Oliver West, one of the original TCA organizers and a well known figure from Boeing. West was a hard taskmaster and shook up his engineers by telling them they had two weeks to solve the noise problem! Like many other things West did at Canadair, this did not endear him to anyone but it certainly got action. However, there could be no two-week solution and the problem was to test Canadair and TCA for several years.

A later experience recalled by Al Lilly shows West's intransigence. Lilly was on a demonstration flight with BOAC personnel when an engine had to be feathered. Coolant was streaming from the engine and West came forward insisting that the engine was on fire and that smoke was pouring from it. After landing, West ordered the aircraft stopped on the runway, then raised a storm claiming that the airport firemen were not on the ball!

Remembering the trip to the west coast years later, Ron Baker wrote, "We enjoyed the trip out west and down the coast. As we checked in over the radio checkpoints there were numerous remarks about our speed, as the North Star was so much faster than most transports. It was a good trip."

C.D. Howe Explains . . .

A concise account of the development of Canadair in the late-war and early postwar years was provided in the House of Commons by C.D. Howe on March 20, 1947. This statement was in response to a series of questions posed by MPs Fraser and Jackman. In particular, it describes the 1946 sale of Canadair by the Canadian Government to the Electric Boat Company of Massachusettes.

Early in 1944, it became apparent that additional orders for PBY types would not be forthcoming. It was imperative that plans be made either for closing the plant and laying off some 8,000 skilled workers then in employment, or alternatively that another type of aeroplane should be undertaken. Requirements for fighter planes for the RCAF were well in hand, but both RCAF and Trans-Canada Air Lines had substantial requirements for long range transport planes. The stated requirement of RCAF for long-range transport was then a minimum of 50 aircraft, and for TCA, a minimum of 25 aircraft. At that time, practically all four-engine long-range transport planes were being built in the United States and, of course, the Douglas C-54 was generally considered to be the best. An extensive study was made by Canadian engineers, in collaboration with Douglas and Rolls-Royce, to determine the type of plane best suited to Canadian requirements. The conclusion was that the commercial adaptation of the Douglas C-54, known as

Cartierville on August 1, 1947, at the peak of North Star M1 production. The main plant with its 657,000 sq.ft. of space can be seen along Laurentien Boulevard. Most of the North Stars seen near it are engineless due to a delivery lag from the UK. The aircraft farther up Laurentien are at the Curtiss Reid flying school, the oldest resident at Cartierville. At the corner of Laurentien and Bois Franc Road is a Canadian Car plant and west along Bois Franc is the big Noorduyn plant, by now involved in Canso and DC-3 conversion work. Land along Laurentien was bought at war's end by Canadair for 6 cents a foot. (Canadair Ltd. via Graham Wragg)

the DC-4, powered with Rolls-Royce engines, would be the aircraft peculiarly well adapted to Canadian service and commercial air line requirements.

The government therefore negotiated with the Douglas Aircraft Company, and secured a licence agreement under which Douglas undertook to furnish plans of its DC-4 type, as well as its DC-6 type then being designed, and to assist in the development, redesign and construction in Canada of an aircraft based on Douglas designs and fitted with Rolls-Royce power plants. In this arrangement, the Rolls-Royce Company cooperated fully. The negotiations were completed in the fall of 1944.

The Cartierville plant had been built and managed for the crown, under a management agreement, by Canadian Vickers Limited, who had additional aircraft facilities at St. Hubert. Later, in 1944, the Canadian Vickers Company was engaged in an extensive shipbuilding program for the allied navies and the merchant service, which involved heavy financial commitments. It was considered advisable to ask Canadian Vickers to divest itself of its management responsibilities related to the Cartierville plant, and to devote its full resources to shipbuilding. The government then entered into a management contract with a group of men who were at that time in charge, for Canadian Vickers, of the management of the Cartierville plant. These men formed a corporation, under the name of Canadair Limited, as a privately-owned management company, charged with the exclusive duty of operating the Cartierville plant as agents of the crown for the production of transport aircraft, as well as to carry out further commitments such as subcontracts for the United States navy for the production of PBY hulls and centre sections. These latter orders engaged the facilities at Cartierville until March 1945. During the same period, engineering and tooling for the Canadian transport plane was advanced as rapidly as possible, having regard to the fact that Canadair, Douglas, and Rolls-Royce were then filling important war commitments.

After the end of the European war in May 1945, United States subcontracts were terminated. Employment at Cartierville was then maintained by accepting contracts from commercial air lines in Canada, the United States, South America, and Europe to meet a desperate peacetime need for transport equipment. Engineering and tooling for our four-engine plane had not reached the point where facilities at the plant could be fully employed, and therefore orders were accepted from more than 16 commercial operators for the conversion of United States Army Dakota type transports to the commercial DC-3 type. The work involved stripping down army planes and practically rebuilding them into transport types, almost the equivalent of building a new aeroplane. This work commenced in the early summer of 1945 and is still in progress. Up to the middle of September 1946, 87 DC-3 aircraft had been delivered, and this work is still continuing. The integration of these various programs enabled a high level of employment to be maintained at Cartierville, and also enabled the plant to retain the services of trained personnel for the production, in due course, of a program of four-engine transport planes.

The first four-engine transport plane, named the North Star, left the assembly line in June 1946, less than two years after the program was initiated. This prototype has performed satisfactorily, and has fulfilled the expectations of the group that planned the new plane. A total of seven finished aircraft have been produced to date, and it is expected that TCA will operate six of these planes in its regular transatlantic service on or about April 1, 1947.

Out of several long-range transport planes planned by manufacturers in the United States and the United Kingdom at about the same time, some of them earlier, only one is in operation to date, namely the Lockheed Constellation. Operational experience with that plane has not been altogether satisfactory, resulting in its being grounded for periods on two occasions. The Canadian plane being the second postwar transport plane, it was felt that exhaustive tests under all conditions of weather should be made before the plane was put in operation. These have been conducted during the winter months, and minor modifications have resulted. It is worthy of note that ours is the second postwar plane to go into operation. Not a bad record for Canada.

Total value of all production at the Cartierville plant up to September 14, 1946, which is the date of transfer of the management to Electric Boat Company, was as follows:

US government contracts and sub-contracts................	$ 54,780,000
RCAF contracts for PBY's......	27,340,000
Trans-Canada Air Lines for DC-3 conversions.................	2,730,000
Foreign customers for DC-3 conversions.................	10,480,000
Inventory on hand at September 14, 1946........................	9,750,000
Total......................	$105,080,000

Employees at Canadair Limited at December 31, 1946, totalled 7,838 persons.

During 1946, the Electric Boat Company, a substantial American corporation, became interested in the Cartierville project. This company, with large financial resources, had examined carefully into Canadian economic conditions, and became convinced that it could wisely and profitably engage some of its liquid capital in Canada. After extensive negotiations, Electric Boat Company acquired substantially all of the stock of Canadair Limited, and has entered into an agreement with the Canadian government, effective as of September 14, 1946, covering the crown facilities, contracts, etc., at Cartierville. The main features of the agreement are as follows:

(a) Electric Boat Company to provide Canadair Limited with $2,000,000 working capital forthwith;

(b) Canadair Limited to waive all claim or right to fees or profits from the beginning of its operations to and including September 14, 1946;

(c) Canadair Limited to buy at book value the assets of the project, as at September 14, 1946, exclusive of plant, tools and automotive equipment, and to assume the liabilities as set out on the balance sheet of the project prepared as of that date;

(d) Canadair Limited to buy such automotive equipment as it might require, at prices to be agreed upon with War Assets Corporation;

(e) Canadair Limited and the government to enter into a fifteen-year lease-option agreement covering the plant facilities and tools under which the company will pay a fixed annual rental of $200,000, and will have an option to acquire all of the facilities, including the tools, at an inclusive price ranging from $4,000,000 in the first year to $2,511,961 in the 15th year;

(f) Of the assets to be purchased by the company, a balance of $3,886,821.10 to be paid for in equal quarterly instalments from June 30, 1947, to December 31, 1948, with interest on the unpaid balance at the rate of 3½ per cent per annum;

(g) The government to assign to Canadair Limited its licence agreement with the Douglas Aircraft Company;

(h) Canadair to agree to sell to the government and Trans-Canada Air Lines 34 RCAF type and 20 TCA type aircraft at negotiated fixed prices of $630,000 and $660,000 each respectively, sales tax extra.

Before entering into this agreement, the Department of Reconstruction and Supply satisfied itself that this company is able to provide capable and experienced management for the plant. Mr. H.Oliver West, an early officer of TCA, who had since become manager of operations of the Boeing plant at Seattle, and had been in responsible charge of the gigantic war production of Boeing, was designated as president of Canadair. The government is satisfied that the present Canadair Company is providing expert and experienced management for the Cartierville property.

Shortly after the war ended, requirements of both the RCAF and TCA were revised downward. The present requirement of RCAF is for 24 aeroplanes and for TCA, 20 aeroplanes. This reduction in requirements has of course added to the cost of each individual plane, although the cost of the planes is still well below the cost of purchasing and importing into Canada planes of competitive type.

Hansard, March 20, 1947.

Pressurization Problems

One troublesome system in developing the North Star in its early days was cabin pressurization. The US cabin supercharger first used proved costly and inefficient. TCA engineer Clayton H. Glenn was eventually able to offer

Air Canada's first M2 North Star appears on the left in this production line view at Canadair in 1947. The aircraft became CF-TFA and bore the name *Cartier* for a short time until TCA abandoned the practice of naming its aircraft. CF-TFA ended its days in Mexico in the early 1960s. (Canadair Ltd. via Graham Wragg)

a solution. In this section he describes the problem and how it was overcome:

From the onset it was apparent that the hydro-mechanical monster called the Douglas Cabin Supercharger was not going to be long-lived in the North Star. There were two of these in the aircraft driven by the outboard engines; and theoretically they could provide a constant flow of air to the cabin regardless of engine rpm, altitude or cabin pressure. They cost about $12,000 and were extremely complicated. I became quite familiar with them when I visited California in February 1947 with Don Long of Canadair, although I had nothing to do with their actual installation in the aircraft. There were no alternative units available in the United States and the only ones available overseas were Roots type superchargers used in the Tudors. These had no flow control and the air output varied with engine rpm and altitude and cabin pressure. It was necessary for the flight engineer to manually spill overboard any excess air to reduce cabin pressure fluctuations. Jim Bain thought that some consideration should be given to this type of unit as manufactured by Sir George Godfrey and Partners Ltd.

It was decided to send a team of Canadair and TCA people to England to investigate the Godfrey type supercharger. Don Long of Canadair and Bill Sansom and F.T. "Bun" Moore of TCA made the trip. In time they turned in a report to the effect that the Godfrey unit was an impossible proposition. They said that the units which would be of the proper size to provide the required amount of air would be too large to install in the small space available behind the outboard firewalls, and, in any event, there was no automatic airflow control system available to guarantee an even flow of air into the cabin. Because the North Star had no flight engineer it would be impossible to work a scheme as they did in the Tudor, even if the superchargers could be fitted into the airplane. The recommendation was unanimous: the Godfrey supercharger could not be fitted.

Nonetheless, after a trip to Boeing to a high altitude symposium in August, I began to see a possible solution to this concern. The symposium covered structural problems, particularly related to pressurization, as well as lubrication, electrical and engine problems.

During the lectures I became aware that there was a critical pressure ratio for a venturi tube through which air was passing, above which there would be no additional flow. At this pressure ratio, sonic speeds would be encountered in the throat of the venturi creating a block. I could see this as a means of designing a flow control for the Godfrey supercharger, and although I later used this principle in my reasoning with Jim Bain, that I could design a flow control mechanism, in actual fact I found out that it was impractical because with the cabin pressure which we were encountering in the North Star, critical pressure ratios were never reached.

After the Boeing symposium, I ap-

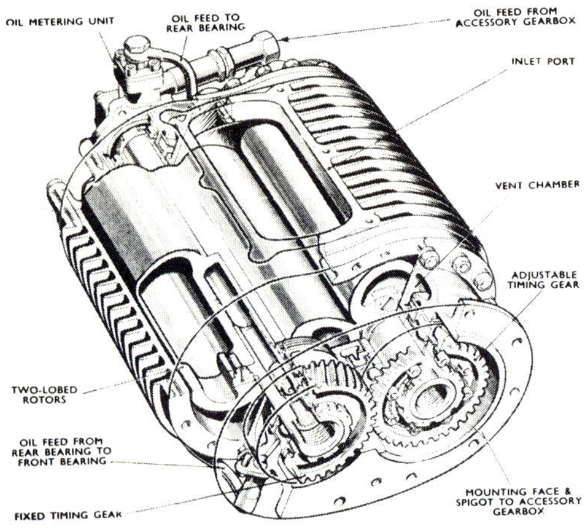

Cut-away of the Type 15 Godfrey blower as used by TCA. The essential simplicity of Roots-type supercharger is apparent. (Flight International)

proached Jim Bain and told him that in spite of the report of the group that had visited Britain, I felt that the Godfrey supercharger could be made to work in the North Star. He jumped at the possibility of having some alternative equipment, particularly if it was British built.

At Godfrey's, Chief Engineer Norm Anderson laid out a proposal for meeting the physical limitations of the supercharger and he suggested that we use a small one and run it at high speed rather than a large one at low speed as was more usual. Norm showed me on the test bed that we could achieve almost 13,000 rpm on a unit designed for 8900. From the air supply point of view it appeared that the small unit could do the job if restressed. There was then the question of driving the supercharger at the desired speed.

I took a set of supercharger and Merlin auxiliary gear box drawings then went up to Derby. I explained to Rolls-Royce what I wanted and in less than a week they had designed me a step-up gear box to fit between the main gear box and the supercharger. With a few modifications, the gear box worked. My problem now was one of airflow control. With this, I went off to Normalair at Yeovil, the only manufacturer at the time working on airflow control devices, but they were unable to offer much assistance.

I returned to Canada and convinced Jim Bain that I could do the job and he allocated $25,000 to do one aircraft. I ordered three superchargers from Godfrey's, three gear boxes from Rolls-Royce, six enormous air silencers from Burgess, a couple of inlet air filters and three AiResearch 3-inch butterfly valves. I then proceeded to design the automatic airflow system.

By this time, early 1948, I was back in Winnipeg. I discovered that my initial ideas would not work out and was in somewhat of a bind. However, I knew that the Douglas cabin supercharger was controlled by the airflow through a diffuser. I could use a venturi tube to produce a pressure differential which would be sensitive to airflow and operate electrical spill valves. I selected a design pressure such that the airflow instrumentation in the North Star cockpit, as used for the Douglas superchargers, could be used. I then had Sam Brown, one of our instrument men, redesign a compression ratio limiting switch to become a differential pressure control switch. This, connected to the venturi, and operating a slow-moving spill valve, was to be my airflow control. I designed into the original circuitry a mechanism so that the system would not hunt as well as a dampening device such that airflow transients would not activate the control mechanism.

I had a very skilled sheet metal worker, Bob Binney, build my venturi tubes and when they were connected to the airflow system on the test stand they produced the exact pressures expected. So far so good!

I could not see developing a cabin supercharger system without involving Bun Moore, who was responsible for air conditioning in TCA. Bun jumped at the opportunity. This turned out to be a good working relationship, though a letter from Engineering didn't help matters. It suggested we stop our crazy project before we blew up a cabin fuselage. I covered for this by installing a second emergency pressure relief valve in the fuselage.

The Douglas superchargers were of the centrifugal type with a wide variety of speed. The supercharger was driven by the auxiliary gear box, and through a system of fixed and variable displacement hydraulic pumps coupled with an epicyclic gear train, a constant air mass flow output could be obtained. My plan was to drive a fixed displacement supercharger, the Roots type with the rotors that look like interlocking figure eights, from the same auxiliary gear boxes, and any excess air that might be produced for aircraft pressurization and air conditioning needs would be spilled overboard. Most of the spilling would be at low altitude where there was lots of air. High up, almost all the output of the supercharger would go to the cabin.

The airflow would be as follows: in through a filter system to the superchargers, out past a branch pipe which would divert air through a spill valve, through acoustic silencers, then through an airflow measuring device which would control the spill valve, thence into the cabin. As designed, no surging of airflow was to be encountered, and the first installation worked out well.

At this time TCA and Rolls-Royce were investigating Merlin engine failures. It was decided to run one of the Merlins in flight at full takeoff power until it failed, or at least long enough to determine the impact of using high powers. Aircraft 215 was selected as the test vehicle. For five hours we suffered unbearably high noise, but the engine finally ran out of oil and was shut down. For us, the flight was a success. At design altitude and engines running at design speed the supercharger installation worked perfectly.

As aircraft 215 had been so modified, the DOT insisted that the person responsible for the new design fly with the aircraft for the first 1000 hours! Thus did Bun Moore and I fly aboard 215 for several months. One or the other of us had to be aboard as long as passengers were being carried.

One of the early flights was to the Caribbean. We hadn't realized how much noise the cabin superchargers made until an approach over Nassau early one morning. The scream of the unsilenced supercharger air being dumped awakened the whole town as we made our approach and we were greeted with quite a delegation of protesters. We now knew that this was a problem that had to be dealt with.

In August 1950 I went to the UK to assist BOAC in designing the installation of the Godfrey supercharger for their Argonauts. They took our drawings and modified them slightly to incorporate the British requirements, but on the whole, their installation was very like ours.

Operating costs for the Douglas superchargers had been as high as $4.80/hour whereas early estimates for the Godfrey replacements were that it would cost about $.65/hour, though ultimately we were able to

achieve overhaul costs as low as $.10/ hour. Over the life of the North Star fleet the overall estimated saving was $2,000,000, so in the end the project was considered a great success. For years it was cited as a fine example of just how the Engineering Branch could keep fleet operating costs down.

Clayton H. Glenn

Early Sales

With the North Star program speeding along, there was great interest in aircraft sales. The Canadian government of course was bankrolling the entire venture and the story goes that when TCA went to C.D. Howe asking for a fleet of 12 North Stars he retorted, "Twelve! That's not enough. We'll order 20 and even that won't do." As one TCA pilot put it, "Howe didn't spend time on any exhaustive studies. He just operated on intuition."

Jim Bain later remarked, "Howe's demand for 20 instead of 12 aircraft was a very good piece of judgement. Every one of the 20 went straight into service immediately as it was received. In fact, before the 15th through 20th were received, we were waiting for them."

Coupled with the order of 20 North Stars for TCA was a 1946 order for 24 North Stars for the RCAF. The RCAF aircraft were soon taking shape on the production line as they were fabricated largely from surplus components Franklin bought from the Douglas plant.

The prototype North Star nears completion at Canadair in June 1946. The initial batch of North Star airframes was built mainly of C-54 components, but the later pressurized ones included: DC-4 empennage, rear fuselage, wing fillet, flaps, ailerons and wing tips; C-54G wing centre and outer wing panel; and DC-6 fuselage centre (minus 80 inches), fuselage nose section, and landing gear. (Canadair Ltd. via J.T. Bain)

On July 14 Lilly and Norm Milne flew 17501; and on August 8 17503 took to the air with Al Lilly and Bill Anglin in the cockpit. The program was shaping up nicely, but there was not the activity in the sales and promotion department that some expected. Jim Bain had expressed his concern about the slack sales efforts in a letter of December 13, 1946, to TCA President H.J. Symington: "I do believe that we presently are unnecessarily losing ground by the delay in finalizing the future of Canadair Limited and I suggest that in the interests of the Canadian government and the costs per aircraft to TCA, some action be taken immediately to publicize the North Star . . . The cancellation of the Tudor II has opened up to us a market which we simply cannot afford to neglect for a moment."

Meanwhile TCA was responding to enquiries from around the world. In one case Bain replied to Qantas on May 27, 1946: "In TCA we satisfied ourselves that the Tudor I or II did not meet our requirements at all." At this time, of course, Bain held senior positions with both TCA and Canadair, so was doubly dedicated to the success of the North Star.

Diana Dudley, who headed up design of North Star interior accommodations and decor for TCA. She was later succeeded on the North Star by Janet Lowe, who recalls, "My work on the North Star was done in close cooperation with the TCA engineers, and liaison people from Canadair. We worked both in Montreal and Winnipeg. I had a number of buying trips to New York, mainly for fabric selection. Materials had to be coordinated in a pleasing design yet be functional, durable, easily cleaned, sound proof and fire proof. . . Lots of amusing things happened before the North Star got airborne, though at the time some of them really weren't so funny. I remember the panic everyone was in when the plane was weighed in — too heavy! Everything had to be cut back, even to such extremes as trimming back curtains and drapes." (Air Canada Archives)

TCA supplied many airlines, such as Sweden's Aerotransport, with copies of its detailed report favourably comparing the North Star with the Constellation in terms of operating costs.

The third North Star seen in a striking air-to-air pose on April 26, 1947. CF-TEL was used by Canadair as a demonstration aircraft before being turned over to TCA. It became the first North Star to be lost in an accident when it crash landed at Sydney, Nova Scotia, on August 12, 1948. (Canadair Ltd. via Graham Wragg)

May 17	Dorval-Gander	4:00 hours
	Gander-London	8:50
18	London-Hucknall	:35
19	Hucknall-London	:30
21	demo	1:20
	”	:55
22	”	1:20
	”	1:00
23	”	:35
27	London-Hurn	:35
	demo	:50
	Hurn-London	:30
	London-Hurn	:50
28	demo	2:05
	Hurn-London	:35
May 29	demo	2:15
	”	:50
	”	:15
	”	:20
30	”	:30
	London-Dunsfold	:15
	Dunsfold-London	:15
	demo	:15
31	”	1:00
	”	:25
June 1	”	:55
	”	:45
2	London-Paris	1:05
3	demo	:50
	”	:55
	”	:10
June 4	Paris-Geneva	1:25
	demo	:50
5	Geneva-Prague	1:50
	demo	:45
	Prague-Geneva	2:10
	Geneva-Amsterdam	1:55
6	demo	:35
	”	:25
	Amsterdam-Copenhagen	1:35
	demo	:40
	Copenhagen-Geneva	3:40
	Geneva-London	2:25

The itinerary of CF-TEL, as extracted from the log of Captain Al Lilly, noting its travels around Europe during its sales and demonstration tour there in 1947. The trip was concluded with no serious technical delays.

Those associated with the North Star European tour, June 1947. Left to right in the back row are Mike Springer, Smokey Harris, Bill Forbes, G. Laverdale, Louis Gauthier, and Lofty Lake. In the centre are Al Ramsay, Gord Stringer, Al Lilly, T.J. Emmert, Clem Christie, Lowell Dunsmore, Ron Baker and "Boo" Paschall. In the front are Ed Ainger, H.O. West and John Shallenberger. (via A.J. Lilly)

CF-TEL overflies the Rolls-Royce engine plant at Derby, England. (Rolls-Royce WN260 via Ron Baker)

Nonetheless only Canada and England were ever to purchase new North Star aircraft. What seems to have been the reason appears in a confidential memo from TCA president G.R. McGregor dated May 4, 1950. Included in it is this comment: "Canadair built the North Stars, and the following version of the aircraft known as the C-4 used by CPA and BOAC, under licence from the Douglas aircraft company. That licence contained a clause which prohibited the sale by Canadair of aircraft built under Douglas licence, except in the UK and Canada, and prohibited the resale of those aircraft by the original purchaser for a period of two years after the initial purchase, or until the aircraft had been flown for 5,000 hours." The reason for this limiting clause was Douglas' desire to restrict competition from the North Star for its DC-4 and (primarily) DC-6 airliners. Though there was keen interest in the North Star from airlines in many countries, some of which were ready to sign contracts with Canadair, the North Star was doomed to a small production run.

Meanwhile development and demonstration flying continued and in early 1947 CF-TEN-X was operating in the southern US where clear runway conditions prevailed for landing tests. Most of this work was done at Pinecastle, an abandoned US Navy air station near Orlando. Besides doing the job at hand, some of those on the project indulged in an unusual pastime during off hours. The airport was infested with snakes and the sport of picking off the slithering fauna with .22s flourished during the North Star's brief stay of February 12-26.

On May 2, 1947, the North Star was first demonstrated to observers from BOAC. Shortly after, CF-TEL was sent on a European tour, May 17-June 6, during which it visited England, France, Switzerland, Czechoslovakia, Holland and Denmark. Flying was intensive and on many nights the crew had as little as three or four hours' sleep.

Back in Canada the North Star was again the news, completing a record transcontinental flight on October 13, 1947. Herb Seagrim writes, "The flight was conducted for PR purposes and carried C.D. Howe, H.J. Symington and many others from government and industry. Crew of this flight were myself as captain, B.A. Rawson as first officer, and Phyllis Harding as stewardess. Our record on the flight from Vancouver to Montreal of 6 hours 52 minutes was not broken to my knowledge until the DC-8 era. Tail winds in the order of 100 knots at 20,000 feet contributed to the record of course. Also, as I recall, the Merlin engines were operated at METO (Maximum Except Take Off) power most of the time. My most vivid recollection is of the cabin noise level, what with short stacks and METO power. However, C.D. Howe was vastly pleased with the flight, notwithstanding the racket."

TCA engineer Clayton Glenn was also aboard the flight: "In October 1947 we had the first pressurized version of the North Star ready to fly. We wanted to put some time on it and a trip with C.D. Howe was organized. We flew to Toronto, then to the Lakehead, which was Howe's constituency, then on to Winnipeg. We overnighted there and proceeded next day to Saskatoon, Edmonton and Vancouver, where Howe did some electioneering." Glenn adds that on the return flight "Jim Bain, who was very anxious to set a speed record, would come into the cockpit and when nobody was looking would push the throttles forward to get a few more miles per hour out of the aircraft. After he had left, Jim Hall of Rolls-Royce would come forward and pull them back." A good time was had by all and the aircraft set down at Dorval with, among other things, a big load of B.C. salmon!

BOAC Buys Canadian

In the early postwar years Britain was striving to restore and expand its commercial aviation network. Initially aircraft such as the Lancastrian, York and Halton were placed on the long-haul routes, but these were strictly interim equipment. Modern commercial transports were sorely needed if the British airlines were to compete on an equal footing with their foreign competitors. Whether to develop such aircraft at home or to buy them abroad was a thorny question for several years in the late forties.

In its July 30, 1948, issue *The Aeroplane* covered Parliamentary proceedings about civil aviation, making the following key points:

- BOAC and BSAA still operating mainly stop-gap aircraft to their detriment.
- BOAC operating Constellations successfully with 8.3 hours/day utilization on the Atlantic.
- Stratocruiser considered the most promising airliner, though not to be forthcoming until 1949.
- Lord Pakenham very optimistic about the upcoming Hermes.
- Great enthusiasm over the Saunders Roe Princess flying boat to be used on South Atlantic.
- News from trials of Tudor II not encouraging, but Tudor IV to be used by BSAA.

The report went on to say: "Twenty-two DC-4M aircraft were to be ordered. The Canadair carried 40 passengers against 32 in the Tudor IV; that was a 25 per cent higher earning capacity. Over long distances with full tanks the relative advantage of the Canadair Four was somewhat greater. In general, the superiority of payload varied according to distance from between 3000-3500 lb. The Tudor IV's undercarriage had large single main wheels which would impose heavy loads on certain Empire airfield runways and constitute a danger to breaking up the surface. The Canadair Four, on the other hand, had four main wheels and a nosewheel, requiring much less strength of runway and having no such operating limitations. It had been estimated that at the very least the country would be saved £5,000,000 during the next five years by buying the Canadairs. . . .The 'Fly British'

C.D. Howe shakes hands with Captain Herb Seagrim on their arrival at Dorval following the record 6 hour 52 minute non-stop flight from Vancouver in CF-TFA. Others, from the left, are stewardess Phyllis Harding, H.J. Symington and Captain Barney Rawson. The date is October 13, 1947. (Air Canada Archives)

Backbone of BOAC's fleet in the immediate postwar years was the Avro York. Developed as a military transport during the war and a spin-off from the Lancaster, the York served BOAC well but was not competitive with modern postwar airliners like the Constellation and DC-4. *Malmesbury* was photographed at London on March 1, 1955. (Peter Keating)

policy remains the government's ultimate objective, but 'Fly Commonwealth' was a pretty good substitution."

From the first demonstrations of the North Star, BOAC was interested in the new airliner. On September 30, 1948, it placed an order for 22 aircraft. In BOAC service these were to be known as Argonauts. Attractive terms were presented to BOAC by C.D. Howe to help finalize the deal and the aircraft were offered on a speedy delivery basis, almost as if Canadair were still producing for wartime needs. The price for each Argonaut as quoted in the letter of intent of July 24, 1948, was Can. $594,000, excluding engines, but later on, in a letter dated September 28 that year, the price appeared as $671,107, with the total cost for the fleet including 15% spares coverage being $16,633,000.

Rather tongue-in-cheek, a senior Canadair management figure from the postwar era claimed recently that the Canadair-BOAC deal for Argonauts had a unique origin. He explained that at the time Britain had two needs: one, for a commercial airliner; two, to divest itself of a particular colony, namely Newfoundland. According to him, the terms of the deal were simple: Canada traded Britain 22 Argonauts for Newfoundland! The first BOAC Argonaut flew in March 1949, while the Canada-Newfoundland terms of union came into effect the following month. "Can't you put two and two together?" asked my source. Stranger bargains have been struck!

On March 6, 1949, Al Lilly and Bill Longhurst took the first BOAC Argonaut, G-ALHC, on its maiden flight. By now BOAC had its people in Montreal preparing for the arrival of the Argonaut fleet. The initial aircrew course was given by TCA in Montreal during February 1949, the course being both technical and flying. In June-July that year BOAC sent an Argonaut on a gruelling tour of the Middle East and Far East. This proving flight was the toughest to which the airline had ever subjected a new aircraft and the Argonaut came through with flying colours. There were no mechanical delays on the trip, which included operations from nearly 40 airports.

June 14	London-Rome	4:30 hours
	Rome-Cairo	5:35
15	Cairo-Bahrain	5:20
16	Bahrain-Bombay	6:45
17	Bombay-Hakimpet	2:15
	Hakimpet-Madras	2:00
	Madras-Colombo	2:00
18	Colombo-Singapore	7:40
19	Singapore-Hong Kong	6:35
20	local test	8:00
21	local test	8:00
23	Hong Kong-Okinawa	6:30
	Okinawa-Nagoya-Osaka-Tokyo	7:00
25	Tokyo-Hong Kong	10:30
26,27	Hong Kong-Manila	3:15
28	Manila-Hong Kong	3:15
28,29	Hong Kong-Hainan-Bangkok	5:20
30	Bangkok-Rangoon	1:40
	Rangoon-Akyab-Chittagong-Calcutta	3:50
	Calcutta-Allahabad-Delhi	4:20
July 1	Delhi-Karachi	3:15
3	Karachi-Tehran	5:50
4	Tehran-Kuwait	2:10
	Kuwait-Abadan	:20
	Abadan-Basra	:15
	Basra-Baghdad	1:25
5	Baghdad-Damascus	2:25
	Damascus-Nicosia	1:10
6	Nicosia-Lydda	1:10
	Lydda-Athens-Bari-Malta	8:35
7	Malta-Nice	3:10
	Nice-Marseille	:30
8	Marseille-Toulouse-Bordeaux-London	4:35

BOAC's intensive route proving flight with the Argonaut, June-July 1949. The crew included BOAC Senior Captains Farnsworth, Peter Cane, Nigel Pelly and Maurice Haddon, as well as Canadair's Al Lilly. The aircraft was G-ALHD, *Ajax*.

Ajax taking off from Kai Tak during its proving flight, June 1949. (BOAC via Peter Cane)

Flaps 40. A BOAC Argonaut just about to touch down at Cartierville after a shakedown flight. (Canadair Ltd. via Graham Wragg)

◁ Captains Peter Cane and Nigel Pelly aboard *Ajax*. (BOAC via Peter Cane)

Lower left

G-ALHG on a pre-delivery test flight near Montreal. Dubbed *Aurora*, it was later to operate BOAC's last Argonaut service on March 8, 1960. (Canadair Ltd.)

Captain Peter Cane uses the sextant aboard *Ajax*. (BOAC via Peter Cane)

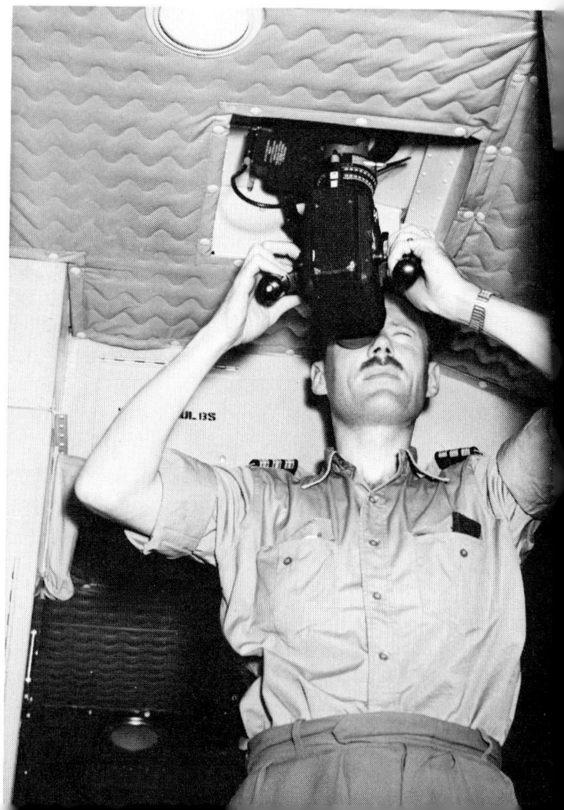

Trans-Canada Air Lines: Service Begins

To get its North Star operation in the air as quickly as possible, primarily on the competitive transatlantic routes, TCA borrowed six unpressurized DC-4M1 aircraft from the RCAF. They were delivered to TCA directly from Canadair, their C-54 cargo doors modified for passenger use. The first, CF-TEK, was delivered on November 19, 1946. The sixth aircraft was delivered on June 17, 1947. In early spring 1947, a transatlantic proving flight was made to the UK, and on April 15 TCA inaugurated North Star service with a flight from Dorval to Prestwick. The aircraft, CF-TEM, left Dorval at 4:00 p.m. local time amidst great fanfare. A large party of VIPs was at the airport, including A/V/M A.T. Cowley, Director of Civil Aviation; H. Oliver West of Canadair; H.J. Symington of TCA; and J.A.D. McCurdy. In 1909 McCurdy was the first to fly a heavier-than-air flying machine in Canada.

Two years later he made the first aeroplane flight between Florida and Cuba and on this occasion commented, "Although not disposed to make any wild calcuations . . . I am seriously impressed with the possibility of making intercontinental transocean flights by aeroplane." The send-off he witnessed at Dorval was a symbolic fulfillment of McCurdy's prophecy. Long-time aviation writer Bob Halford was to write years later of this same milestone, "Its introduction to commercial airline service marked the achievement of maturity by Canada's aircraft industry."

Crew aboard CF-TEM for the TCA inaugural were Captain G.W. "Gil" McLaren, Captain J.L. "Lindy" Rood (flight operations superintendent), F/O D.E. McInnis, Navigating Officer H.T. Warkentin, Radio Officer A.J. Blackwood, and Purser-Steward Douglas V. Miller. The 19 passengers included C.P. Edwards, Deputy Minister of Transport; Jim Hall of Rolls-

An exceptional air portrait of a North Star flying low over the Montreal waterfront in October 1947. (Air Canada Archives)

Civil Lancaster and M1 North Star operations as noted in the 1946 log book of Captain J.L. Rood of TCA just as the North Stars were undergoing their proving flights before entering service in April 1947. Other pilots aboard and the aircraft fleet numbers are noted.

▽

Nov. 23	Lanc 103	Dorval-Dorval	H.W. Holland	1:19 hours
25	Lanc 107	Dorval-Dorval	J.H. Jones	:45
30	DC-4 84	Dorval-Dorval	H.W. Seagrim	1:00
Dec. 4	DC-4 84	Dorval-Dorval	J.G. Haslett	3:48
6	DC-4 84	Dorval-Dorval	J.A. Wright	3:36
7	DC-4 84	Dorval-Dorval	J.G. Haslett J.A. Wright	3:18
11	Lanc 105	Dorval-Prestwick	J.H. Jones	12:43
13	Lanc 104	Prestwick-Dorval	J.H. Jones	14:11
17	DC-4 84	Dorval-Dorval	J.G. Haslett J.P. Laskoski B.J. Trerice	3:15
20	DC-4 84	Dorval-Dorval	H.W. Seagrim	1:15
23	DC-4 84	Dorval-Dorval	R.M. Smith	1:00

Personnel involved with TCA's North Star inaugural of April 15, 1947: J.A.D. McCurdy presents Captain Gil McLaren with a photo of the Silver Dart; Stewardess Marguerite Brezinski; and Captain McLaren in the cockpit. (Air Canada Archives)

Captain Gil McLaren during his bush flying days ▷ along the Mackenzie River. Many of TCA's early pilots came to the company from the bush and brought with them a wealth of practical flying experience. (via G.W. McLaren)

◁ One of the Mackenzie Air Service Norsemen which Gil McLaren and his mechanic Sherman Found flew in the Arctic. Both men became senior captains with TCA/Air Canada, logging thousands of North Star hours. Their "PhDs" were earned in the school of hard knocks north of Edmonton as far down as the Arctic coast. (via G.W. McLaren)

CF-TEM, which made TCA's first transatlantic scheduled flight, is seen here at London Airport in April 1948. (Flight International via Stephen Piercey)
▽

Radio operator's station aboard an M1 with R/O Harry Beardsell at work. (TCA via G.W. McLaren)

◁ TCA M1 crew Captain Gil McLaren, F/O E.E. Jones, Radio Operator Harry Beardsell and Navigator Bill Henderson. (TCA via G.W. McLaren)

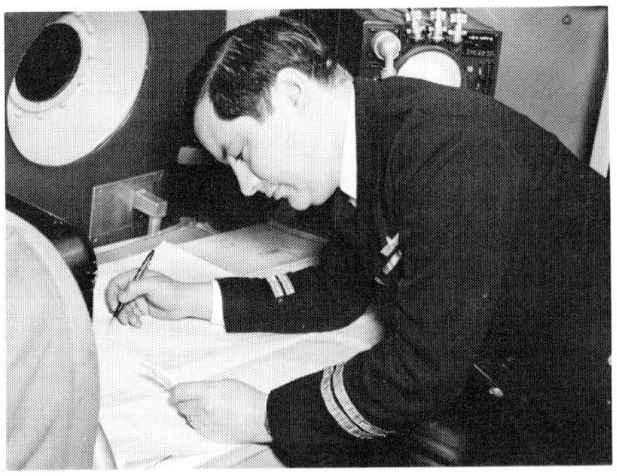

Bill Henderson at work at his navigator's table. (TCA via G.W. McLaren)

◁ "Now we are six" reads the personal caption on the back of this snap of a sharp-looking TCA crew and their M1 North Star. From the left are Captain Gil McLaren, F/O Ernie Jones, Navigating Officer W.K. Henderson, Stewardess Betty Hemingson, Radio Operator Harry Beardsell and Purser Alex Stebner. (via G.W. McLaren)

CF-TEP, later 17522, on the tarmac at Dorval with another M1, TCA Lanc CF-CMT and a BOAC "Connie." (Air Canada Archives)

Crew from a TCA transatlantic familiarization flight of 1948 at the dinner table in London. At the table in the foreground and looking clockwise from the bottom left are: Kent Davis, Bob Smith, Wendy Reid, Bill Graham, Ralph Leek, Colin Campbell and Joe White. (*Montreal Gazette*/Public Archives of Canada PA129312)

TCA North Star crew of Captain George Lothian, Purser Don Greavey, Stewardess Val Day, unknown, Captain Bob Bowker and Radio Officer Fred Dixon in 1947. (Air Canada Archives)

The met man briefs the TCA crew who include, from the left, going clockwise: unknown, Norm Ramsay, George Campbell, Herb Russell, Kent Davis, Ross Stevenson, Whitey Kelsey, Chuck Kenworthy, Bob Penrose, Bruce Bowden, Ralph Leek, unknown, Joe Crispin and Doug Holland. (*Montreal Gazette*/Public Archives of Canada PA129318)

Crew at TCA's Dorval ground school at a Merlin engine lecture. From the left in the back row are: Ralph Leek, Bob Penrose, Joe Crispin, Bruce Bowden, and Pat Roy, while in front are George Campbell, Kent Davis, Norm Ramsay, Herb Russell, Ross Stevenson, Whitey Kelsey, Chuck Kenworthy and Doug Holland. (*Montreal Gazette*/Public Archives of Canada PA129319)

The familiarization crew listen to Captain Bob Smith at this London session. Clockwise from, the bottom left are: Stu Ruddick, Gordy Kay, unknown, Ross Stevenson, Joe Crispin, Don Barnett, Norm Ramsay, Herb Russell and Doug Holland. This series of photos was taken by *Gazette* photographer Glay Sperling. (*Montreal Gazette*/Publc Archives of Canada PA129313)

Royce; G.R. McGregor, F.M. McGregor and J.T. Bain of TCA.

Other inaugurals soon followed the introduction of the pressurized DC-4M2 North Stars. The first of these, CF-TFA, had initially flown August 17, 1947. The type made its first trip overseas on September 4. On that flight, pilots Al Lilly, Bill Anglin and Ron Baker made a record transatlantic crossing in 10 hours 40 minutes. Inaugurals of note in 1948 were to include Montreal/Toronto-Bermuda (May 1), first domestic service, Vancouver-Calgary (May 1), Toronto-New York (May 14), Toronto-Chicago (June 1) and Montreal-Toronto-Nassau-Kingston-Port of Spain (December 2).

Captain Gil McLaren recalls a couple of adventures from the early TCA North Star days. The return part of the transatlantic inaugural was flown April 20. McLaren writes, "The fuel flow to No. 3 engine got very rich. We had to shut it down and turn back to Prestwick; but we were not authorized to dump fuel at that time and could not land overloaded. We proceeded to London but in the meantime two other engines began running rich. To make a long story short, we flew all night over London burning enough fuel to get down to our landing weight. We were up for 9 hours 9 minutes. The passengers later reminded us that that was about long enough to have gotten us across the ocean!"

On another occasion McLaren was flying from Barbados to Toronto with a planned refuelling stop in Bermuda: "We were advised that a hurricane was forecast for our arrival in Bermuda, so we requested clearance to divert to Nassau. We continued along, working radio with Pan American when their dispatcher advised us to land at San Juan, figuring that with the fuel figures we had given him we'd run dry before reaching Nassau. Well, as I had two extra pilots and two extra navigators on the flight, we worked the fuel on board over and over, estimating we had enough. Pan Am was just as sure of their calculations. We pushed on and made Nassau, then found out that Pan Am had been figuring with US gallons and we with Imperial. After this incident we used pounds of fuel instead of gallons."

One of TCA's early transatlantic North Star operations was the post-war refugee airlift between Europe and Canada. The Merlin engine, that had not much earlier droned over

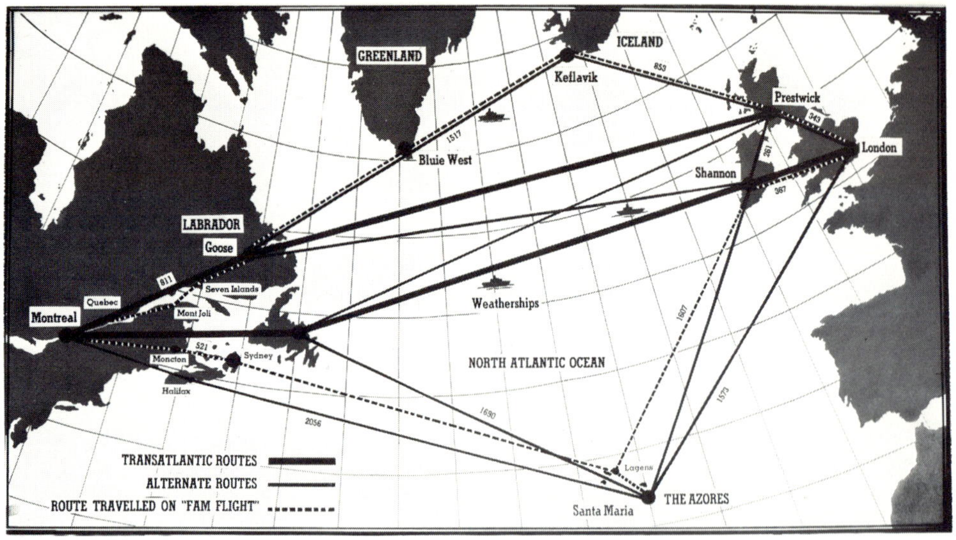

North Atlantic airways, including one followed on a TCA familiarization flight in 1948. (Public Archives of Canada PA129320)

Purser Frank Davyd, radio operator Fred Dixon, mechanic Mack Caverhill and Captain Sherman Found of TCA in front of Hitler's peace monument in Munich during the TCA airlift of refugees, January-March 1949. (Air Canada Archives)

Munich bringing Lancaster bombers to devastate the city, now brought North Stars there to collect loads of the homeless. The January to March 1949 shuttle between Prestwick and Munich carried over 1200 adult refugees and many children. In part, the TCA record of the airlift notes: "The operation was conducted with remarkable regularity; often aircraft were on the ground at Munich only a scant 25 minutes for loading. The North Stars flew high to avoid the Berlin air-lift traffic, their passage interrupted only by periodic radio contacts with control regions for enroute clearance. Although the continental airways control system is similar in many respects to our own, to Canadian ears the multitude of different languages over the airwaves sounded like the Tower of Babel."

First Mishap

On August 12, 1948, North Star CF-TEL, on the transatlantic service, was landing at Sydney, Nova Scotia. Just before it crossed the threshold, its wheels dug into a mound of earth left after some excavation and the landing gear was driven upwards, rupturing fuel tanks. A fire erupted as the aircraft slithered to a stop. Quick work by Purser Jack Triggs and Stewardess Rita Meyers got all 11 passengers out before the aircraft burned up. The rest of the crew got out safely as well, these being Captain Doug Holland, F/O Bob Penrose, Navigator Gil Evans and Radio Operator Bob Wright.

All-pressurized Fleet

TCA was anxious to replace its M1 North Stars as soon as possible, and its first M2 was delivered October 1, 1947. Delivery of the remaining 19 aircraft came very quickly, the last being turned over June 4, 1948. This permitted the speedy return to the RCAF of the remaining five M1s. It is of interest that the M1s were used by TCA without cost. The understanding with C.D. Howe over airline use of the M1 specified, "TCA will not be required to pay rental to RCAF for aircraft while in TCA possession,"

TCA's first three North Star M2s seen lined up at Malton on May 14, 1949. (Jack McNulty)

CF-TFD in an unusual gear-down photo over Montreal, October 20, 1947. (via Graham Wragg)

Mar. 13	Vancouver-Vancouver	H.W. Seagrim	Special	1:30 hours
14	Vancouver-Dorval	R.J. Baker H.W. Seagrim	Shakedown test	9:54
19	Dorval-Dorval	E.G. Dundas	Test	1:53
20	Dorval-Dorval	R.W. Chadwick	Test	2:29
24	Dorval-Goose Bay	R.M. Smith	Test	3:52
25	Goose Bay-Dorval	R.M. Smith	Test	5:23
31	Dorval-Dorval	H.W. Holland	Test	:35
Apr. 2	Dorval-Dorval	A.W. Pavey	Test	:34
3	Dorval-Toronto-Dorval	S. Albulet G.K. Edward	2051-3	2:58
8	Dorval-Toronto-Dorval	J.A. Wright	Courtesy	2:54
8	Dorval-Toronto-Dorval	J.A. Wright	Courtesy	3:08
9	Dorval-Ottawa	G.B. Lothian D.E. McInnis	303-9	:36
9	Ottawa-Dorval Dorval-Ottawa Ottawa-Dorval	G.B. Lothian D.E. McInnis	Special 89-9 Courtesy	2:42
9	Dorval-Winnipeg	G.W. McLaren K.G. Main	Test 89-9	5:49
14	Dorval-Dorval	R.M. Smith	Test	2:10
15	Dorval-Prestwick	G.W. McLaren D.E. McInnis	200-15	12:08
16	Prestwick-London	S.S. Albulet G.S. Quinn	200-15	1:38
19	London-Prestwick	S.S. Albulet G.S. Quinn		1:35
19	Prestwick-Dorval	S.S. Albulet G.S. Quinn	201-17	12:48
22	Dorval-Prestwick	J.R. Bowker C.S. Buchanan	200-22	12:47

Early North Star operations with TCA from the 1947 log book of Captain J.L. Rood. Other pilots on each trip are noted.

but it was TCA's responsibility to reconvert them to RCAF specifications. This job was completed at Winnipeg and Clayton Glenn recalls: "It was decided that the conversion . . . would be carried out in Winnipeg during the winter of 1948-1949. The airplanes were not identical from the point of view of equipment and it was necessary to install the military type radio, their standard instrument panel, as well as minor modifications to the engines. It was decided that the cabin interior would remain untouched, that is, it would be an airplane with a full complement of seats and galley, although we had to modify the front end to accommodate extra RCAF crew.

"I was assigned to the project as was Bill Bird, and along with John Sutherland and people such as Rennie Selkirk, we formed a team that would do the work. We took over the corner of No. 1 hangar and had three or four draughting tables and several desks. This became my headquarters for about a year. The five aircraft along with electronic equipment rolled in one at a time. My contact in Ottawa was A/V/M James and we often communicated over the supply situation.

"Because the aircraft had been built to carry passengers the wiring was not exactly compatible with that of the military version. At the time I defied

North Stars, DC-3s, a C-47 and an Anson adorn the ramp at Dorval (Air Canada Archives)

anyone in future years who had any trouble with the airplane's electrical or radio system to find out how it was put together. The modification was a patchwork arrangement, but everything seemed to work and the RCAF accepted it without too many complaints.

"To make the project easier for us, the RCAF flew in one of their standard C-54GMs to be used as a sample. It was parked in the corner of the hangar for several months and Bun Moore and I used it as sleeping quarters on occasion."

TCA in 1948

The TCA annual report for 1948 shows the rapid growth of its operations, reflecting the impact of the new fleets of DC-3s and North Stars. The report lists the following achievements:

- Carried 27% more passengers on all services than in 1947.
- Flew 106% more ton miles of express and air cargo.
- Began the first nationwide transport of unsurcharged first-class mail by air in any American country.
- Introduced 20 North Star aircraft, which increased carrying capacity by 60% and established new standards of travel comfort.
- Reduced transcontinental flight time between Montreal and Vancouver to 13 hours.
- Began air service between Canada and Bermuda and between Canada and the British West Indies.
- Brought 6000 new Canadian citizens from the British Isles and Europe in one of the largest mass movements ever undertaken by civil aviation.
- Carried 9000 persons and 600,000 pounds of essential cargo across the Rockies in emergency service during the British Columbia flood crisis.

Domestically, 532,555 revenue passengers were carried on North American routes, up 24% from 1947. The average distance flown per domestic passenger was 469 miles. For the year, mail carried was up 86% and revenue from it was up 22%. On the Atlantic, Ontario and Dominion government contracts accounted for 175 westbound charters with immigrants. The report also noted, "Travel and shipping time between Canada and the West Indies has been reduced from seven days by ship to six hours from Toronto to the Bahamas, nine and a half hours from Toronto to Jamaica and fifteen hours from Toronto to Trinidad." This,

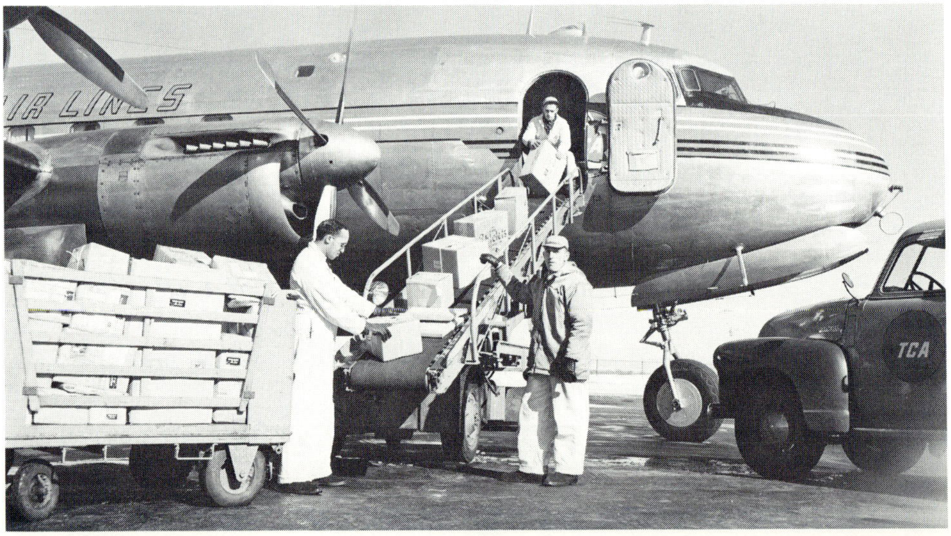

Showing its open stacks, a North Star takes on freight at Dorval in 1949. (Air Canada Archives)

An M2 North Star photographed over Dorval in the late 1940s. The open stacks of its Merlins show that the photo predates installation of crossover exhausts. (Air Canada Archives)

thanks to the new North Stars.

Overseas traffic included 32,821 passengers for 1948, 23,429 of them on scheduled flights. Bermuda and Caribbean passengers totalled 2951. Most Atlantic passengers were westbound. Only during the summer months was there any heavy travel from Canada to the UK. Under "Property and Equipment" on the Atlantic the report states, "Nine pressurized North Star aircraft were assigned to Atlantic and Bermuda service on June 1. These, together with the six unpressurized North Stars that had been used since April 1947, comprised the overseas fleet during the busy summer months. The latter aircraft were retired from TCA service during the autumn. Two additional North Stars were later assigned to provide for the Caribbean service."

Major overseas routes as noted in the report were:

Montreal—London
via Sydney, Shannon
and Glasgow.......... 3540 miles
Montreal—Bermuda
via Toronto 1462
Montreal—Port of Spain
via Toronto, Nassau
and Kingston.......... 3273

In closing, it was reported: "The Company was well satisfied with the performance of its new fleet of four-engined aircraft. The Canadian-built North Stars were subject to many of the usual difficulties that accompany the entry of a new type into commercial service, but they also demonstrated that they are among the finest aircraft flying today."

TCA North Star Inaugurals

April 15, 1947	Transatlantic Service, M1
Oct. 13, 1947	Record flight, Vancouver-Montreal, 6 hours 52 minutes
May 1, 1948	Montreal/Toronto-Bermuda, M1
May 1, 1948	First M2 introduced, Vancouver-Calgary, all subsequent inaugurals being of M2s
May 14, 1948	Toronto-New York
May 28, 1948	Arrival in Vancouver from Montreal of first official North Star flight
June 1, 1948	Atlantic and Bermuda
June 1, 1948	Toronto-Chicago
July 1, 1948	Non-stop Montreal-Winnipeg (all up mail), to August 23, 1948
Dec. 2, 1948	Montreal-Toronto-Nassau-Kingston-Port of Spain
May 1, 1949	Saskatoon, Edmonton on non-stop service Edmonton-Vancouver
Dec. 2, 1949	Barbados added on Caribbean service
April 1, 1950	Montreal-New York
April 2, 1950	Tampa added on Caribbean service
April 1, 1951	Paris added on Atlantic service
April 27, 1952	Montreal-Moncton-Sydney-St. John's
Nov. 5, 1952	Dusseldorf added on Atlantic
April 26, 1953	Regina, Lethbridge added on transcontinental service
Nov. 1, 1953	Toronto-Cleveland
Jan. 16, 1954	Tampa-Mexico (to November 1, 1955)
Feb. 1, 1954	Saint John, Halifax, Stephenville, Gander added
May 2, 1955	All cargo transcontinental service
Nov. 1, 1955	Montreal-Quebec City
June 15, 1956	Boston

TRANS-ATLANTIC SERVICE

Light Face Figures Indicate A.M.
Dark Face Figures Indicate P.M.

TABLE 1 READ DOWN EASTBOUND

Flight No. →			200
MONTREAL	(EST)	Lv	4.00
•SYDNEY	(AST)	Ar	8.20
SYDNEY	"	Lv	9.20
GLASGOW	(GMT)	Ar	11.30
GLASGOW	"	Lv	12.30
LONDON	"	Ar	2.30

TABLE 2 READ DOWN WESTBOUND

Flight No. →			201
LONDON	(GMT)	Lv	7.30
GLASGOW	"	Ar	9.30
GLASGOW	"	Lv	11.00
•SYDNEY	(AST)	Ar	7.25
SYDNEY	"	Lv	8.25
MONTREAL	(EST)	Ar	11.00

•Flights will stop at Sydney for passenger traffic and cargo, subject to availability of space and providing 24 hours advance notice is received by the Company. Stops will be made at Goose Bay, Labrador, on approximately the 1st and 15th of each month.

TCA's exclusive LONDON VISITORS' BUREAU is at your SERVICE—Hotel Accomodation, Sightseeing, Shopping, Amusements, Recreation, etc.

PAGE ONE

TCA's transatlantic North Star service as of May 1, 1947.

MONTREAL - TORONTO - WINNIPEG - REGINA - EDMONTON - CALGARY - VANCOUVER

Read Down / Read Up

41	31	★1	343	43	★303	★3	45	★5		TABLE 2	★6	30	★300	40	★4	32	42	★2	44
7.00	8.50	11.40	12.55	.....	5.00	5.45	9.00	10.30	EST	Lv MONTREAL .... Ar	10.15	11.55	2.20	.....	4.45	7.45	9.30	10.45	11.55
7.50	9.40		1.45	.....	—	—	9.50		"	Ar OTTAWA Lv	↑	11.10	—	.....	—	7.00	8.45	↑	11.10
8.00	9.50		1.55	4.55	—	—	10.00		"	Lv OTTAWA .... Ar		11.00	—	2.20	—	6.50	8.35		11.00
9.45	11.35	1.30	3.40	6.40	6.50	7.35	11.45	12.20	EST	Ar TORONTO Lv	8.35	9.30	12.40	12.50	3.05	5.20	7.05	9.05	9.30
	51		53										52						50
.....	8.00	1.50	4.00	.....	.....	7.55	.....	12.40	EST	Lv TORONTO .... Ar	8.15	.....	11.25	.....	2.45	.....	.....	8.45	9.15
.....	10.25		6.25	.....	.....	—	.....		"	Lv SAULT ST. MARIE ... Lv		.....	9.30	.....	—	.....	.....		7.20
.....	12.55		Ar 8.40	.....	.....	Dly Ex Mon Tue	.....		"	Lv FT. WM.-PT. ARTHUR Lv		.....	7.30	.....	Dly Ex Tue Wed	.....	.....		5.20
.....	—		Dly Ex Tue Wed	.....	.....	—	.....		CST	DULUTH (See Table 8)		.....	Dly Ex Wed Thu	.....	—	.....	.....		—
.....	2.40	5.35		.....	.....	11.40	.....	4.25	"	Ar WINNIPEG Lv	3.05	.....		.....	9.35	.....		3.35	1.40
			71		65								66				72		
.....			6.15		5.00				CST	Lv WINNIPEG .... Ar			11.59	.....	.....	.....	3.05		
.....			—		6.00	.....			MST	Ar YORKTON (See Table 9) Lv			9.20	.....	.....	.....	—		
.....			8.25		.....	.....			"	Ar SASKATOON Lv			.....	.....	.....	.....	11.25		
.....			8.40		.....	.....			"	Lv SASKATOON ...... Ar			.....	.....	.....	.....	11.10		
.....			10.55		.....	.....			"	Ar EDMONTON Lv			.....	.....	.....	.....	9.20		
				63	61		65					66				62			64
.....		5.55		6.05	.....	.....	5.00	4.45	CST	Lv WINNIPEG .... Ar	2.45	11.59	.....	.....	.....	.....	.....	3.15	10.55
.....	.....		.....	7.15	.....	.....	—		"	Lv BRANDON (See Table 9) Lv		—	.....	.....	.....	.....	.....		10.00
.....	.....		.....	8.00	.....	.....	6.55		MST	Ar REGINA Lv		8.25	.....	.....	.....	.....	.....		7.30
.....	.....		.....	.....	4.45	.....	.....		"	Lv REGINA .... Ar		.....	.....	.....	.....	2.30	.....		.....
.....	.....		.....	.....	5.55	.....	.....		"	Lv SWIFT CURRENT Lv		.....	.....	.....	.....	1.30	.....		.....
.....	.....		.....	.....	7.05	.....	.....		"	Lv MEDICINE HAT Lv		.....	.....	.....	.....	12.25	.....		.....
.....	.....	8.20			7.55	.....		7.10	"	Ar LETHBRIDGE Lv	10.45	.....	.....	.....	.....	11.35		11.15	.....
83	85		87	89			81					88		86	84		82		80
11.25	3.45		8.50	10.55	.....	.....	7.40		MST	Lv LETHBRIDGE .... Ar		10.10	.....	8.05	3.10	.....	10.40		7.00
12.25	4.45		9.50	11.55	.....	.....	8.40		"	Ar CALGARY Lv		9.10	.....	7.05	2.10	.....	9.40		6.00
12.35	4.55		.....	12.05	.....	.....	8.50		"	Lv CALGARY .... Ar		9.00	.....	6.55	2.00	.....	9.30		.....
1.50	6.10		.....	1.20	.....	.....	10.05		"	Ar EDMONTON Lv		7.45	.....	5.40	12.45	.....	8.15		.....
.....	.....	8.40	.....	.....	.....	.....	.....	7.30	MST	Lv LETHBRIDGE .... Ar	10.25	.....	.....	.....	.....	.....	.....	10.55	.....
.....	.....	10.05	.....	.....	.....	.....	.....	8.55	PST	Ar VANCOUVER Lv	7.15	.....	.....	.....	.....	.....	.....	7.45	.....

★ Flights 1, 2, 3, 4, 5, 6, 300, 303 — 40 Passenger North Star Skyliners ★

BERMUDA *NON-STOP* SERVICE

TORONTO - MONTREAL - BERMUDA

TABLE 10

READ DOWN — CANADA - BERMUDA

Flight Numbers →		Saturday ★600	Sunday ★602
Lv. MONTREAL	EST	10.30	.....
Lv. TORONTO	EST	—	10.15
Ar. BERMUDA	AST	4.20	4.30

BERMUDA - CANADA — READ DOWN

Flight Numbers →		Sunday ★601	Saturday ★603
Lv. BERMUDA	AST	6.00	6.00
Ar. TORONTO	EST	—	10.15
Ar. MONTREAL	EST	9.50	.....

Domestic and Bermuda North Star service published in TCA's June 1, 1948, timetable.

One of TCA's DC-3s over a Prairie landscape. CF-TDS was rebuilt for TCA by Canadair in 1946. The DC-3 replaced the airline's aging fleet of Lockheeds. (Air Canada Archives)

Merlin Problems

The initial experiences with the Merlin in service seemed to confirm the expectations of the doubters. Among other problems, it was a gas guzzler and very noisy. The North Star was no sooner in the hands of TCA than it began suffering from engine shut-downs. Though some of these were precautionary, the result of false fire warnings, others were mechanical. On the Toronto-Winnipeg run one month, 20 shut-downs occurred. With experience, power charts were modified and, instead of cruising at 1100 bhp, pilots set the throttles at 950 bhp and most of the serious trouble stopped.

These two RCAF North Star incidents typify the Merlin saga. One involved 17515 on September 9, 1950. W/C C.H. Mussells and F/L K.T. Wallace were piloting the aircraft north from Tacoma. The incident report notes: "In flight, No.4 coolant loss indicator commenced to flicker and the engine was feathered. Approximately three minutes later No.1 coolant loss indicator commenced to flicker and this engine was also feathered. Gas was jettisoned and a normal two-engine landing was made at Sandspit. No.4 engine leak was repaired and the aircraft returned to McChord on three engines." Another incident from the litany of North Star engine failures occurred October 2, 1950. Aircraft 17510, piloted by S/L H.A. Morrison and F/O M.D. Broadfoot, was involved: "During flight, coolant fluid in No.2 engine was lost. Engine was feathered and aircraft returned and landed at Wake Island." The phrase "successful three engine landing carried out" occurs repeatedly in the RCAF's flight safety records for the North Star, though as the aircraft settled into squadron service these incidents dropped off.

BOAC too was having teething problems with its Argonauts. One of its pilots, Captain Leslie Gosling, recalls: "Out of my first nine crossings of the South Atlantic on Argonauts I arrived four times on three engines. These were teething troubles with a new type and mostly due to faulty warning lights, but in one instance there was an injection pump failure."

Other early problems are described

Merlin maintenance in the TCA shops. Tommy Daniels and Joe Fay talk shop by an inverted engine, while Brian Wood adjusts a part on another engine. (Air Canada Archives)

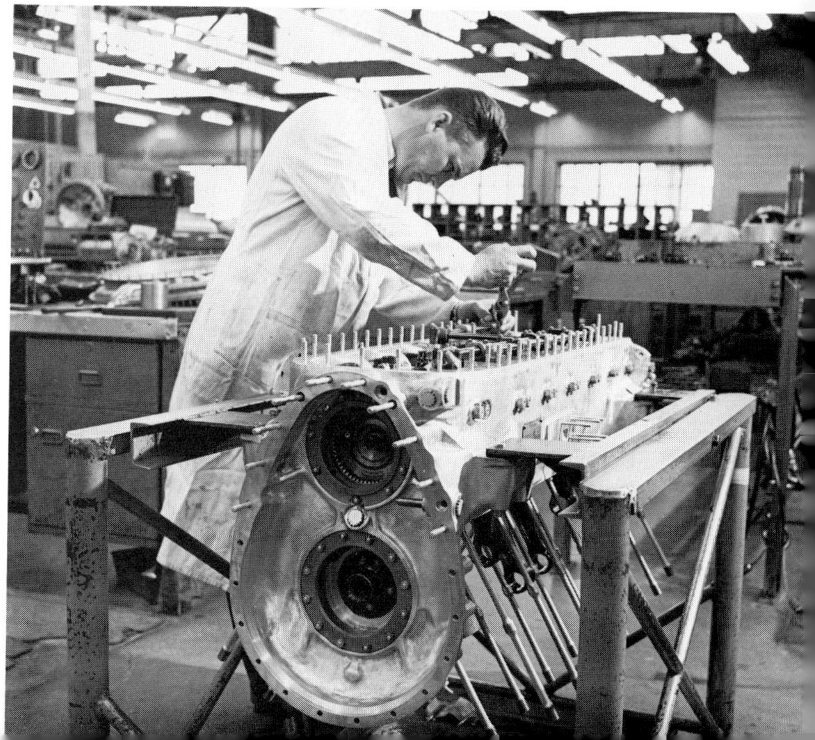

"Oscar 2," TCA's Merlin test rig at Dorval in 1954. Ken Dawson, A.J. Leblanc and Jules Fiorini of TCA were responsible for this interesting concoction. So far as we know, Oscar never got off the ground. (Air Canada Archives)

by TCA's engineering test pilot, Ron Baker. He describes the problem of surging engines: "This wasn't apparent until we started flying at higher altitudes with the pressurized M2s. It was unpredictable. At certain times, one or all four engines would start wild swings in rpm and manifold pressure which were difficult to stop. I did a lot of flight testing on it, with engineers from TCA and Rolls-Royce. We had difficulty duplicating the problem but realized it had something to do with compressor design, along with speed and atmospheric temperature and pressure. We found that we could stop it best by applying hot air to the intake and/or pulling the throttle back. I had a smart performance engineer in my group, Ian Macdonald, who went into it quite deeply. We sent him to England to talk with Rolls-Royce, Frank Whittle and others. As a result, Rolls-Royce agreed that the problem was in compressor design, something we simply learned to live with by establishing restrictive limits of operation."

Baker goes on to explain: "We also had some cylinder problems and on one test I was to run an engine in flight at takeoff power, throttle wide open. Engines are normally rated for a maximum of five minutes at this setting. We ran this engine for three hours, fifteen minutes, until the oil level was getting low and we had to reduce power. I remember being on pins and needles all this time as I was afraid any moment that the engine would fly to pieces. Nothing happened and on tear-down the engine was found to be in perfect condition. I don't think any other engine at that time could have survived such a rigorous test!

"There was also a certain amount of propeller trouble, including failures. After one incident, we took a North Star to Hartford where Hamilton Standard had their operation. The propellers were instrumented with strain gauges and tests were carried out with the aircraft at various angles to the wind (furnished by a Northrop P-61 Black Widow fighter aircraft). As a result, there was a propeller redesign along with some restrictions in operation on the ground."

Because of these teething troubles, the North Star came in for a hard time from the press. Articles appeared regularly reporting engine failures and bluntly stating that the North Star was unsafe. The paranoia reached a peak in May 1949. Even when a North Star was seen taxiing on two engines, not an uncommon practice, this was reported in the press as another case of engine failure! In this vein, veteran TCA pilot Herb Seagrim remembers: "Louis St. Laurent, Prime Minister at that time, wanted to fly from Montreal to St. John's, Newfoundland, using a chartered North Star to accommodate his staff plus a number of the press, some 40 passengers in all. At the time the North Star was having quite a few problems and was the target of attack by the Conservatives, headed by George Drew, in the age-old business of trying to discredit the Liberals who had sponsored the airplane. The media, of course, were having a ball publishing all the alleged failings of airplane and engines.

"I was captain of this flight. We took off but before reaching cruising altitude one of the engines started to surge rather badly. This was a problem having to do with supercharger design and under certain atmospheric conditions the offending engine would have to be shut down. In this instance, however, with all these newspaper vultures waiting for something to happen, I was not about to give them any satisfaction by returning to Montreal, shutting the engine down or even announcing any kind of trouble. Instead I throttled the troublesome engine to a fast idle and carried on to St. John's, effectively on three engines. With the propeller still turning, no one except the crew knew what was going on and in no time the newspaper gang was full of free booze and food. We were 25 minutes late, but I attributed this to headwinds. Safety was not a factor on the flight as the faulty engine could have been used to some degree if necessary."

Merlin Warranty

On TCA use of the Merlin engine, Jim Bain has written, "We had had experience with Merlins in the Lancasters but that operation had not been the equivalent of what we anticipated with the North Star, and I had real difficulty in figuring out the basis for a warranty. In one discussion with Jim Pearson of Rolls-Royce he remarked that he was sure that TCA would not be sorry for choosing Merlins, so I asked him, 'Why not make that the basis for a contact warranty?' and he agreed. The engine warranty which was approved by Lord Hives, chairman of Rolls-Royce, was written on a single sheet of paper and became known as the 'Won't Be Sorry' contract. And we weren't—Rolls-Royce lived up to it completely."

The exact wording of the TCA/Rolls-Royce "Won't be sorry" agreement, as originated by J.T. Bain of TCA and J.D. Pearson of Rolls-Royce, reads:

Under the terms of the purchase contract between Trans-Canada Air Lines and Rolls-Royce Limited specifically outlined in Mr. C.D. Howe's letter of July 28th, 1944, Rolls-Royce Limited agree to give Trans-Canada Air Lines a special warranty on power plants and engines, which will protect them against excessive cost of operation. Such warranty will be in addition to the normal warranty given by aero engine manufacturers, covering minimum life expectancy of engine components, and detailed as in Appendix I.

Definition of Power Plant and Engine: For the purpose of this warranty the power plant and engine is defined as the demountable part of the power egg in front of the firewall, excluding the propeller, accessory gear box and all other items not supplied by Rolls-Royce Limited.

The special warranty will cover the engine and power plant for a life of 6000 hours or a period of three years, from date of delivery, whichever is the least.

Rolls-Royce Limited under the terms of the special warranty, agree to compensate TCA with the supply of free replacement parts, should the operational cost of the engine and power plants become excessive for any reason associated with their design or manufacture.

As it has not been found possible to determine what the "normal" operating costs of the engines and power plants should be, the point at which the operational costs are considered to have become excessive will be left for discussion and agreement between the Rolls-Royce and TCA engineers. Should such discussion have to take place then the necessity of Rolls-Royce having to maintain goodwill would be a guarantee for TCA.

Montreal, Que.,
November 3rd, 1945.

Of engine problems Bain explains, "Certainly we had our share of teething troubles but we only had one really difficult problem with the Merlin. That was the excessive wear of the pistons, rings and cylinder linings. This problem proved to be one of lubrication and was cured by switching to a compounded oil supplied by Texas Oil Co. which came up to Dorval to research the situation. The cost of correcting the problem was high, but was completely carried by Rolls-Royce which stood firmly behind their 'Won't Be Sorry' warranty agreement."

Merlin Efficiency

In an April 20, 1948, technical paper, Rolls-Royce reported on the record of the Merlin engine in the first year of North Star operations. It notes that "during this period TCA effected 957 single crossings or 2.6 crossings each day of the year (April 15, 1947-April 15, 1948). The average utilization per aircraft was 7 hours 42 minutes, and it is interesting to note from the attached tables that the individual utilization of the six M1 aircraft is within 30 minutes of each other. . . .

"Analysis shows that TCA 184 and TCA 185 flew 1187 and 1093 hours respectively without a single engine failure.

"The engine performance tables indicate that two engines completed 2000 hours without failure and seven completed 1750 likewise.

"It will be seen that the average Time Expiry life was 654 hours and the average period of service for all engines, taking into account failures and repairs, was 557 hours. . . .

"The average North Star M1 operating conditions over the past year have been to fly at constant power at 7000 to 9000 ft. in M.S. gear both east and west at 2300 rpm plus 40 in. Hg boost, charge temperature 47°C, fuel flow 380-400 lbs/hr, 875 to 900 BHP at an IAS of 210 mph approximately. Plus 70 in. Hg takeoff boost being used always.

"The time taken for an east and westbound crossing is approximately 15 hours in an M1 aircraft, whilst the M2 (pressurized), flying at 19,000 ft. have done the trip in 13 hours approximately."

Noisy North Star

The unacceptably high noise level in the North Star was evident from the first few test flights in 1946. Now that passengers were travelling on the airplane, something had to be done to rectify the problem. That the main source of noise would be the inboard exhaust stacks had been foreseen in the design stage and because of this there was a warranty clause in the TCA/Canadair contract making noise suppression a joint responsibility. Canadair was obligated to develop and pay half the cost of any modification needed to reduce cabin noise to an acceptable comfort level. It has been reported that earlier on it had been in the Canadair specification that noise level would not exceed that of US counterparts like the DC-4 or Constellation, but when Electric Boat took over Canadair this detail was deleted.

Initially Canadair spent about a year and a half trying to develop a quieter exhaust system but it finally dropped this pursuit, claiming the responsibility lay with Rolls-Royce. Meanwhile, Rolls-Royce was working on the crossover scheme adapted from Canadair's, and this was eventually proven and successfully put to use on the Argonaut. TCA was not completely happy with this sytem and set to work on its own. The project was centred at Winnipeg and Canadair and was under the direction of Merlin "Mac" MacLeod.

MacLeod was known as a technical genius from boyhood and during the 1930s he worked as a mechanic with Canadian Airways. That period of his life is replete with stories of his ingenuity. He joined TCA in 1937 and was responsible for many innovations which greatly improved existing aircraft performance.

In 1947 MacLeod began work on an exhaust gas crossover manifold for use on the North Star. Working in the Canadair shops, he tinkered, hammered, bent and cogitated on a "hands on" basis. This drove those around him to distraction as he would come in each morning with something new to try out. To some at Canadair, this seemed inefficient and costly. Even so, the system took shape and was proven on the North Star. Simply put, the final product was some piping that connected both sets of exhaust outlets to one crossover manifold that would dump exhaust gases and noise outboard of the engines, away from the cabin.

Success didn't come easily. The main problem was that the convergence of two streams of hot, fast-moving gases was more than the manifold could bear and it would crack. MacLeod's answer was to place the exhaust dump at the point where the two streams converged. This permitted the gas to escape outboard of the engines without sacrificing ejector

Inboard views of the TCA crossover exhaust system, cowled and uncowled. (Air Canada Archives, K.M. Molson Collection)

thrust. Noise was reduced by 6 decibels overall and by 15 decibels in the speech range. Passengers could now hear themselves think, though the North Star continued to be rather noisy compared with other airliners.

TCA test pilot Ron Baker notes of the development of the MacLeod system: "The airborne noise tests of the system indicated a big improvement as far as cabin noise was concerned. But as any loss in engine performance was a key factor, we had to do a complete certification and cruise/climb performance tests. The certification involved takeoff and en route climb performance on four and three engines. We did this at Montreal, Winnipeg and Calgary to get a range of altitude for takeoff and climb out."

On one flight, the crossover on No.3 engine burned through. Everyone was disappointed and the flight returned to Winnipeg. Nobody else wanted to deliver the news to Mac MacLeod, so Baker was elected as he was the one who had done the damage. He got MacLeod on the phone and said, "Mac, we have a bit of trouble."

"I know, the exhaust broke," replied MacLeod. "I thought it might happen. It's too flat on top and it's panting. I'll be right out." The next exhaust MacLeod delivered had more curvature at the top and worked much better. In time the system was used on all TCA North Stars with great success. Ron Baker sums up the development of the crossover this way: "It was a remarkable undertaking for a small organization, when major engineering companies wouldn't touch it."

Between January and August 1952 the MacLeod system was in continuous use on a TCA North Star plying the Montreal-New York route. When finally removed for inspection no serious defects could be found, even though the shortest time for any of the four crossovers on the aircraft was 1300 hours. Meanwhile a tender was put out by TCA for the manufacture of 120 sets of crossover manifolds. Lowest bidder was MacDonald Brothers of Winnipeg; second lowest was Canadair, at about half a million dollars. On the basis of quickest assured delivery, Canadair was given the contract in January 1952. As it turned out, delivery was slow and much haranguing followed between TCA and Canadair. The first set of crossovers was delivered in April 1953 and by year's end half the fleet was equipped. Finally in March 1954 the contract was complete.

The TCA crossover in practice was

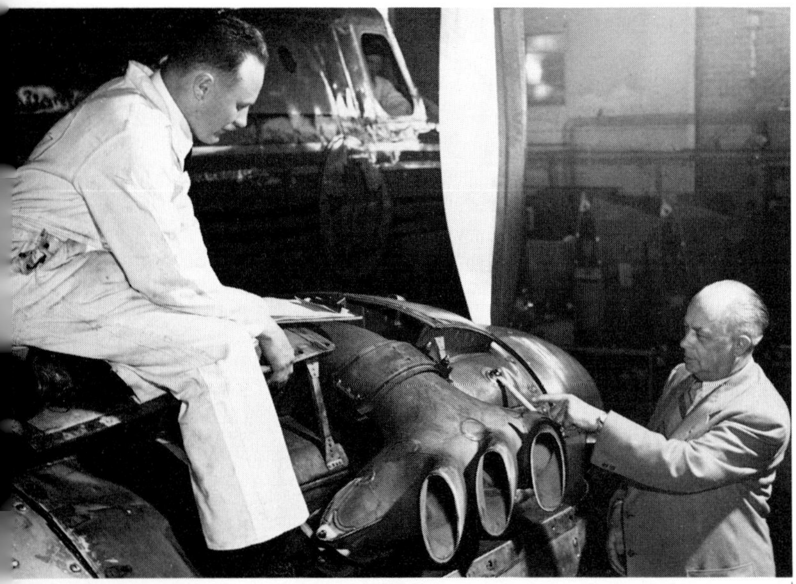

TCA crew chief Gordon Phillips and M.W. "Mac" MacLeod admire the "MacLeod Crossover." (Air Canada Archives)

Outboard view, cowled. (Air Canada Archives)

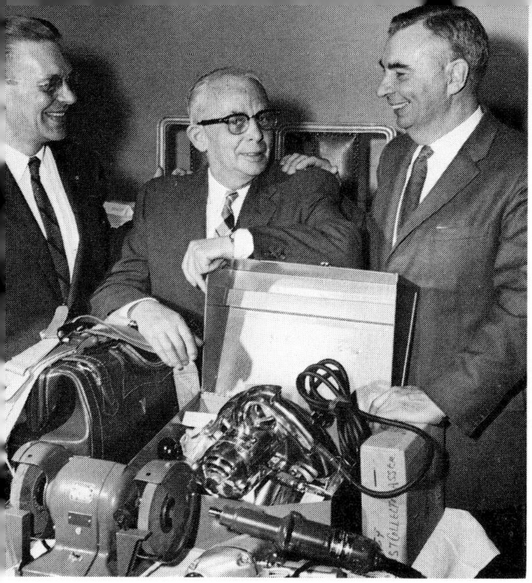

Jim Bain on the left and Jack Dyment on the right chat with "Mac" MacLeod on the occasion of his retirement from TCA in February 1959. (Air Canada Archives)

not a vast improvement to the ear. This irony was made embarrassingly clear on the press flight set up to introduce the system. Jack Dyment was forced to shout his announcement to the assembled reporters that the crossovers had solved the North Star's noise problem! However Captain Ron Baker, who was flying the airplane, remembers, "I flew at low power to keep the noise to a minimum, bearing in mind that the press had already made up its mind that the flight would be noisy. If Jack had to speak up it was likely to make himself heard above the din of the noisy media people!"

TCA President Gordon R. McGregor aboard a North Star with Captain Lowell Dunsmore. (Air Canada Archives)

On the Argonaut, beyond making the aircraft a bit quieter, the BOAC system, which evolved from Canadair's research, actually gave an added bonus: overall aircraft performance was increased by 4%. Captain Leslie Gosling of BOAC recalls the Argonaut's crossovers: "With their stub exhausts the Argonauts were initially very noisy. Then crossover exhausts were fitted to the inboard manifolds of the inboard engines, and this helped a little. Later all four engines were fitted, which made the aircraft acceptably quiet. *Atalanta* was the first to be fitted with crossovers on all engines, it being used on Royal flights."

In his book *Adolescence of an Airline*, Gordon R. McGregor, President of TCA between 1948 and 1968, presents his own account of the Merlin engine and how TCA came to terms with it:

Since each of its four engines had six open exhaust stacks blasting directly at the cabin windows, and since sound insulation was a poorly developed art at the time, it was not long before our sturdy and otherwise satisfactory aircraft were dubbed the "noisy North Stars." I had spent several hundred hours sitting close behind Merlin engines, and therefore was not unused to being about 4-1/2 feet away from the business end of 12 open exhaust stacks. Even with this background, I was impressed with the noise level in the cabin of a North Star, particularly abreast of the engines. There was a logical explanation for this unfavourable comparison, because in fighter aircraft our ears were covered with radio-telephone receivers mounted in thick pads. It was obvious something must be done, especially with respect to the blocks of cylinders exhausting toward the aircraft cabin.

The company was fortunate in having in its Winnipeg shop a practical mechanical genius by the name of "Mac" MacLeod. (A strange coincidence and a well kept secret was the fact that his first name was "Merlin.") He, working so far as I know single-handed, developed an exhaust manifold for the Merlin engine which must still be regarded as a minor miracle. In effect, it conducted the exhaust gases from the inboard block of each engine into a large arch shaped duct to the outboard side of the engines without appreciably increasing the size of the necessary cowling. Furthermore, this was managed without creating exhaust system back pressure which, if it had occurred, would seriously have deteriorated the engine efficiency. The other great problem in the exhaust manifold design was to avoid structural stresses due to dimensional changes brought about by the wide range in temperature to which the manifold would be exposed. Mac achieved this by an ingenious slip joint inserted between the manifold and the exhaust port of each cylinder.

Canadair, whose contractual responsibility it was, had long since given up the problem, and therefore it was particularly galling that for economic reasons that company should get the contract for the manufacture of the crossover exhaust manifolds, to the MacLeod design. The crossover exhaust did not reduce the cabin noise level to anything like modern standards, but did greatly improve it.

The original Merlin exhaust system had provided passengers with yet another reason for comment. Orange exhaust flame, fresh out of the cylinder ports, is completely invisible in sunshine. At the time, TCA was operating a transcontinental flight that, at one season of the year, arrived in Winnipeg at about sunset. It stayed on the ground about half an hour, and many through-passengers stayed on board. It would then start up, taxi out, and take off at full throttle, with all four engines apparently enveloped in flame, now completely visible due to the advancing dusk. The alarmed passengers knew perfectly well that the flame had not been there before, and were convinced that the stop in Winnipeg had done the poor North Star no good whatever. When the crossover exhaust system was installed in 1953 and relieved the passengers of a substantial proportion of exhaust noise, it also deprived them of the thrill of suspected multiple engine fire.

The next problem with which the North Star confronted the company was the high level of cost of engine replacement parts. The Merlin had been designed as a high performance military engine. Its weight per horsepower was therefore extremely low. Also, most of its flying time had been under war conditions, in both fighters and bombers. The rate of aircraft casualties dictated that few engines would achieve a long life in terms of

CF-TFC is refuelled at Dorval while handlers prepare CF-TFB for its next flight. Both aircraft are in the M2-3 configuration. (Air Canada Archives)

durability. As Lord Hives, the then head of Rolls-Royce was later to say: "Having built thousands of them during the war, we never learned anything about the Merlin until you airlines started to fly them."

Under these circumstances it is not surprising that Merlin engines proved to be voracious consumers of expensive replacement parts at time of overhaul. This applied particularly to the valve mechanism and to a lesser extent to the pistons and their rings. More massive components such as crankshaft, main bearings and reduction gearing stood up comparatively well.

One of my more unpleasant duties as president of TCA was to call on Rolls-Royce management at Derby, England, armed with a detailed record of the cost of replacement parts for Merlin engines per hour of operation. This dicussion was called for because a far-sighted term of the contract for purchase of Merlin engines contained a clause under which there was manufacturer's warranty that such costs would not exceed those established by air-cooled radial engines of comparable output.

Discussion began with references to this "never be sorry" clause, and to my figures, which showed a most unfavourable comparison with the experience of US airlines using big radial engines. I explained to Lord Hives that TCA, which was then in a serious deficit position, could not continue under the burden of operating costs for which the Merlin engines were largely responsible. (At the time I was quite sure that TCA was equally incapable of bearing the cost of repowering the North Star fleet.) To my astonishment, Lord Hives did not seem unduly perturbed over my tale of woe. He merely brushed some ashes off his waistcoat and said with a cheerful smile that Rolls-Royce would make good on the terms of the contract to which I had referred. I then felt in duty bound to give him my estimate of the warranty payment from Rolls to TCA which would be involved. This, in those days, looked to me astronomical.

Lord Hives then asked me if I would like to have a lawyer draft a formal agreement, or if a letter from him to me would be acceptable. I felt that, dealing with a man of the calibre of Lord Hives, a request for legal services was both unnecessary and bad mannered. Accordingly, we repaired to his board room, where most of his department heads were assembled, and he, in their hearing, dictated a letter to me which committed Rolls-Royce to furnishing TCA with replacement parts at a cost per operating hour not greater than the leading radial engine of equivalent power, which at the time was a Pratt & Whitney product. Throughout the dictation, the faces of the Rolls-Royce Air Division management team grew more and more horrified, because they knew, perhaps better than Lord Hives and certainly better than I did, what the letter was letting them in for. However, within an hour it was typed and signed and in my pocket.

In discussing the matter afterwards with me, Lord Hives said that the Rolls-Royce reputation for contractual integrity could not be weighed in terms of Pounds Sterling, and that while the clause in the original purchase contract, in the light of later developments, was probably injudicious, it was going to be honoured. "Rolls-Royce engines should not cost more to operate than the best produced by any other firm," he said, and after an amazingly short time, they came very close.

TCA was never without a substantial number of Rolls-Royce engines in its fleet from that date, although in recent years, with the phasing out of Viscounts and Vanguards, the proportion of Rolls engines to total dropped sharply.

The airline's experience with the Merlin engine on a continuing basis is worth recording. The first full year of operation after the Hives/McGregor meeting was, as forecast, extremely expensive to the Rolls-Royce company in the matter of parts warranty claims, but many things occurred all of which proved advantageous. Rolls sent representatives to Canada ostensibly to make sure that we were not discarding parts which still had useful life in them, but in actual fact I think to make sure that the airline was not mistreating the engines in operation.

Parts items which were failing after the shortest period of operating time were singled out and redesigned by Rolls-Royce. These "bottlenecks" as they were called, shifted around the engine in an amazing manner, as the total time between overhauls gradually could be extended, and within three or four years the warranty claims under the replacement parts agreement disappeared. But I think it is true to say that Rolls engineering staff never stopped improving the design and performance of the Merlin engine throughout the long period of its operation by TCA and BOAC, and I would not be surprised to learn that some of the original Canadian Merlins are still in service today.

I suppose no airline is ever completely satisfied with the performance of its engines, and on that subject they do as a regular thing compare notes with one another. BOAC and TCA, being the only two major airline operators of Merlin type engines, compared notes continually, and it is an astonishing fact that seldom was it the case that the difficulty experienced by one airline was also a source of trouble to the other. For example, at one time TCA was having great trouble with what are called piston lands, the narrow partition between the piston ring grooves. BOAC had never heard of this trouble, but were grumbling loudly about reduction gear failure, which TCA had never experienced. This grew to be such a mystery that we even exchanged lubricating oil with one another to try and arrive at some logical pattern, without success. It would be wrong to drop the subject of the Rolls-Royce Merlin engine without saying that it was a great engine, manufactured by a great and completely ethical company, but basically a racehorse called upon to pull a heavy cart, when it fell into the hands of the airlines.

The Merlin, however, had other headaches in store for both flight and maintenance people. It was equipped with the most sophisticated fire-warning and control devices which had been installed on any aero engine up to that time. In practice, a detection device which activated a red warning light for each engine above the normal instrument panel proved so sensitive that the faintest hint of trouble touched it off. There was, for example, a case at Winnipeg Airport when a mail pick-up truck engine was left idling close to a North Star, and its exhaust produced a fire warning signal on two adjacent engines on the side nearest to the truck.

The procedure to be followed by a pilot when the engine fire warning light came on, was to press a related button which "blew the bottles." The bottles contained a fire suppressant, methyl bromide. One of these bottles in each engine nacelle discharged directly into the air intake; thus all the induction side of the engine and the cylinder were suffused with the gas, and since combustion was brought to a halt the engine completely lost in-

CF-TFN showing short-lived underwing markings. (Air Canada Archives)

terest in continuing to run. One unfortunate pilot found two fire warning lights burning on the same side of his aircraft during the final stage of a night approach to a landing at Stevenson Field. Being very close to touch down, and loath to complete the landing with power on only one side, since no fire was visible from the cockpit, he completed the landing normally. Immediately after, both fire warnings went out.

Somewhat later, I was on the flight deck of a North Star which was delayed in Saskatoon with a sheared scissor bolt in its undercarriage. The passengers were picked up by the diversion of a following flight, and at about sunrise the next morning the aircraft took off to complete its flight to Vancouver. There was a long plume of thin smoke from, I believe, an incinerator chimney visible in the sky ahead. In turning from the runway heading onto its planned course, the aircraft flew through this, and all four fire warning lights came on simultaneously. The captain, by this time fairly well acquainted with the built-in practical joking ability of the system, simply groaned, "Oh, no," in well-justified disbelief, and kept on; but had only one light shown, he might have taken the warning seriously, discharged the bottles, and necessitated a return to the Saskatoon Airport.

Eventually the sensitivity of the warning system was cut down to deal effectively only with fire in an engine nacelle, and not react to sudden changes in temperature often encountered at low altitudes near airports, just when full power is most desirable. Incidentally, TCA was the first airline to switch from CO_2 to one of the bromides for fire fighting, and to have the ability to direct a second bottle into the first engine, in case the fire rekindled.

During the disastrous Winnipeg flood of May 1950, TCA North Stars were chartered by the Ontario Department of Lands and Forests to rush sandbags from Montreal to Winnipeg. Sandbags are seen here being loaded. (Air Canada Archives)

Another Merlin idiosyncracy was one which caused many delayed departures from Vancouver for too long a period, before its cause was correctly diagnosed by an ex-Spitfire pilot. A westbound North Star would go into Vancouver in the evening or night, land, and report all four engines functioning normally. On the pre-flight check-up the following morning, all four engines would start with difficulty, and then run on varying numbers of cylinders, but never on all 12. Eventually, two and two were put together, and the answer came to light. At the time, westbound flights maintained 12,000 feet at least, across the Rockies, and this necessitated a fairly fast letdown to approach altitude at Vancouver Airport. During this period of reducing altitudes practically no engine power was required, and they were accordingly throttled back. Since little power was being developed, but the forward speed of the aircraft remained high, engine temperatures dropped sharply, allowing substantial deposits of lead to be made on the spark plugs. The cure, of course, proved to be a short period of fairly high-powered flying prior to final approach at Vancouver, and it is possible that many passengers wondered why, with the airport in clear view, the pilots did not seem to know where it was and took wide swings out over the water prior to heading for the runway in use. However, morning start-ups changed from the headaches which they had been, to a quite normal operation.

G.R. McGregor

Away from his memoirs and dealing with the facts of corporate life, McGregor was hardly nostalgic about the North Star, as this 1950 memo reveals: "The aircraft power plants have proved exceptionally costly to maintain, with a high replacement parts consumption, relatively high prices for each replacement part, a short overhaul interval (450 hours for the cylinder blocks and heads), and fuel consumptions in excess of those originally forecast by the manufacturer. . . . The North Star has proved definitely uneconomical on trans-Atlantic operations. Due to fuel requirements and all-up weight limitations, only 32 of its 40 seats may be forecast as available for sale on winter trans-Atlantic flights, and only 36 may be booked in advance for summer flights. None of the aircraft are equipped with sleeperette or sleeper accommodations, a condition which is going to put TCA at a competitive disadvantage with Stratocruisers and Constellations, which either are or shortly will be authorized for Montreal trans-Atlantic service by BOAC, KLM and Air France. The cabin noise level in the North Star, due primarily to engine exhaust, is higher than any other four-engined aircraft operating in North America, and is a constant source of passenger complaint." These remarks, though, were tempered with the fact that "maintenance costs have recently shown an encouraging decrease, due partially to an agreement negotiated with the Rolls-Royce company last October, and partially to greatly reduced occurrence of mechanical failure which has followed a reduction in operating horsepower, the good effect of which has been augmented by the progressive introduction of design modifications."

North Star Under Attack

Technical problems with the North Star gave muckrakers what they thought was a grand campaign issue as the 1949 federal election neared. The Liberals under Louis St. Laurent were seeking re-election and were vigorously opposed by the Conservatives under George Drew. To Drew and company, the North Star looked like just the Liberal project to level some broadsides at. Their approach was to challenge and embarrass the Liberals for having purchased a flying death trap for TCA and the RCAF. Drew's mouthpiece was the Toronto *Globe and Mail*, headed by arch-Conservative George McCullough. In May 1949 the paper began a campaign to discredit the North Star by "exposing" its supposed flaws and alleged collusion between government and big business in the sale of Canadair to Electric Boat.

Globe and Mail headlines like these began appearing: "UK Secret Report Says North Star Wings Crack," "Mystery Report: First of North Stars Built of Surplus Parts Bought Cut-Rate in U.S.," "Super North Star for Prime Minister Bars Merlin Engine," "Failures of North Stars Cited in TCA Bulletin," "Two Planes Lost Is Cost of Trying Out Merlins for Long-Range Flying," and "Canadair Sold to Arms Ring by Howe for Fifth of Its Cost." There was, of course, a grain of truth in each headline, but rarely was a smear campaign so ill conceived. The paper's facts were largely incorrect or taken out of context to deceive the reading public about an airplane that was having less developmental trouble than most of its contemporaries in the US and Britain.

The "Two Planes Lost" story, for example, attempted to link the loss of two CGTAS Lancasters to the current North Star woes. No such connection could be made by anyone aware of the facts. The entire *Globe and Mail* campaign was aimed at duping the public into accepting a fantasy, yet it defended its claims with editorials headed, "Let the Facts Come Out" and "Establish the Truth." It railed, "The *Globe and Mail* is doing an objective job on reporting on this North Star mystery, and will continue to do it. Nobody who has read the articles on the subject already published . . . can have the slightest doubt that the complaints against the engine are well founded."

Meanwhile, the *Toronto Star* was backing the North Star: " 'Preposterous' Howe Says of Reported TCA Engine Switch" and " 'Treasonous' Says TCA, Engines in North Stars

CF-TFE on the ramp at Malton on May 14, 1949, dandelions in full bloom. In the background is 17520, one of 412 Squadron's VIP North Stars. (Jack McNulty)

Resplendent in Corporation trim, Tudor I *Elizabeth of England* was photographed on January 1, 1947, en route to London Airport to be christened. (via Graham Wragg)

World's Best Developed." At a party nomination meeting C.D. Howe said, "Now the engine, the civil Merlin is, of course, the outcome of a military type engine. As a matter of fact, there is no large aircraft engine in service in the world today that has not been first developed as a military type. Further, I doubt if the engine manufacturers in Great Britain and the US would ever have produced a high-powered aircraft engine if it were not for the development work sponsored and paid for by the governments of those countries in the interests of national defence . . . TCA's experience has been no worse than that of any other airline, and in most respects has been far better. I, at least, do not intend to follow the practice of the Georges by throwing mud at other people's fine achievements." His statement about engine development was essentially correct. The Pratt and Whitney R-2800 came as much from US military engine development in World War 2 as did the Merlin.

This election issue was dealt with in rather cavalier fashion by the Liberals. They were not to be ruffled by such nonsense, not publicly at least. In fact, behind the scenes even C.D. Howe was a bit rattled. He wrote to G.R. McGregor, "I am very much afraid that current criticism will hurt patronage on TCA aircraft." McGregor replied explaining what dirty tricks the Drew-McCullough team was using to get information and added, "So far it has been possible to conclusively attribute 17 Toronto cancellations . . . to uneasiness raised . . . by the press statements of aircraft unreliability." Howe then requested of TCA technical records of Pratt and Whitney engine failures with TCA. The records showed: November 1946-April 1949, 142 TCA Merlin engine failures; May 1938-October 1942, 211 TCA Pratt and Whitney engine failures. This was ammunition to help the Liberal cause in the House of Commons. The newspaper war raged for several weeks, then the election was held and the Liberals were solidly returned. As an issue the story faded and development and operation of the North Star continued as before.

The Avro Tudor V. An ungainly-looking tail dragger, this North Star contemporary was 105 ft. 7 in. long with an all-up weight of 80,000 lb. This one was the only Tudor registered outside the UK. It operated briefly in Canada during 1952, hauling freight between Toronto and Nassau. In this photo, it is seen at Sept Iles, Quebec. (Morrison-Knudsen Co.)

British Competitors

While the North Star had its problems during its developmental years, its contemporaries were not exactly "shining stars." Most large aircraft at the time were brought to service only after prolonged difficulties. With some designs, major, even disastrous flaws were carried into the operational phase. Some aircraft were such overall headaches that they failed to make the grade in service and took early trips to the scrap yard. Viewed in such a perspective, the North Star was not the hopeless old crock it's sometimes depicted as being.

The Avro Tudor was conceived in Britain during the war for use by BOAC on the Atlantic. It was to be an interim airliner to fill the gap until the arrival on the commercial scene of the Brabazon and Comet airliners. The prototype, powered by Merlin 102s, made its first flight on June 14, 1945. After a troublesome development period, the design was rejected by BOAC in early 1947. Avro pressed on, however, introducing the Tudor IV as a 28-32 passenger aircraft and this entered service with British South American Airways. After two disastrous crashes in 1948-1949, the Tudor IV was withdrawn from general passenger service. The crash of a Tudor V in the UK in 1950 took another 80 lives. Hereafter most of the 36 Tudors of various marks that had been completed were relegated to hauling freight. Altogether, the Tudor was not one of the great successes in airline history. Neither was the Handley Page Hermes.

The civil version of the RAF's Hast-

A Hermes seen at London Airport, August 14, 1954. (Peter Keating)

ings transport, the Hermes first flew on September 2, 1947, over two years after the go-ahead for its development had been given. First flight of the production Hermes IV took place twelve months later but more than a year passed before the first aircraft was accepted by BOAC—with some reservations because of overweight problems. By January 1951 all 25 aircraft ordered by BOAC had been completed and the 40-passenger type had been in service for about six months, supplanting Yorks and Solents on routes between London and Johannesburg.

When it became apparent that the Hermes would not live up to expectations, BOAC planned to operate it on an interim basis, then replace it with the Argonaut. The Hermes itself had stood in for the unsuccessful Tudor but it was not to last due to short spar life. By late 1953 the Hermes fleet had been mothballed, though four re-entered service briefly upon the grounding of the Comets, to be finally withdrawn from BOAC use in December 1954. By this time the Argonaut had taken over completely from the Hermes and had established itself as the most successful airliner in the BOAC inventory.

American Competitors

Airliners developed in the United States for the postwar market also had their shortcomings, but these designs enjoyed much greater success than did the Tudor or Hermes. The Lockheed Constellation faced many difficult hurdles in its growth. Between

A Braniff Airlines Douglas DC-6 about to land at Minneapolis, August 20, 1963. A larger aircraft with more passenger appeal than the North Star, over 500 DC-6s of all versions were eventually produced. (Larry Milberry)

1943 and the end of 1949, 178 "Connies" had been built, though only 14 of these before 1946. Yet of these 178, 17 were lost in serious accidents with great loss of life between 1945 and the end of 1949. Operators involved included TWA (six lost), PAA (four lost), KLM (three lost), Eastern Airlines (one lost), Air India (one lost) and the USAF (one lost). Included in the Constellation troubles were deadly fires erupting in the supercharger and induction systems. During this period, the CAA lifted the Constellation's Certificate of Airworthiness for a month.

The Douglas DC-6 also suffered fatal crashes, including a fire on June 17, 1947, which led to the crash with the loss of 43 lives of a United Airlines flight. On October 18 that year an American Airlines DC-6 caught fire in the air and was severely damaged but disaster was averted by a hasty forced landing.

Another much-plagued airliner of the North Star era was the Boeing Stratocruiser, powered by P&W R-4360s. As recorded by US aviation historian John Whittle in *Flypast*, "Long delays in getting Stratocruisers into production were partly caused by engine problems. Even so, when the aircraft went into service, the engine was far from snag free. . . . Early engines suffered valve problems to such an extent that cylinders had to be changed after a mere 150 hours. Oil consumption was also a problem, and white smoke trailing behind a taxiing Stratocruiser was the norm rather than the exception.

"Failures were frequent, and aggravated in the early days by diluting engine oil with fuel on shutdown to

Stratocruiser *Canopus* at Dorval in the company of North Stars. (Jack McNulty)

facilitate the next start in cold conditions. The fuel, however, freed a lot of engine sludge which blocked oil lines on the next flight, causing a rather precipitous engine seizure.

"In January 1953, BOAC began to experience seizures on newly overhauled engines and grounded its fleet for about two weeks until the problem was traced to a crankshaft bearing lubrication problem.

"The propeller was perhaps the Achilles heel of the Stratocruiser. . . The record contains many instances of propellers and even complete engines shed. Propellers were directly involved in at least three major Stratocruiser accidents."

Whittle observes that, of only 57 commercial Stratocruisers built, nine were destroyed in major accidents. Several others were written off in serious incidents, mainly upon landing. Not an enviable safety record. The fact is that in its early years, the North Star faced development problems no more serious than did other contemporaries. It outshone the Tudor and Hermes, and in safety, for the 70 aircraft put into service between 1947-1949, established a laudable record. The North Star, though as cantankerous as other four-engined piston-pounders, has been given an unfairly bad press. It was in most ways a typical airliner of the day.

The Canadair C-7

During 1950 Canadair proposed a new aircraft as a replacement for TCA's North Stars. This was the C-7, a somewhat larger airliner powered by the Pratt and Whitney R-2800-CA15 engine. At first TCA was interested in the C-7, especially as Canadair was keen to market its North Stars abroad if TCA purchased the new plane. But at least two key factors went against TCA acquiring the C-7. For one, the aircraft, at $850,000 each, would be too expensive. The airline's budget simply couldn't stand it, as the resale value of the North Stars would be in the $500,000 area. Fleet-wide re-equipment was thus not possible. TCA considered acquiring a few C-7s for improved transatlantic service but realized that it would be complicating matters by having two major types to service and maintain. Then there was the question of investing in a new fleet of piston-engined aircraft. G.R. McGregor was not prepared to buy C-7s as the plan would commit TCA to piston-engined aircraft for the next eight to nine years. Observed McGregor, "The development of turbo jet and turbo prop aircraft is proceeding at a pace which threatens the piston-engined aircraft with an advanced state of obsolescence within five years." Throughout May 1950 the C-7 was news around TCA and Canadair but the airline dropped the matter quickly, leaving Canadair with little choice but to do the same. As it happened, the airline soon did opt for a piston-engined aircraft for its long-haul routes, purchasing a fleet of Super Constellations while it awaited developments in the turbine field. The Connie, being a larger and faster aircraft than the C-7, seemed the most promising as an interim type.

Canadair DC-4M2 Compared with Contemporary Aircraft

	Canadair DC-4M2 North Star	Handley Page Hermes IV	Avro Tudor II and V	Lockheed Constellation	Douglas DC-6	Boeing Stratocruiser
Span	117' 6"	113'	120'	123'	117' 6"	141' 3"
Length	93' 10"	96' 10"	105' 7"	95' 1"	100' 7"	110' 4"
Height	27' 6.3"	30'	24' 3"	23'	28' 8"	38' 3"
Wing area	1452 ft^2	1408 ft^2	1421 ft^2	1650 ft^2	1463 ft^2	1769 ft^2
Empty weight	49,950 lb	55,350 lb	46,300 lb	57,000 lb	51,460 lb	83,500 lb
All-up weight	79,575 lb	86,000 lb	80,000 lb	102,000 lb	95,200 lb	145,800 lb
Engines	Merlin 724, 1760 bhp	Bristol Hercules 763, 2100 bhp	Merlin 621, 1770 bhp	Wright R-3350, 2500 bhp	P&W R-2800, 2400 bhp	P&W R-4360, 2800 bhp
Max. speed	325 mph	355 mph	295 mph	350 mph	365 mph	375 mph
Cruise speed	250 mph	261 mph	235 mph	328 mph	313 mph	340 mph
Range	3000 mi	2000 mi	2330 mi	3000 mi	3000 mi	4600 mi

Note: These figures should be regarded as approximate. Weight and performance figures vary from one source to the next, and would indeed change through the production run of each aircraft.

TCA's first Super Constellation lifts off at Burbank for its delivery flight in February 1954, Captain George Lothian at the controls. TCA bought 14 of these beauties. They went straight into service on the Atlantic. The Super Constellation carried more passengers, faster, farther and in greater comfort than the North Star. Even the stately "Connie," though, had its detractors. Some referred to it as "the best three-engined airplane flying the Atlantic"! (Lockheed Aircraft)

Other Replacements

In the early 1950s TCA evaluated many aircraft in the search to update its fleet. As early as 1949 the National Research Council had made preliminary calculations on the Dart, Tay and Orenda as North Star replacement power plants. An engineering memo of August 29 on this subject noted that "we are presently losing a lot of money every time a North Star flies on intercity or transborder operations . . . the urgency for correction is well nigh desperate . . . we must leave no stone unturned to achieve the necessary corrections."

In 1951 the company studied many aircraft for domestic use. The list included the Viscount 700, Bristol 175, Comet, Hermes, Ambassador, Apollo and Heron from the UK. A DC-3 converted to the Mamba turboprop engine was also being marketed and Canadian Car and Foundry at Ft. William proposed it to TCA. Types like the Heron and Apollo were quickly eliminated, while the Comet was seen as having some domestic role in Canada. Of the Bristol 175 (later known as the Britannia) TCA noted, "potentially outstanding as a trans-Atlantic or transcontinental aircraft." Of all British proposals it was the Viscount that most interested TCA. Of American aircraft, the DC-6B, Super Constellation, Convair 340 and Martin 404 were of much interest. For domestic use, TCA was greatly impressed by the latter two. When it came to a decision, the Viscount would be the TCA choice though the Convair was a close second.

Also of interest was the all-jet Avro C.102 Jetliner. TCA had been investigating turbine power even before the war had ended. Rolls-Royce informed TCA in 1945 of its new jet engine, the AJ65, and assured it that the engine would be made available for commercial use. A TCA delegation then went to Avro at Malton and met with Walter Deisher, Fred Smye and Joe Morley to talk over the possibilities for a commercial jetliner. They, along with TCA, became very excited about the prospects, and under Jim Floyd of Avro, detailed specifications were prepared. Soon Avro's parent company in Manchester was in on the act and Sir Roy Dobson, the company's president, came over and met with C.D. Howe.

Dobson was not pleased with Howe's reluctance to provide funding for the new aircraft. Howe saw it as a peacetime project and expected Avro to bear much of the financial load. Soon after this meeting, the AJ65 engine was unexpectedly placed back on the "secret" list and, with Dobson's approval, Avro at Malton drew up specifications for an airliner powered by four small Derwent engines. This was a shock to TCA. Jim Bain notes, "The power output of the four engines was about the same as for two AJ65s, but there the similarities stopped. Weight and balance, payloads, fuel consumption, takeoff performance, altitude performance and cruise speed all deteriorated." In spite of TCA's objections, Dobson went ahead with the modified Jetliner.

The Jetliner flew in August 1949, North America's first jet transport. It began a rigorous flight test program, and conducted some record-breaking intercity flights in Canada and the U.S. It was demonstrated to a series of US airlines as well as to the

Figures taken from Trans-Canada Air Lines route studies made around 1950. These compare the North Star with various contemporary types as applied to two of the company's routes.

Montreal-New York, 340 miles (summer conditions)

	North Star	DC-3	Martin 404	Viscount 702	Jetliner	Convair 240
Block time	1:55	2:21	1:51	1:27	1:18	1:46
Payload, lb	12,420	3920	9395	10,780	12,540	10,175
Passengers	40	21	40	40	40	40
Fuel used, lb	3440	1250	1855	3490	7585	1850

Montreal-Moncton, 455 miles (summer conditions)

	North Star	DC-3	Martin 404	Viscount 702	Jetliner	Convair 240
Block time	2:17	2:50	2:10	1:50	1:28	2:06
Payload, lb	12,420	3995	9445	11,150	12,600	10,220
Passengers	40	21	40	40	40	40
Fuel used, lb	3940	1505	2155	4450	9275	2155

military. But as far as TCA was concerned, it no longer met the required specifications. A lot of bickering went on between Avro, the government, and TCA over the Jetliner. In the end the Jetliner program fizzled and the sole aircraft, veteran of hundreds of successful flights, was scrapped, by then surpassed by more advanced technology in the form of such aircraft as the Boeing 707.

Weight Watching

Weighing nearly four tons more than the standard C-54, the North Star had a more critical payload situation. And since payload was the name of the airline game, North Star operators were constantly seeking ways to reduce basic weight to make room for more payload. A TCA memo of May 16, 1951, details the company's struggles to make the North Star pay its way: "Since delivery of the fleet of North Star aircraft from Canadair in the fall of 1947, over 2100 lbs have been reduced from the operational empty weight of the Oceanic M2s and almost 1800 lbs from the Domestic M2s. . . . The payload on the Domestic Service at the end of the 1948 decreased from 12,300 to 12,140 lbs due to modifications applied and fluctuations in removeable equipment and food. Similarly the North Atlantic payload decreased from 11,175 to 10,880 lbs and the Caribbean/Bermuda from 11,385 to 11,045 lbs. . . . By the second half of 1949, the weight increase from modifications was sharply reduced as a result of such weight savings as the removal of aircraft flares (-43 lbs); removal of the astrodome (-17 lbs); and the introduction of the fibreglass carpets (-51 lbs) and the plastic air-filled mattresses on the crew bunks (-56 lbs). Total -167 lbs.

"The principal weight increases resulted from the installation of a fairing over the fire extinguishing system in the rear cabin (+19 lbs); an intercooling mod (+31 lbs); smoke mask equipment (+15 lbs); and the introduction of miniature liquors (+30 lbs) on the Caribbean Service . . . The payload on the Domestic Service decreased from 12,500 lbs to 12,405 lbs and on the North Atlantic Service from 11,450 lbs to 11,370 lbs. . . .

"During the first quarter of 1951, the payloads on all services showed the greatest increase since acceptance of the North Stars from the manufacturer in 1947-1948. The M2s were all reweighed between October 26-April 23 and produced an average unaccountable weight difference of minus 300 lbs between the weighings carried out in Winnipeg in 1948-1949 and the recent programme . . . No.2 and No.3 auxiliary fuel tanks are being removed from the Domestic M2-3s for a weight reduction of 358 lbs. . . . The brass/copper radiators are being replaced with the aluminum types on the Oceanic M2s at a weight saving of 688 lbs per a/c. . . . The payloads have increased as follows: Domestic Service from 12,650 to 13,060 lbs; North Atlantic from 11,850 to 12,430 lbs; Caribbean from 11,935 to 12,595 lbs. It is pointed out that the present payloads are greater than at the time of delivery from Canadair. . . ." Weight savings since delivery are noted as: Domestic, 760 lbs; Oceanic, 1255 lbs; and Caribbean, 1550 lbs. Upcoming weight saving schemes were noted as: Domestic, 1134 lbs; and Oceanic, 641 to 731 lbs.

Avro Canada's C.102 Jetliner. First flown in 1949, the world's second jet transport to fly, this futuristic design excited the aviation world, but Canada did not exploit the concept. Only one Jetliner was built and the project died quietly as other nations took over the field of jet transport. (Hawker Siddeley Canada Ltd.)

Conversion Programs

The commercial airline market was changing rapidly in the early 1950s. More and more of the travelling public were flying and airlines were having trouble accommodating increased traffic. One interim solution was to increase seating aboard existing aircraft. It seems that near miracles resulted. In the case of the North Star, passenger capacity rose from 36-40 in the beginning, past 50, past 60 and ultimately, with other operators, to 78!

By late 1952, TCA had decided to standardize its fleet of 23 North Stars as follows: 36 passengers (6 aircraft, Atlantic only), and the rest either 40, 48 or 55 passengers, with all floors stressed for 55 seats. In this TCA bulletin of March 1955, J.L. Rood, Director of Flight Operations, provides the details of another North Star cabin facelift. The memo is dated March 11, 1955:

In the interest of providing increased passenger accommodation on our North Star aircraft, a conversion program was embarked upon some time ago, further details of which, together with the cargo conversion and the spar capping modification, are outlined below.

Initially, the conversion was planned to provide accommodation for 56 First Class seats or 67 Tourist seats, but this was subsequently changed to provide 52 First Class seats or 62 Tourist seats.

The plan for our 56 First Class-67 Tourist Passenger Conversion was essentially as follows:

The forward cargo area above the floor between the main electrical distribution panel at Station 181 and the flight compartment to cabin bulkhead at Station 260 was to be replaced by the addition of two rows of passenger seats. This area was to be refinished with windows, baggage racks, ceiling, lights, etc., and necessitated also the relocation of the inverters and elec-

Technician Ted Heard uses his electronic kit to get an accurate weight for a TCA North Star. The system used for the annual North Star weigh-in produced a figure guaranteed within 50 lb of the airplane's true weight. (Air Canada Archives)

CF-TFV at Malton in the early 1960s. Formerly CPA's *Empress of Auckland*, this aircraft was sold to TCA in 1951, then converted to a freighter in 1955. Payload on these aircraft was 16,500 lb. (K.M. Molson)

tronic rack—the latter change forcing a complete new radio rack installation at the existing Radio Operator's Station. Thus, all baggage and cargo space was limited to the forward and rear belly compartments.

We had completed three of these aircraft when the realization came about that our cargo and air express commitments were developing beyond expectations. Several proposals were investigated with a view to enabling mail and cargo to be carried in the new forward seat area when required. The end result was the decision to temporarily hold the conversion, install cargo hampers in those three aircraft converted, and also redesign the front end to cope with the addition of one seat row only. Thus, the final conversion of all North Stars used as passenger carrying aircraft will be 52/62, that is, 52 First Class seats or 62 Tourist seats. To accomplish this end, the forward cabin bulkhead at Station 260 will be moved ahead by forty inches to Station 220. The area between Stations 220 and 181 will be used for cargo with the exception of a small coat room area on the left hand side.

Coincident with the 52/62 conversion, we are also increasing the aircraft landing weight 70,000 to 72,000 pounds, and accomplishing project "Spar Cap."

The increase in landing weight involves a major rework to the flaps

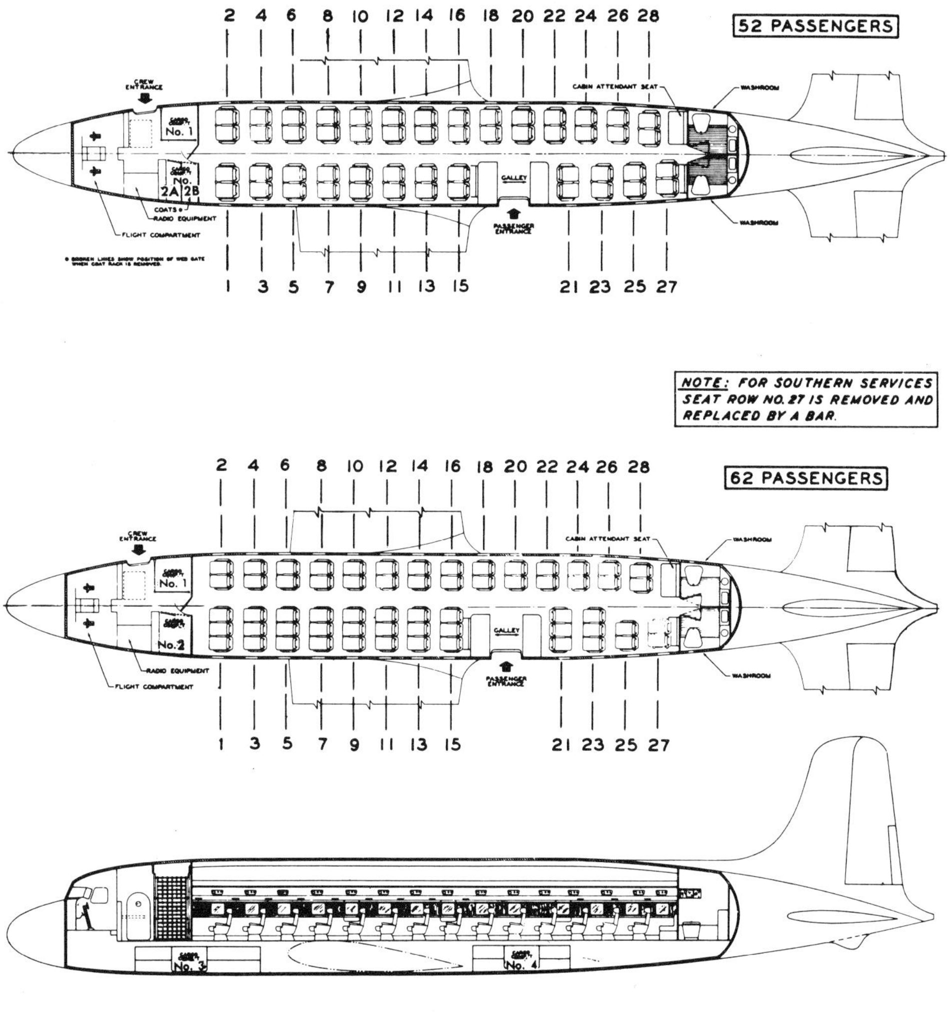

Cargo and seating configuration TCA North Star, 52 and 62 seats.

Two views of North Star freighter CF-TFB. Loading cargo at Dorval in November 1957, and landing at Malton on November 7, 1959. (Air Canada Archives, Larry Milberry)

and the replacement of the flap hinge and actuating linkage.

Project "Spar Cap" involves a major rework to the aircraft centre section area and consists of the reinforcement of the front spar and the centre section bottom skin panel between front and rear spars.

The planned North Star cargo conversion is essentially as follows:

The C-4-1 aircraft (Nos. 221 and 222) will be stripped of all passenger cargo facilities during their forthcoming Maintenance "5 Checks" in March and April of this year. This involves removing seats, baggage racks, galley, etc., installing protective panelling over all bulkheads, sidewalls and windows, and installing cargo tie-down fittings and equipment. No further alterations are planned until the end of the 52/62 Passenger Conversion Program, at which time these two C-4-1 aircraft will be routed to the conversion hangar for the following modifications:

A. Increased landing weight modifications.
B. Project "Spar Cap."
C. Cockpit standardization (new radio rack, inverter relocation, etc.)

Prior to the final clean-up of the two C-4-1 cargo airlift, present plans call for a third all-cargo aircraft to be converted from an M2-48/57 passenger aircraft.

The entire program should be fully completed by January 1956.

Such conversion work was, of course, costly. In 1950, for example, Canadair had quoted TCA a price of $34,700 each to convert 14 North Stars to 48 seaters.

By October 1955 the TCA fleet of North Stars stood as follows:

5 @ 62 seats.
7 @ 52 seats (all domestic aircraft except #201 and 218 in 52/62 seat configuration).
2 @ 48 seats.
2 as operational spare.
4 @ 52/62 seats, Caribbean aircraft, i.e. with navigator's station, these being #213, 214, 216, 217
2 @ 46/48 seats, Caribbean aircraft, these being #219 and 220.

Aircraft #210, 221 and 222 slated for cargo conversion.

One of TCA's Bristol Freighters. The airline purchased this type to inaugurate all-freight service across Canada but the Bristol proved uneconomical and was phased out as soon as TCA could convert some North Stars to freighters. (Air Canada Archives)

Bogged down. North Star CF-TFB sits in the overrun at Vancouver. TCA engineer Earl Gerow writes, "The aircraft was attempting a takeoff with about two inches of slush on the runway. The nosewheel threw up the slush like a bow wave into the radiators and fuel system air intakes of No. 2 and No. 3 engines causing them to fail. Takeoff was aborted and the North Star overran by about 200 feet." Later, TFB was started up, turned around and taxied back towards the runway. But it hit a soft spot and sank in the mud. After being winched out of this predicament, it was towed to the hangar, checked over, and placed back on the line. (R.E. Gerow)

"Immediately Following the Bang . . ."

"Double check your information because there are many prejudices against the North Star and the Merlin, particularly among the flight group. Some of the stories you'll hear stray considerably from fact and should be taken with a grain of salt." So I was warned by a senior TCA captain as I researched the history of the North Star and Argonaut. There are, of course, some who insist that the aircraft was a mistake from the beginning. But not everyone totally condemns it. Even those who were unhappy with TCA's choice in 1944 realized that, properly maintained and flown by competent crews, the North Star was a reasonable airliner. Others, especially those who came onto it in the 1950s after many of its bugs had been removed, look back with real fondness upon the aircraft.

Many who flew the North Star in its early days remember the frequent engine failures. At the time they were cruising the airplane at 1100 bhp. This gave a fast cruise speed, but engines were very prone to breakdowns and time between overhauls was a mere 400 hours for blocks and 800 hours for engine beds. Before long it was found that the North Star cruised nicely at 1000 bhp and problems were fewer. Then the cruise setting was cut back further to 950 bhp and mechanical failures were reduced to a reasonable level, and TBOs (Time Between Overhauls) for the Merlins were eventually extended to 1250 hours, quite a respectable period.

One captain recalls that the three-bladed North Star performed a little better than the four-bladed ones, and that the three-bladed propellers were standard on the Atlantic. He also recalls a few of the North Star's idiosyncracies. When light, he found it nose-heavy on landing so that, on rounding out, the pilot was very busy juggling power, main controls and trim to effect a smooth touchdown. Power synchronization was provided for the engines, but on takeoff it took a while for everything to be humming along properly. Rolling down the runway, one engine could be surging ahead of where the pilot wanted it while another could be lagging. On landing, as the pilot throttled back, the same thing would happen and a smooth power reduction was not assured. On the whole, though, this pilot remembers the North Star as a good all-round airplane in the hands of a competent pilot. As a money-maker, he points out that with its high overhead it was marginal. He also recalls how TCA North Star pilots used to feel real jealousy as they parked on the ramp beside DC-6s and Connies back in the early fifties.

Several interesting landing incidents occur in North Star lore. One pilot recalls a landing at Winnipeg into an 80 mph wind, which hadn't been forecast en route. Once the aircraft was on the ground, it could not be turned off the runway as the winds were so severe. The same pilot remembers flying from Dorval to Malton the night of Hurricane Hazel in 1954. En route, the four radiators were bashed in by hail and glycol was lost from all engines. The North Star reached the ground all right, but with four overheated engines. Not recommended, as glycol loss from even one engine meant a shut-down and usually a return to takeoff point.

Another such incident occurred on a Winnipeg departure. A North Star had been de-iced on the ramp but had been parked in a slightly tail-low attitude. The aircraft departed for its

Facts and Figures: TCA North Star Operations 1949

Average monthly salary, captain	$850.00	Propeller repair, overhaul/sched. mile	$ .0145
Average monthly salary, first officer	$398.33	Instrument repair, overhaul/sched. mile	$ .0086
Fuel consumption/sched. mile	1.1834 gal.	Radio repair, overhaul/sched. mile	$ .0107
Oil consumption/sched. mile	.0117 gal.	Aircraft depreciation/sched. mile	$ .1150
Cost 100 octane fuel, Toronto	$ .3050/gal.	Engine depreciation/sched. mile	$ .0482
Cost 100 octane fuel, Montreal	$ .2740/gal.	Annual utilization per North Star	2520 hrs.
Servicing/sched. mile, labour and supplies	$ .043	Daily utilization per North Star	6.9 hrs.
Repair, overhaul/sched. mile, labour	$ .0627	Turnaround time Vancouver	60 min.
Repair, overhaul/sched. mile, material	$ .0610	Turnaround time Toronto	45 min.
Engine repair, overhaul/sched. mile, labour	$ .0938	Turnaround time Montreal	45 min.
Engine repair, overhaul/sched. mile, material	$ .2258	Taxi time per North Star flight	15 min.

Wintry scene at Malton during the early 1950s: CF-TFJ and CF-TFV on the ramp with TCA DC-3s. (Air Canada Archives)

next destination but no sooner was it airborne than one engine overheated. It was shut down but the three others immediately overheated as well. The captain turned around and made an emergency landing. It seems that ice melted during de-icing had drained back to refreeze against the oil coolers. While taxiing out, everything seemed normal as the still-cold oil was bypassing the coolers, but as the North Star took off and the oil heated up, it was rerouted into the oil coolers, all of which had suffered ruptured lines due to ice. The oil simply spilled out of the system. Only quick action in the cockpit prevented an accident.

On another Winnipeg landing in severe winds, pilots Crocker and Bosma at the helm, the result was real embarrassment. As the TCA North Stars had no reverse pitch capability and as maximum braking was required, the captain decided on touchdown to hit the idle cut-off switches. The aircraft rolled to a stop but all power, even lighting in the cabin, was lost. The crew called for a Hobart, a ground starting unit. Someone misinterpreted this as "tow bar" and, shame of shames, the North Star ended up being towed to the ramp.

Some pilots have stated that the North Star was as much airplane to handle as any two-man crew could want. This was perhaps especially true on TCA's Maritime routes where runways were usually short and weather more often poor than elsewhere in Canada.

A typical day could begin at Dorval as Flight 850. It was slated to be airborne for Moncton at 0140 for an 0450 touchdown. The Moncton-Halifax leg appeared on the timetable as 0520-0610. Push on after 20 minutes for Sydney and be there at 0740. Leave for Stephenville at 0810 to arrive at 0950(NDT). Theoretically, be at Gander at 1130, and finally reach St. John's at 1255. Some aircraft would then turn around and return to Gander. That would be after seven landings, so long as everything had gone well.

But Maritime Canada's weather being what it is, Flight 850 was often a much more complicated affair. With airports down to minimum weather conditions, three, four or five landing approaches were sometimes made before getting down, and these could be white-knuckle experiences. Sometimes the North Star had to give up and go somewhere safer. One notorious place to get into was St. John's airport, Torbay. A landing from over the water towards the west headed the airplane at a 925-foot cliff. The airport was in a trough with rising terrain towards the west. The

CF-TFG bearing rarely seen double registration—on the tail and fuselage. This North Star was sold in Mexico in 1961 and remains the last M2 in existence. For many years it has sat at Mexico City airport as a café. This photo was taken at Malton on October 22, 1949. (Jack McNulty)

wind would sweep down into the trough then rise up and over the cliff. Captain Maurice Labine recalls being on final and hitting severe subsidence created by this phenomenon. With full power on, the North Star would still drop and pilots would find themselves staring straight at the cliff. As climbing was too difficult, a hasty 180 was the prudent manoeuvre. Go around and try again. The passengers must have been thrilled!

In fact, the passengers usually accepted the rigours of North Star flying to Newfie. After a particularly gruelling trip, they would often give the crew a cheering round of applause when the aircraft finally plunked down at Torbay. Of course, the same weather is there today and no doubt the DC-9 and 737 crews occasionally get the same applause.

For a while TCA operated a weekly service from Dorval to Goose Bay. Goose was another airport with fast-rising terrain nearby. Taking off there one day at all-up weight, Captain Labine suffered an engine failure. A piston disintegrated in No.3 and the engine began to fall apart. The ceiling was 400 feet and, knowing that those hills were not far away, Labine stayed visual, did a quick racetrack turn without even bothering to dump fuel, and was back on the ground in less than five minutes.

Rigorous training programs saved many a pilot's biscuit during these TCA years. Standards were extremely high. Out of Dorval, landing practices were sometimes done over the long, straight Soulanges Canal. People below must have wondered at times as North Stars, perhaps on two engines as they practised emergency situations, descended quite low along the canal. Many a pilot also remembers the check pilot feathering an engine on takeoff. Taking off from Dorval one hot day, with the North Star at maximum weight, the F/O selected flaps instead of wheels up. The captain was suddenly stuck with an airplane on the stall and couldn't even take a second to reach over and retract the gear. Staring him in the face was a water tower, getting closer and closer. Training and experience paid off, thank goodness, and the North Star made it.

One practice in North Star days that would not be acceptable today was the windmill start. Starters were notorious for failing and sometimes a Merlin would not turn over. The routine here was to disembark passengers and taxi out, roar off down the runway and let the prop windmill to start the engine. The pilot would then brake, taxi in, and re-board the passengers with the engines still running.

Unusually rough landings occasionally greeted North Star passengers. One occurred at Saskatoon on June 30, 1953. CF-TFG, with 35 passengers

Detailed Comparison of DC-4M2-3 and DC-4M2-4

	M2-3	M2-4
1. *Engines*	Merlin 622/722	Merlin 624/724
2. *Propellers*	4-blade, 12' 8" diam.	3-blade, 13' 1" diam.
3. *Max. certified weight*	79,400 lb	80,200 lb
4. *Max. weight, zero fuel*	63,000 lb	63,000 lb
5. *Operational empty weight*	51,842 lb	51,528 lb
6. *Max. payload (4 - 5)*	11,158 lb	11,472 lb
7. *Max. disposable load (3 - 5)*	27,558 lb	28,672 lb
8. *Increase in disposable load, M2-4 over M2-3*	1114 lb	
9. *Increase in permissible payload, M2-4 over M2-3*	314 lb (equals 2-1/3 passengers on transatlantic)	
10. *True air speed (still air, 70,000 lb)*	197 mph/10,000' 226 mph/20,000'	203 mph/10,000' 225 mph/20,000'

A view of the M2 galley aboard the CF-TFC, April 1948. (Air Canada Archives)

◁ Cabin view of a TCA North Star in February 1947. Interior decor was by TCA designer Janet Lowe. (Air Canada Archives)

Wider view inside an M2 in 1948. Years were to pass before airline safety would consider the hazards of overhead luggage. (Air Canada Archives)

CF-TFE running up on a blustery winter's day at Malton. (Larry Milberry)

North Star CF-TFA being towed at Vancouver after throwing No.3 propeller on September 27, 1954. (Gordon Peters)

The errant propeller just before it was stuffed into the garbage can to the left. (Public Archives of Canada C117896)

aboard, rounded out short and landed on the grass 70 feet short of the runway. It bounced through a mound of earth left behind by some excavators, then carried on down the runway. Nobody thought too much of the incident and TFG went about its business at the terminal then taxied out. As they turned onto the runway, the pilots commented about the job they had done rearranging the turf near the end of the runway, then took off for Toronto. Once there, the North Star was inspected and found to have suffered serious structural damage, particularly at the wing centre section adjacent to Nos. 2 and 3 engines. In his report on the incident the captain observed, "The bump was not as bad as a normal poor landing, more like a car on a washboard road."

CF-TFA experienced an unusual incident at Vancouver on September 27, 1954. Rolling down the runway on takeoff, a bang was heard. The TCA report notes that, "Immediately following the bang . . . the first officer turned in time to see from his cockpit window the propeller and reduction gear of No.3 power plant cartwheeling away to starboard." A DOT inspector aboard added, "At the same time a flash was seen to pass over the starboard wing and a great belch of smoke came out of No.3 engine. The power was cut immediately and the aircraft taxied back to the ramp, still smoking very badly and with gas pouring out over the starboard aileron." The inspector seemed a bit put off that the pilot parked the North Star right at the airport gas pits! CF-TFA had earlier experienced two other incidents. It had encountered severe turbulence over Turner Valley, Alberta, on April 12, 1950,

Ramp scene at Vancouver in 1959: North Star CF-TFF, a Super Connie and three DC-3s. (Air Canada Archives)

leaving several passengers shaken up. On May 12, 1953, it suffered an engine fire.

On April 10, 1957, CF-TFB, with 15 aboard, was landing at Sydney at about 0230 local time. The weather was down and a return to Halifax might have been prudent. Nonetheless the captain persevered and got the wheels down on his second try. The problem was, he landed about half way down an icy runway. At about 35 mph TFB sailed off the end of the runway and headed cross-country. It came to rest in a ravine, its tail sticking up in a less than flattering pose. Another Sydney event involved CF-TFK on October 24, 1960. Soon after landing, the nose gear inadvertently retracted, leaving the North Star in a rather unnatural pose.

A takeoff incident took place December 19, 1948, as CF-TFL was about to depart Winnipeg with 36 passengers. Takeoff weight was 71,606 lb. The runways were covered with 4 inches of wet snow. It was the captain's intention to abort his takeoff half way down the 5000-foot runway if performance was unsatisfactory. At that point he had 100 mph IAS (indicated

CF-TFB after it overran the runway at Sydney. Damage was substantial but the North Star was soon back in service. (Public Archives of Canada C117896)

TCA North Stars were never radar equipped. Flying on the southern route one day, Captain Bob McRae called Miami for an update on some weather. When asked why he didn't simply refer to his radar he replied, "We don't have any. They issued us with St. Christopher medals instead"! (Air Canada Archives)

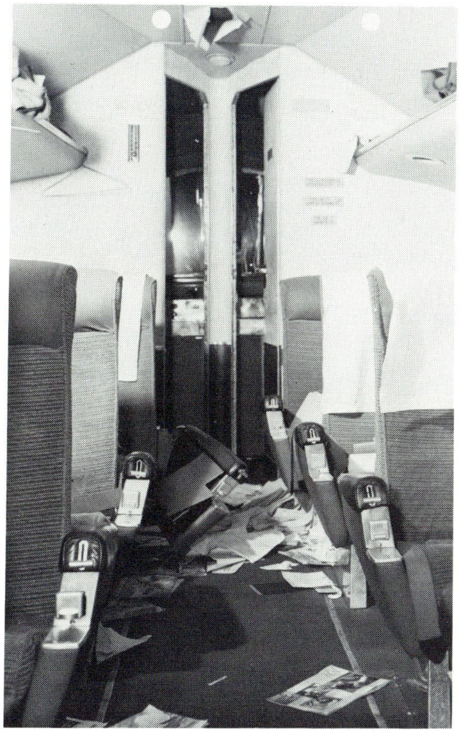

The interior of CF-TFU after the North Star ran into turbulence on January 24, 1953. (Air Canada Archives)

air speed) so carried on; but the nosewheel didn't lift until 3300 feet. At 4000 feet speed was only 103 mph and the aircraft showed no interest in leaving the ground. At this point the captain aborted, but too late. CF-TFL ran out of runway and carried on 700 feet into the overrun. Causes of the incident included: engine intakes almost totally clogged with snow, slush covering much of the wing, wheel wells plugged with slush, leading edge of right stabilizer crushed over most of its breadth by flying slush, and drag caused by slush on the runway.

As incidents like the rough landing at Saskatoon illustrated, the North Star was an exceptionally rugged airplane. That is one feature that has never been challenged. Captain Gus Bennett gives further testimony to the fact. On January 25, 1950, he was in command of a North Star near Chicago when the aircraft entered severe turbulence. The buffeting was unbelievable, yet the crew didn't lose faith in their airplane and they came through unscathed. In a famous event in the annals of Northwest Airlines, one of its Stratocruisers in the same weather didn't do so well. It shed a complete engine, as the Stratocruisers were wont to do from time to time when the going got rough. Captain Bennett has few complaints about the North Star!

Another in-flight "irregularity" involved CF-TFU on January 24, 1953. It was bound from Toronto to Idlewild at 7000 feet with 18 passengers. At Elmira it was cleared to 9000, then at Allantown, Pennsylvania, New York Centre cleared it to begin descent. Passing 5500 feet there were a few seconds of moderate turbulence then a violent downdraft with TFU losing 700 feet. Captain Lennie closed the throttles immediately and recovered with no further trouble. Everyone aboard was shaken up but none seriously. Though undamaged structurally, the North Star sustained considerable damage to its interior.

Another occurrence from time to time with the TCA North Stars was the burning through of crossover exhausts. This happened at Chicago to CF-TFI on November 10, 1953. No.4 was shut down on takeoff "as a result of visible indication by the flight crew of fire at the top of the inboard cowling." The flight proceeded to Toronto where a 4 x 6 inch hole was found burned through the cowling. A series of such happenings led to a directive to inspect all crossovers every 25-35 hours of use.

One problem with the North Star was its marginal takeoff performance in hot weather. It climbed laboriously and the situation was aggravated by the big rad flaps which added drag. The TCA manual recommended climb speed after takeoff at 135 knots. In hot weather conditions this would usually leave the aircraft hanging precariously in the air. One captain found that by lowering the nose a bit and picking up speed, the climb-out was much more satisfactory, with coolant temperatures holding in the safe 40 degrees Centigrade area. The captain regularly used this unorthodox procedure, but the company eventually got wise to it and laid on a check pilot one day. As it happened, the departure was from Idlewild on a very hot day. Takeoff needless to say was by the book with the check pilot in the right seat. As the North Star crept into the air, the captain could see him out of the corner of his eye, busily adjusting the rad flaps and appearing to be breaking into a sweat. "Don't worry," advised the captain. "I've got a nice clear stretch of the Hudson River picked out for when everything boils over." The check pilot finally relented and down went the nose a bit, up came the speed, and the temperature gauges came into line. Rumour has it that sometime later, TCA North Star operating manuals were amended to include the once-frowned-upon practice of faster climb-outs.

With the North Star in service for so many years, lighter stories of every kind abound. All was not flinging propellers across the infield or burning through cowlings. Captain John Bosma remembers a flight where among the cargo in the forward hold, adjacent to the cockpit, was a large, noisy tomcat cooped up in a wicker cage. After putting up long enough with its yowling, the captain ordered the beast let out. It immediately leapt up onto the glare shield where it sat contentedly for the rest of the flight, except for landing. At that point, it retired to the floor, somewhere down amidst the rudder pedals.

Glen Cawker served TCA in various capacities—steward, ground school instructor, and eventually second officer supervisor on Lockheed 1011s and Boeing 747s. As a flight attendant on the unpressurized M1s he recalls that on flights above 10,000 feet one purser and one steward, often with 44 passengers, had to join a series of 4- or 5-foot sections of oxygen hose together with bayonet fittings to make enough to be able to walk through the cabin to inspect the passengers. He adds, "I did one trip out of Dorval, through Sydney, where we climbed straight to 19,500 feet from Sydney and had the seat belt sign on all the way to London due to turbulence. I stepped on the long hose and it separated. I wandered about for 30 minutes in a state of euphoria before I realized what was wrong!

"On another trip with Purser Bill Grant we were demonstrating the life vest. As we had no PA, this was done by individual seat row. There was a lady on board who had been complaining about everything since we left Dorval. As Bill explained to her about the oral inflation tubes for the jacket she complained, 'Surely you don't expect us to be able to blow that thing up by ourselves!' Bill replied, 'I don't know about the rest of the passengers, M'am, but I'm sure you'll do all right!' She laughed and was quite pleasant for the rest of the trip.

"Once we had a cargo fire warning light come on and the F/O had to come back to inspect as the light stayed on after the bottle had been discharged. I remember how in the dark of night he had to go through a hatch in the aisle floor, crawl forward, open a bulkhead and have a look. I lay on the floor grasping his ankles, ready to extricate him if need be. Fortunately the passengers slept through the whole procedure.

"On one downwind landing at Prestwick onto a wet runway and coming in too fast the F/O later said that the worst thing about the landing was seeing the captain pass him! Referring to the sideways attitude of the North Star as it slithered down the runway!

"Going through the Azores westbound, we arrived to find another North Star stuck with a bad engine. That crew took our aircraft and we waited for a ferry flight back to Dorval, captained by Norm Ramsay. He didn't have clearance from Lisbon so we had to wait another 24 hours. Armed guards, always prevalent in the Azores, wouldn't let us near the aircraft till the clearance came through. We finally got away, but lost an engine, forcing a delay at Gander. Finally, after three days, we got to Montreal, some 48 hours late."

TCA Vancouver passenger sales manager Laurie Manchee recalls an experience when he escorted the illustrious C.D. Howe and one of Howe's friends aboard a North Star. Manchee ushered his two VIPs to the rear of the passenger cabin where noise would be at a minimum.

After a congenial chat, Howe's companion asked, "Mr. Manchee,

Prime Minister Louis St. Laurent about to board a TCA North Star, January 16, 1951. Stewardess Mary Ann Charbonneau welcomes the PM. (Air Canada Archives)

when are you going to get rid of these noisy North Stars?"

Manchee had had plenty of experience explaining to the travelling public about the superb Rolls-Royce engine married to the proven Douglas airframe and so on *ad nauseam*, but this was different. As he tells it, "With the boss of all we beheld, including both aircraft and airline, this seemed a bit too hot for me to handle."

Laurie Manchee replied, "Here's the Father of the airline. Ask him."

And without hesitating Howe came back abruptly, "If he doesn't like the airplane, we'll throw him off the damned thing!"

A Pitch for Publicity!

Laurie Manchee was also TCA's traffic representative in Toronto in 1947. One episode he recalls from those days involved the Toronto Argonaut football team: "The first postwar football season was about to get under way. The Argonauts were considering a charter flight to Winnipeg, which would be a 'first' in Canada. I got a rate for them on the North Star but it turned out to be more than the Argonauts' budget. They wanted to know the publicity potential of such a

Oct. 9	CF-TFP	C.L. Skelding	Dorval-Gander	3:30 hours
10	"	"	Gander-Prestwick	7:30
	"	"	Prestwick-London	1:56
	"	"	London-Orly	1:26
	"	"	Orly-London	1:31
11	CF-TFM	"	London-Shannon	1:54
	"	"	Shannon-Keflavik	3:46
12	"	"	Keflavik-Gander	8:20
	"	"	Gander-Dorval	5:02
21	CF-TFO	J.R. Bowker	Toronto-Bermuda	5:01
	"	R.M. Smith	Bermuda-Toronto	5:15
26	CF-TFR	G.A. Smith	Dorval-Gander	4:08
	"	"	Gander-Prestwick	9:07
	"	"	Prestwick-London	2:15
27	"	S.R. Found	London-Prestwick	2:00
	"	"	Prestwick-Keflavik	3:27
	"	"	Keflavik-Goose Bay	7:12
	"	"	Goose Bay-Dorval	4:06

Excerpts from Ron Peel's log for October 1951. Peel was chief TCA navigator at the time. The log names the captains on the flights. Aircrew certainly worked for a living in the "good old days!" Like other veteran captains of the day, S.R. Found passes off with a bit of a smile the claim that the North Star flew "across" the Atlantic. The last trip noted here shows what was going on: the North Star flew "around" the Atlantic as a rule, not "across" it, especially on westbound crossings.

charter in the hope that this might bring the charter rate down to something more in line with their budget.

"I agreed to make a pitch along these lines and proposed to head office that we could gain much-needed publicity and that giving the team a 50 per cent discount would prove self-liquidating. The idea was accepted. I chased up the three Toronto dailies and got reasonable assurance that, as a first in Canada, it was a story worth covering.

"The flight left Malton on a Friday evening amidst pressmen and flashing cameras. It was a joyous ending to a busy week. Still, I was anxious to see the Saturday evening papers, hoping to see a generous spread of pictures and type covering the event. As it turned out, the 'ink thieves' basically ignored my scheme, choosing instead to cover the Canadian National Exhibition World Swim in Lake Ontario, with endless pictures and page after page of accounts. In a small squib on the sports page of each paper was the account of the North Star carrying the Argonauts to Winnipeg. Needless to say, I was greatly relieved that no hell-raising from head office followed."

Foolish Young Men . . .

With the millions of miles of droning across land and sea, it was inevitable that the TCA North Stars would cross paths with the CF-100s and Sabres flown by the RCAF. Those who drove the Clunks, as some irreverently called the CF-100, were generally a sober crowd, at least while on the job. They were all-weather jocks, strictly no nonsense. The Sabre boys, though, were a different breed—glory bound, move-over-Charlie-I'm-comin'-through-spirited lads. As a rule, Clunk drivers and their sidekick navigators lived longer. The gung-ho if sometimes foreshortened lives of the Sabre jocks have been immortalized by one of their number, R.J. "Chick" Childerhose. In one of his books, *The F-86 Sabre*, he describes an incident with a North Star.

In their what's-done-is-done view of life, the Sabre quartet seemed to have no idea of the madness of their action, nor much feeling for the North Star crew who obviously had a different definition of professionalism:

Another autumn day, glowering cloud but no rain, ceiling perhaps 2000 feet. The clag extended to 25,000 feet, also perhaps. Mohawk Black section of four was returning to base and were directed to hold on top.

"We have a TCA North Star inbound at the present time."

"Okay," said Black leader with aplomb. "We'll let-down visual."

"You'll what!"

"Black, dive brakes go."

And down and down they went. Four Sabres in finger formation. Lovely formation. They held it all the way down through 23,000 feet of alto-stratus. They held it right to that instant when they broke out of cloud and beheld, larger than life itself, a red and white North Star.

"Break!" screamed Black leader.

He said that after everyone in Black section had pulled for their lives. It was a sort of downward Prince of Wales manoeuvre. Sabres going in every direction. But mostly downward.

Eventually the four Swords congregated, assembling themselves in a shaky semblance of a formation. And there was only the trace of a tremor in the voice that announced the news:

"Black is three out with four."

"You're cleared initial, Black. Call the pitch."

The scene—or 'skit' as these things were known—was being related with gestures and shrieks in the flight room, when the telephone interrupted. It was the airport manager. He desired the presence of the pilots of Mohawk Black section in his office. At once.

Four G-suited pilots filed into the office. Standing beside the desk of the senior citizen who was the manager, were two ashen-hued and angry Trans-Canada pilots.

"You nearly rammed me!" said the TCA captain. In essence that is what he said.

The airport manager was a calming influence.

"Hold it, Ted. Hold it. Let's have the story again."

And the TCA captain repeated his horror tale of a sky full of Sabres going in every direction in their efforts to avoid collision with his sacred vessel full of sacred passengers.

So emotional had been the captain's address, that not a soul in that office was unaffected. The airport manager was quite pale as he turned to Black leader and asked, gravely, what he had to say for himself.

"I didn't see any North Star," said Black leader. "Did you guys?"

No one had.

"I was just flying formation," said Black Two.

"What altitude were you flying?" asked Black Three.

No answer. Indeed, the good captain seemed incapable of answer.

"I think I saw a North Star," admitted Black Four. All eyes were on him as he scrunched his bushy eyebrows in the effort of remembering. "But he was a long way off. Two,

North Star CF-TFF in Bermuda, April 1948. At this time TCA's promotional material urged, "Ask about TCA's Bermuda Triangle Tour. Tickets good for passage from Montreal-Bermuda-Toronto-Ottawa-Montreal in either direction. Tour may originate and terminate at any point of call on the Triangle."! (Air Canada Archives)

Captains Lothian and Bowker at work in an M1. (Air Canada Archives)

three miles." He turned to the airline captain. "Were you flying just below the cloud?"

It was two against four. Four Sabre 'captains' against one TCA captain and his assistant.

I agree. We were very foolish young men.

Chick Childerhose

Bowker's Airline

Captain J.R. "Bob" Bowker was one of the early pilots on TCA's southern services. He had started flying in 1929 in the RCAF and spent several years mainly on government photo survey work with Fairchild 71s, Bellancas and Vedettes. In 1938 he spent several months with American Airlines, training at Newark as copilot on Stinson As and DC-3s. Later that year he joined TCA and was put on the Vancouver-Seattle run with a Lockheed 10A. During the war he made many an Atlantic crossing in the CGTAS Lancasters, something which he enjoyed as a rule. One feature of this operation he recalls is something called "Atlantic-itis"—too much flying crammed into too few days.

In 1948 Bob Bowker, Ken Main, Roger Smith and a few other pilots were assigned to the new Bermuda and Caribbean routes flying the M2. To fly the Bermuda-Trinidad service it was decided to post one captain for three months in Bermuda. Before long Bowker was living there permanently and running what soon became know unofficially as "Bowker's Airline." Bowker enjoyed the climate and the flying and to this day resides in Bermuda, though he retired from the airline in 1968.

Bowker has many a fond memory of his North Star years. As a rule he covered the Bermuda-Barbados leg in six hours. The cockpit was often a sweltering place so he would at times turn off some of the heat-producing electrics and do without fuel gauges and other amenities. He recalls that at Barbados there was a very keen TCA mechanic who waited all week for the North Star to arrive and prayed for snags. Bowker learned not to report trouble to the mechanic. If he did, the mechanic would go straight to work and within five minutes have the airplane in pieces all over the tarmac. The trick Bowker learned was to keep quiet about things then push on for Trinidad, where there was a night stop. That leg Captain Bowker got down to a predictable 41 minutes each time.

For years Bowker used the same flight plan trip after trip. After a while he pretty well forgot how to write one up and was a bit taken aback one day when George Lothian came down to check the route and Bowker found himself temporarily baffled by the paper work.

CF-TFD in flight near Montreal prior to delivery to TCA, carrying the name *Selkirk*. (Air Canada Archives)

Bowker recalls the first time he saw the North Star. It was at the official roll-out at Cartierville in 1946. He was pleased with what he saw — real washrooms, hot water. Compared to the Lancs, this was heaven. To him the North Star was "a beautiful aircraft to fly." He never had an engine quit, though on three different legs of one trip had to use fire extinguishers because of false fire warnings. Occasionally an engine would run rough, but it was often hard to tell which one it was. This could sometimes be determined by gently placing a finger on each throttle and feeling for the vibrations. As for the Merlin, Bowker says, "It would keep on running with more things wrong with it than any other engine of the time."

Captain Bowker left the North Star for the DC-8 and in 1968 retired from Air Canada with over 25,000 hours. He dabbled a bit in flying, first with a DC-7 freighter operating between Norfolk and Bermuda, then as copilot on a Bermuda-based Lear. He finally give it up for full retirement on his Atlantic paradise. In all he had spent some 10,000 hours flying the North Star.

First North Star Disaster

The first of TCA's North Star accidents occurred over Moose Jaw on April 8, 1954. The aircraft involved was CF-TFW, TCA's Flight 9, Montreal-Vancouver via Toronto, Winnipeg and Calgary. It took off from Winnipeg at 08:58 CST flight-planned along Airway Green 1 maintaining 6000 feet. Aboard were Captain I.A. Bell, First Officer D. Guthrie, Steward L. Penner, Stewardess M. Quinney and 31 passengers. The flight reported progressively over Portage, Rivers, Broadview and, at 10:50, Regina. The final transmission was, "Over Regina at 10:50 CST at 6000 feet. Visual. Estimate Swift Current at 11:30 CST." Thirteen minutes later CF-TFW crashed disastrously into a house on 3rd Avenue N.E. in Moose Jaw, killing all on board and one person on the ground. It had collided with RCAF Harvard No. 3309 flown by P/O T.A. Thorrat, a trainee from No. 2 Flying Training School, RCAF Station Moose Jaw. The Harvard had been flight-planned to pass Airway Green 1 at 7500 feet. In fact it hit the North Star at 6000 feet. Wreckage from the North Star and Harvard was scattered over a wide area. The loss of CF-TFW was the first case of loss of life in a North Star or Argonaut during eight years of intensive flying.

Crash on Mount Slesse

CF-TFD also met sudden disaster. On December 9, 1956, while operating as TCA's Flight 810, Vancouver-Montreal via Calgary, Regina, Winnipeg and Toronto, it crashed into rugged Mount Slesse near Hope, B.C. Aboard were 59 passengers and the crew of Captain A.J. Clarke, F/O J.C.T. Boon and Stewardess Dorothy E. Bjornsson.

CF-TFD had departed the ramp at 18:00 PST. As noted in a company report, "The weather conditions were poor, but not unusually so for that time of year. It did not give any particular concern to Flight 810 in its climb out from Vancouver or to TCA Flights No. 4 and No. 7 which were in the area about the same time."

The flight reported reaching 19,500 feet by 18:52 but then reported a possible fire in No. 2 engine. The engine was immediately shut down and the pilot turned back towards Vancouver. At 19:10 he reported by Hope radio beacon, presumably at 14,000 feet. The DOT-approved minimum en route IFR altitude for this section of Green Airway No. 1 was 9100 feet. Eleven minutes later Vancouver Centre tried unsuccessfully to raise Flight 810. One minute more and it contacted the RCAF rescue centre.

A long and fruitless search began at 20:36 when a CF-100 from Comox flew some lines along Green Airway No. 1 near Hope. Hereafter military and civil aircraft flew hundreds of hours searching vainly throughout this rugged area of the Cascade Mountains, where weather conditions are frequently impossible. Okanagan Helicopters scoured the lower regions while fixed-wing spotters probed the higher reaches of the mountains. Vancouver-based Aero Surveys photographed 600 square miles of the Hope area. All efforts to turn up the wreckage failed for several months. In an interim report TCA noted: "It would seem, even with only three engines operating, that the flight was in a favourable position over Hope at 14,000 feet having a downhill run home. Also, it is reasonable to assume that had ice or further engine trouble become a factor, a radio report would have been received from the flight stating such conditions."

On May 14, 1957, the wreckage of CF-TFD was discovered by mountaineers. It had struck the mountain near its peak at an altitude of 7600 feet, killing all aboard. An official report concluded: "The cause for the aircraft being at an altitude low enough to strike Mount Slesse is undetermined, but there is a high probability that the aircraft while flying on three engines encountered either severe icing, turbulence, subsidence, or a combination of all three, or suffered some other difficulty of such a sudden or dire nature that the crew were unable to communicate with any agency or control the aircraft."

AIR TRAFFIC CONTROL RADIO LOG - VHF 119.7 mcs. From	To		Pacific Stand. Time	TCA RADIO LOG VHF 128.3 mcs. From	To	
810	ATC	This is Flight 810. We have just lost No. 2 engine. We're holding 19000. We're endeavouring to maintain 19000. We would like a clearance immediately to get down if we can. We're losing altitude quite fast here.	1857			
ATC	810	Roger 810. Maintain 14000.				
810	ATC	And is that via Cultus Lake				
ATC	810	Via Green 1, and you should be past Flight 4 shortly. He's just coming up on Hope at the present, maintaining 14000 via Green 1.				
810	ATC	Maintain 14000. We are cleared to descend then go ahead again.				
ATC	810	Maintain 14000.				
810	ATC	Confirm 4 is holding 14000				
ATC	810	4 is holding 13000				
810	ATC	Thank you. We are leaving 19000 at 57 on Green 1				
ATC	810	Okay, and report by Hope on this frequency.				
810	ATC	Roger, Wilco				
ATC	4	Flight 4, ATC go ahead	1901			
4	ATC	Just passing Hope 13000				
ATC	4 & 810	By Hope 01 at 13000 and 810 what's your estimate on Hope				
810	ATC	Another 5 minutes roughly, I believe				
ATC	810	Report if visual by Flight 4. Flight 4 by Hope 01 13000				
ATC	810	Will you be able to hold 14000 okay				
810	ATC	I think so				
ATC	810	We will have further for you by Hope				
ATC	810	810 make your altitude read 14000 or above	1903			
ATC	810	810 Vancouver Center report	1907	ATC	810	Report 810
810	ATC	Vancouver 810, you broke off. What do you wish us to report		810	ATC	You broke off after report.
ATC	810	- - - Report. You by Hope yet		ATC	810	Roger. Are you by Hope yet.
810	ATC	Negative	1908	810	ATC	Negative.
ATC	4	Okay. Flight 4, Vancouver Center				
4	ATC	Flight 4 - over				
ATC	4	What altitude do you require after Princeton				
4	ATC	By Princeton 08 13000 requesting	1908	ATC	4	What altitude do you require after Princeton
ATC	4	Say again your altitude		4	ATC	15000
4	ATC	(unreadable)		ATC	810	What altitude does he (Flt 4) want?
ATC	810	What altitude does he want		810	ATC	Requesting 15000 one five
810	ATC	Requests 15000 - one five		ATC	4	Maintain 15000 approved Flight 4 is acknowledged.
ATC	4	Flight 4 approved maintain 15000				
4	ATC	Flight 4 maintain 15000				
810	ATC	Flight 4 checks				
			1909	4	810	Confirm 15000 looks good to you?
				810	4	Looked alright when we're going up there - right now starting to get some more precipitation, don't know if too good or not.
				4	810	(Unreadable transmission)
				810	4	Didn't get quite up to it - but sure slowed down on way back.
810	ATC	By Hope at 10 request descent down to 10000	1910	810	ATC	By Hope at 10 requesting descent down to 10000
ATC	810	Roger. 810 now cleared Vancouver Range Station, cross Vancouver 8000 or above and remain this frequency		ATC	810	Cleared to Vancouver Range Station, cross Vancouver 8000 or above. Remain this frequency - 810 acknowledged.
810	ATC	Roger. Range cross 8000 or above. Will remain.				
		3 calls to 810 by ATC and no reply	1924			
			1921 to 1934			Repeated calls from ATC, Tower and TCA, but unable to raise Flight 810.

All communication agencies made repeated calls and thereafter issued blind weather and airport information until 2145.

A. RANKIN
Flight Operations Superintendent
Western Region

◁ The last few minutes of radio transmission involving Flight 810.

TCA North Stars Replaced

In early 1961 TCA began to withdraw the North Star from service and by July the fleet was in storage at Dorval awaiting disposal.

Once TCA's North Stars were out of service, the airline was anxious to dispose of its fleet. Under veteran check pilot R.E. Hadfield a sales program was initiated and many possible purchasers were approached. TCA wanted to dispose of the North Stars as a package deal — aircraft, spares, tools, simulator, to a single buyer. This proved difficult for, although the price was right for a fleet of 21 big airliners, the turbine jet airliner was taking over and there was a glut of four-engined piston pounders on the market at bargain basement prices.

In mid-1961 a deal was finally struck with Ron Myhill of Overseas Aviation in the UK, for 15 North Stars plus spares for $660,000. Eleven aircraft were delivered to the UK, but in July 1961 Overseas suddenly went broke and ceased operations.

Meanwhile CF-TFA, CF-TFB and CF-TFH were acquired by Lineas Aereas Unidas SA in Mexico. LAUSA had been formed by Cuban exile and former Cubana official Jose Lopez Vilaboy. He put at least one of these aircraft into service on the Poza Rica-Tampico run and had agreed to pay TCA at the rate of $10.00/flying hour towards the balance owing for the aircraft. He had also acquired CF-TFG and CF-TFI as spares. Four aircraft were delivered to Mexico by November 1961.

With the collapse of Overseas, TCA made a new deal with John Gaul in the UK. For $100,000 cash he took over the 11 ex-TCA North Stars in the UK. For another $65,000 he acquired a large quantity of spares, FOB Montreal. The TCA record for the various North Star deals as of November 3, 1961 is as follows:

Cash from R. Myhill, 15 a/c + spares	$ 77,700
From J. Gaul to liquidate Overseas	$100,000
From J. Gaul for spares	$ 65,000
International Air Freighters, one aircraft	$ 40,000
Offer from World Wide Airways, four aircraft	$ 40,000
Vilaboy, down payment on five aircraft	$ 31,500
Vilaboy due	$ 76,000
Asking for simulator	$ 15,000
Asking for shop equipment	$ 15,000
Total on hand or anticipated	$460,200

On October 30, 1960, TCA operated its last North Star passenger flight in western Canada. The crew included Captain R.J. Smallman, F/O I.M. Woytaz, and Stewardesses M.P. Angus and G. Ryf. (Air Canada Archives)

April 30, 1961, TCA's last ever North Star △ passenger flight, with the crew of Captain Maurice Labine, Stewardesses Judy Turnbull and Joan Teron, and F/O K. Charabin, was Flight 7751, Sydney-Montreal. (Abbass Studio via Air Canada Archives)

△
On July 1, 1961, the last transcontinental North Star freighter operated. Captain Don Orr and F/O R.D. Johnston brought the aircraft back into Montreal bearing the autographs of TCA personnel along the way. (Air Canada Archives)

Middle left

Beginning in 1955, the Vickers Viscount began appearing on TCA's domestic and US services. With this aircraft, TCA became the first North American airline to introduce turbine power. An outstanding success, the Viscount replaced TCA's DC-3s and, on routes like Toronto-New York, North Stars. (Larry Milberry)

The newly introduced Vanguard. This turboprop was a big step forward for TCA, which purchased 22 as North Star and Super Constellation replacements. The Vanguard carried up to 108 passengers at over 400 mph. Vanguard CF-TKI comes in to land on Runway 23 at Malton, June 15, 1961. (Larry Milberry)

CF-TJD, TCA's earliest DC-8, does a touch and go at Malton on April 10, 1960. The aircraft had recently been delivered. Its arrival, and that soon after of the Vanguard, spelled the end of TCA's North Star and Super Constellation fleets. (Larry Milberry)

Another harbinger of the end of the piston airliner age was the outstanding Lockheed Electra turboprop. It offered speed and passenger appeal with which aircraft like the North Star could never compete. Here American Airlines' *Flagship Baltimore* approaches to land at Malton, August 22, 1960. (Larry Milberry)

By this point TCA was almost rid of the North Star fleet and equipment (for exactly how much is difficult to determine) and a comment in a company memo seems to suggest a sigh of relief. It refers to the Vilaboy deal: "If things do not work out, we have at least disposed of five airplanes under the best conditions that we have been able to find." TCA was much more interested in its fleet of new Vanguards and other developments and was relieved to get the North Stars off the books.

As the Vanguard entered service on TCA's domestic, US and Caribbean routes, the North Stars were sold off. Here CF-TFR is readied for delivery to World Wide Airways in December 1961. Like most ex-TCA North Stars, it was not to see much use. Its registration was cancelled within six months and it was scrapped. (Air Canada Archives)
▽

Mexican Excursion

As word spread that the TCA North Star fleet was for sale, there were enquiries from around the world. An Argentine operator wanted North Stars and sent two pilots to Dorval where Captain Ralph Leek spent two weeks with them, giving them a ground school and checking them out in the airplane at least to the point where they could get it around the circuit without hurting themselves. The aircraft were to be delivered via Brazil, but financing difficulties cropped up and the deal fell through.

Next a possibility arose with the Mexican Air Force, which was shopping for a new transport. Captain Leek, F/O Gene Rooze and about 15 others flew down to Mexico City in a North Star to conduct a demonstration and, it was hoped, sell some airplanes. Upon arrival at Mexico, the Canadians were standing around the North Star when a man in full military regalia arrived. He was an air force general and insisted on seeing the pilot — nobody else. Right away, Captain Leek had to take the general up for a ride. He directed the captain towards a tall volcano, saying that if the North Star could overfly it on three engines, he had a deal. This was no problem: Captain Leek feathered one engine, idled another and soared over the volcano. "Bueno! Bueno!" the general kept saying. "You must come and fly with us."

"What's the pay like?" asked Leek.

"Not very good," replied the general, "but you can have a uniform just like mine."

Enthusiasm doesn't pay for airplanes and the TCA personnel were soon back in Montreal. No North Stars were sold to the Mexican air force. Little further was heard from Mexico but at least the general had had his ride.

Canadian history in Mexico. Formerly CF-TFG, this North Star has been operated as the Wings Cafe at Mexico City airport since 1965. It is the only M2 known to exist. (J.M. Ramsden via Stephen Piercey)

Still looking relatively intact, the North Star at Mexico is "the" place to be seen for the local kids. (J.M. Ramsden via Stephen Piercey)

High and dry, CF-TFE/YV-C-LBU being scrapped out at Miami. (Hal G. Martin via Peter Keating)

CPA Enters the Big Times

Canadian Pacific Airlines formally began operations on July 1, 1942. Over the previous few years, the Canadian Pacific Railroad had been laying the groundwork for an airline operation and in 1940 had begun acquiring a number of bush flying operations in the North, most of which had fallen upon hard times. The new airline was busy during the war, involved with Ferry Command, the aircrew training program, aircraft overhaul, and its bush flying operations, but a wartime order-in-council stated that at war's end it must divest itself of its airline activities.

CPA's wartime president, L.B. Unwin, stated, "While the airplane today is an integral part of our wartime mechanism, it should not be overlooked that commercial aviation is destined to play a major role in reshaping a postwar world. In this regard, the Canadian Pacific, as a long-established worldwide transport system, will undoubtedly through the extension of its air operations take its natural place in the development of world trade and travel routes to meet the almost unpredictable demands of the coming air age." This was a bold statement for an airline that was still mainly equipped with aircraft like the single-engined Junkers and Norseman, and it didn't show any fear of a postwar shut-down.

Late in the war, Unwin, W.M. Neale, Grant McConachie and Hugh B. Main of CPA had a series of meetings with C.D. Howe. They suggested that in the interest of developing better transportation services CPA begin to reduce its bush operations and start developing north-south routes using larger, wheeled aircraft. Routes like Whitehorse-Vancouver and along the north shore of the St. Lawrence would profit greatly from this improvement. The idea appealed to Howe as a means to get peacetime air transport going and help provide jobs for ex-servicemen. Howe was also very impressed with the job Unwin had done through the war as CPA president.

One of CPA's DC-3s in flight over Montreal. The DC-3 served CPA for nearly 30 years. (CP Air)

Under H.B. Main, CPA began a major review of its licences and began applying for new routes. One of these was Vancouver-Calgary via a number of stops through the mountains. Howe at first balked at this, but as TCA was not interested in serving the small centres along the route, he went along with CPA. By this time CPA had acquired several DC-3s and was converting them for airline use at Winnipeg. These were badly needed to replace the company's Lodestars.

In early 1947 Grant McConachie became president of CPA. A long-time bush pilot, McConachie had been head of Yukon Southern Air Transport, one of the firms bought up by CPA early in the war. He had long dreamed of establishing an air route to the Orient by flying the great circle route across the north, and he believed the postwar period was the right time to establish such an air link. Once again, C.D. Howe was not excited about the idea, but as TCA was not planning any Pacific service he agreed to it. Such a service would complement the long-established Canadian Pacific steamship routes between Vancouver and the Orient.

McConachie, Main, "Wop" May and others began organizing for rights to fly into various Pacific nations. In Tokyo they had a successful meeting with General Douglas MacArthur, who opened the door for CPA to fly into Japan. There were already good relations with China through the CP steamship tradition there. In Australia a cool Labour government would not commit itself until, at a dinner, the conversation turned to railroads and the prime minister and air minister, both old railroad men, realized that McConachie and Main too had railroading in their blood. That clinched the deal.

Meanwhile CPA had decided on the purchase of four Canadair Fours to use on its Pacific routes. One story states that this purchase was required by C.D. Howe if CPA wanted his permission to fly the Pacific. Another story states that McConachie offered to buy the Canadairs first to butter up Howe. From the more practical side, the Canadair Four (never officially called North Star by CPA) was thought the best choice by CPA, which required a pressurized airliner, and there was nothing else available on short notice which would suit the airline's budget. CPA paid over $800,000 for each of its C-4s, considerably more than did TCA or BOAC.

In 1947 CPA sent a team of representatives to Canadair while the aircraft were being built. These were Ian Gray, Harry Rolfe, Reg Sargeant and Ferdie Vachon. Gray, as director of engineering, was resident at the plant and involved with every detail as the aircraft took shape, including the development of cruise control charts, "how goes it" curves, cabin furnish-

Empress of Hong Kong in a classic air-to-air view. While in CPA service the C-4s were not equipped with crossover exhausts. After being sold to TCA in 1952, this aircraft was lost in a mid-air collision over Moose Jaw. (CP Air via Jan Stroomenbergh)

Grant McConachie, with his famous smile, sits in the captain's seat of a CPA Canadair Four. (CP Air)

McConachie rose to head CPA but the road was long and uphill all the way. Problems plagued him for years as he struggled to keep his bush operations going. Here one of his Fokker Universals is seen in trouble at Prince George in November 1933. (G.W. McLaren)

ings, etc. In charge of the cabin was Reg Sargeant, who had been hired from BOAC. He patterned his plans for the new airline operation on the tradition of the famous CP Empress ships and his own experience as a senior airline steward. Sargeant was single-handedly responsible for the CPA style of passenger service which was to become world famous. Also at this time, McConachie hired Captain C.A. Pentland from BOAC to set the standards of long-range flying for the airline.

In May 1949 CPA accepted its first Canadair Four and the aircraft was turned over for delivery to Captain North Sawle. This was a big event in CPA's history. The Canadair was its first large airliner. It was also its first new airliner and was the basis for CPA's first formal contract with a major aircraft producer. With the Canadair, CPA had risen to the status of a major international airline. It was now in business "for real."

Proving Flights

The story of CPA's initial North Star proving flight across the Pacific and some other operations is detailed here by former CPA Captain Archie Vanhee. Vanhee began his flying career in 1928, taking flying lessons at the Montreal Flying Club. He flew in the bush for a number of years and during the war held several postings in the RCAF, including one as OC, 160 Squadron, a coastal patrol Canso squadron. After the war he flew with CPA and several other companies. Over 50 years after starting his flying career, Archie Vanhee was still flying as training captain with Timmins-based Austin Airways.

The survey flight using a borrowed RCAF North Star departed Vancouver on April 19, 1949, for Anchorage, Alaska, on the first leg of our trip to Hong Kong. Next day we left Anchorage on a flight inside Alaska to visit all the possible alternates such as Northway, Big Delta, Fairbanks, etc. on April 21 we took off for Shemya where we did local training before departing for Tokyo on April 23. Here again we spent one day practicing approaches and landings at Tokyo's Haneda Airport. April 25 saw us on our way again to Shanghai via Osaka, Hiroshima and Fukuoka, landing at Shanghai in a heavy downpour. The civil war was raging on the outskirts of the city, communication was virtually impossible and weather information was unobtainable. To be on the safe side we filled up with fuel and used Manila as our alternate just in case. Weather was still bad when we took off 3 hours later but cleared as we passed abeam of Amoy. It remained excellent all the way to Hong Kong. We landed after visiting Canton where we did simulated approaches as this was to be our regular alternate for Hong Kong once we started our scheduled flights. Since the communists won the war and Canton fell into their power, I doubt if Canton was ever used again by CPA.

We stopped at Okinawa on April 28 on our return trip to Canada and remained in Tokyo overnight. April 30 found us airborne again for Misawa, which was another of our designated alternates, then we continued on to land at Adak and Anchorage. We arrived back in Vancouver on May 1, having overnighted at Anchorage. On this trip the following systems of navigation were used: Dead Reckoning, Radio, Loran and Astro. Every pilot acted alternately as captain, first officer and navigator. It is believed that this was the first North Pacific crossing for a Canadian aircraft of civil registry and the first to land at Tokyo, Shanghai and Hong Kong.

More training and familiarization flights were done in Canada and the United States before undertaking the proving flight to Australia. In the meantime Charles Pentland and I took delivery of one of the C-4s, CF-CPR *Empress of Vancouver*, and flew nonstop from Montreal to Vancouver in 11 hours 11 minutes. This provided a good test of the aircraft's range capability. I remember taking 14 hours 45 minutes a year or so later between San Francisco and Honolulu.

The survey flight to Australia departed Vancouver on May 28, 1949, in CF-CPR. On board were Captain Charles Pentland (director flight operations, CPA), Captains Potter, Stevenson, Lee and myself. After a refuelling stop in San Francisco we landed in Honolulu that same day. Next day we did flight training. On

June 1 we arrived at Nandi, Fiji, having visited Palmyra Island and refuelled at Canton Island. We did local training at Nandi on June 3 (local date having crossed the international date line) and then went on to Sydney, Australia, that same day. On June 6 we visited our registered alternates of Williamstown, Brisbane and Dubbo. On June 7 we were off to Melbourne and back to Sydney.

We visited Auckland, New Zealand, and Ohakea on June 8. On June 9 we

CPA crew who flew aboard CF-TEP on the South Pacific survey flight which left Vancouver on April 19, 1949. From the left are Captain J.K. Potter, Captain C. North Sawle, Captain Louis C. Stevenson, Captain M.D. Lee, Captain George W. Knox, Captain Len Fraser, Navigator P.D. Roy, Captain Robert Goldie, Captain Archie Vanhee, Navigator Frederick Wicker. (via Archie Vanhee)

Bottom left

"Wop" May, Grant McConachie and "Punch" Dickins prior to departure of CPA's pre-inaugural Canadair Four flight to Australia. This photo was taken at Vancouver on July 10, 1949. (CPR)

Captain Archie Vanhee works on a navigation exercise aided by Fred Wicker. Intensive crew training preceded the introduction of the C-4 on CPA's newly won Pacific routes. (CPR)

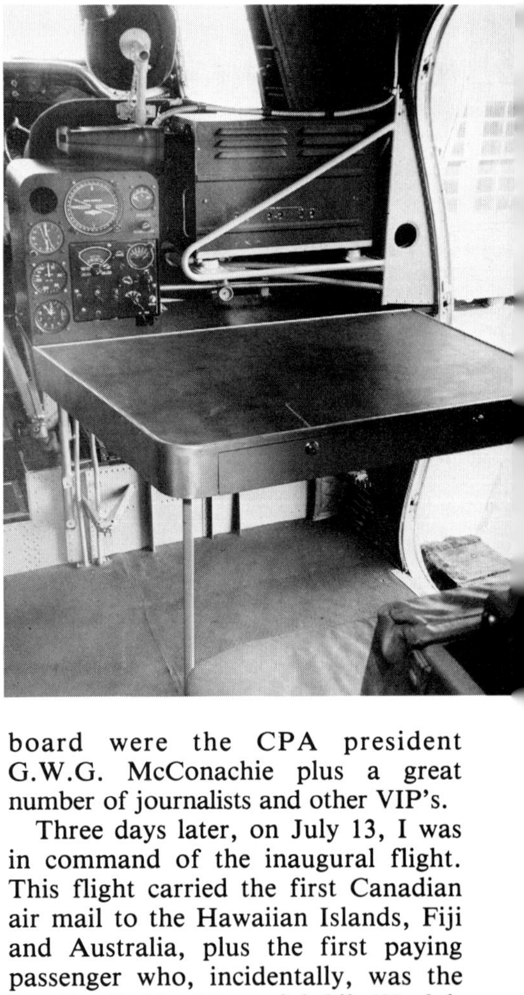

Three interior details of the CPA Canadair Four: the radio officer's station, the navigator's station, and the pedestal showing throttles, radio tuning panel, trim wheels, etc. (Canadair via Graham Wragg)

departed for Nandi, Tongatabu, Canton Island and Honolulu, landing here on June 10, and returning to Vancouver on June 11 after a refuelling stop at San Francisco. Again on this flight the pilots took turns in the captain's, first officer's and navigator's seats. Once again I am confident that this was the first South Pacific crossing, with landings at Honolulu, Canton Island, Fiji, Sydney, Melbourne in Australia and Auckland and Ohakea in New Zealand, by a Canadian civil aircraft.

More training and familiarization flights were carried out to alternates within Canada and the United States, one as far south as Los Angeles, before starting our regular service to Australia. I doubt if any group of pilots had ever received more or better training prior to commencing a regular air service anywhere. At any rate we were ready, primed up and looking forward to the job ahead.

With Captain Potter in command, the pre-inaugural flight got under way to Australia on July 10, 1949. On board were the CPA president G.W.G. McConachie plus a great number of journalists and other VIP's.

Three days later, on July 13, I was in command of the inaugural flight. This flight carried the first Canadian air mail to the Hawaiian Islands, Fiji and Australia, plus the first paying passenger who, incidentally, was the late Pat Reid of Imperial Oil. We followed the same route as the pre-inaugural, i.e. San Francisco, Honolulu, Canton Island, Nandi and Sydney. We had to go via San Francisco because the C-4 lacked the necessary range to fly direct to Honolulu from Vancouver. Thinking of flying direct was out of the question until September 29, 1949. This was the company's seventh crossing and my fourth. After rest at Honolulu my crew and I were ready to resume our flight to Vancouver via San Francisco when the Honolulu weather office informed us that the wind pattern was favourable for a direct flight to Vancouver. At first I was skeptical but we made two different flight plans, one to San Francisco and another direct to Vancouver. The one to Vancouver proved to be shorter in time. Hence we took the direct route. It was an uneventful flight and we made it in 9 hours 30 minutes, thereby setting a record and achieving the first nonstop in the direction Honolulu-Vancouver. In any event it was the first for a Canadian aircraft.

Archie Vanhee

CPA crew during the company's South Pacific pre-inaugural. Left to right are Chief Pilot North Sawle, Chief Navigator Fred Wicker, Supervisor of Passenger Services Reg Sargeant, Radio Officer R. Phipps, Superintendent of Maintenance Fernando Vachon, Captain Bud Potter and Stewardess Helen McCracken. (CPA via F. Vachon)

Crew on CPA's South Pacific inaugural, July 13, 1949. From the left are Stewardess Barbara Aycliffe, Captain Archie Vanhee, First Officer L.C. "Craig" Stevenson, Second Officer Bruce Thompson, Navigator Brian Hellings, the Consul General of Australia, radio officer Freeman and Stewardess Margaret Linehan. (via Archie Vanhee)

In its September 1949 edition, *Canadian Pacific Spanner* described the introduction of the new CPA Canadair Four. On July 11 a pre-inaugural VIP and press flight was conducted along the route from Vancouver to Sydney via San Francisco, Honolulu, Canton Island and Fiji, and three days later the service was formally begun.

CF-CPI took the 32 special guests on the pre-inaugural junket, chief tour guide en route being CPA president Grant McConachie. The guests included A/V/M A.T.N. Cowley (Department of Transport), Clarence Campbell (President of the National Hockey League), Roy Thomson, Jack Webster and various other business leaders, CP and government officials and journalists. Crew included captains J.K. Potter and C. North Sawle; navigator F.D.P. Wicker; radio operator R. Phipps; engineering/maintenance crew A.S. Boon and Fernando Vachon; and cabin crew Helen McCracken, Pamella Hookham and Reg Sargeant. All along the route the passengers were treated to superb CPA service with many extras included as they visited places like Hawaii, Fiji and Australia. In Sydney CF-CPI was christened *Empress of Sydney* by Mrs. Drakeford, wife of Australia's Minister for Air. This strictly first-class junket terminated July 21 when the aircraft touched down back at Vancouver.

McConachie's Dream

Captain L.C. Stevenson, in the following section, gives a detailed account of his experiences during CPA's Canadair Four era:

Sometime in 1948 all of us were aware that CPA would be entering the Pacific arena as the designated Canadian carrier on both the North and South routes to Hong Kong and Sydney respectively. Subsequently, the Captain crew requirements were posted by the

Air travel luxury 1949 style—a view down the passenger cabin in a CPA C-4. (CPR)

company and the successful bidders awarded the positions. I was among the group. Other selected were Captains L.G. "Len" Fraser, J.K. "Bud" Potter, A.D. "Archie" Vanhee, R.N. "Bob" Goldie, W.G.M. "Mel" Knox, M.D. "Mel" Lee, and H.O. "Hump" Madden. Captain North Sawle was appointed Chief Pilot, Overseas. Captain C.A. "Charlie" Pentland became Director of Operations, Overseas. He was a Canadian from Winnipeg who had had a long and distinguished career with Imperial Airways and BOAC dating from the mid-1930s.

A course in air navigation was begun in late January of 1949 which was to last well into the month of April. This course was conducted by an ex-BOAC crew member, F.D.P. "Fred" Wicker, ably assisted by navigator Pat Roy on loan from TCA. After sweating out this course, we began pilot training on two North Star M1 series aircraft borrowed from the RCAF. Our own North Star C-4s were in the late stages of construction at Canadair.

After this training was completed, the first survey flight was organized. Booked onto it was an almost full load of pilot and navigator trainees under the direction of Captains Pentland and Sawle, and navigators Wicker and Roy. During this flight we all had a go at navigating as well as taking turns at the controls. The major airports surveyed were Anchorage and Shemya in Alaska; Tokyo, then known as Haneda; Lungwha Airport, Shanghai; and, of course, Hong Kong. Many other routes along the way were surveyed for alternate and refuelling purposes. This operation took 10 to 12 days to complete. We carried a spare engine in mid-cabin due to the fact that there were no spare Merlins in the Orient at that time.

An interesting feature of the flight was our visit to Shanghai around April 25. Due to the impending Communist takeover of mainland China, we refuelled there as quickly as possible, took on some packages of Canadian diplomatic papers, held a conference with the Canadian Attaché, and left for Hong Kong rather hurriedly. The Communist troops were reportedly only ten miles from the airport and no one expressed any real desire to stay there, though most regretted not being able to have a look at the famous city. The evacuation of Shanghai and other parts of China was just beginning and Kai Tak Airport in Hong Kong was plenty busy during the four days we spent there. Time logged on CF-TEP for the extensive Vancouver-Shanghai-Vancouver trip was 63 hours 16 minutes.

After returning to Vancouver, more training was carried out, this time on our own North Stars. There was a world of difference between the earlier M1 aircraft and the new C-4s. The M1 had no pressurization; no noise suppression insulation; had open exhaust stacks; and was without doubt the noisiest and most uncomfortable aircraft it was my misfortune to be associated with. By comparison, the C-4 was a veritable "Queen of the Air," as it was pressurized and had reasonably good insulation against noise.

However, the C-4s still proved to be too noisy for comfortable travel. Also, as they were limited in payload and range for the long Pacific routes, all flights had to be routed through San Francisco, instead of direct Vancouver-Honolulu. A few such flights were made, but only with ex-

One of Canadian Pacific's C-4 fleet, *Empress of Vancouver*. This aircraft had been in service less than a year when it was written off in an accident in Tokyo. (CP Air)

ceptional tailwinds.

The North Star's maximum seating capacity was 36, and this made it a very uneconomical aircraft for the Pacific. These were the reasons the North Stars were dealt off to TCA as soon as possible. The only advantage of the sparse seating was that there was at least 36 inches of leg room for each seat and the seats were almost fully reclining, a feature which Grant McConachie made full use of in advertising promotion.

The CPA pilots were in agreement concerning the unsuitability of the North Star for the airline's use and welcomed the "new" DC-4s purchased from Pan Am in 1951-1952 as interim aircraft until delivery of our DC-6Bs. Although the DC-4 was unpressurized, it was more suitable for our route structure, having much better range and payload carrying capability.

In the spring of 1949 a North Star survey flight was conducted to Honolulu, Fiji, Australia and New Zealand. The first scheduled flight left Vancouver July 13 for Sydney via San Francisco, Honolulu, Canton Island, and Nandi in Fiji. On this flight myself and Captain Vanhee alternated as first officer under the able supervision of Captain Pentland. An earlier pre-inaugural flight had gone ahead two days earlier with Captains Sawle and Potter in charge. McConachie was aboard this flight along with other

Flight schedule for CPA's C-4 service between Vancouver and Hong Kong. Service on the 6500-mile route was inaugurated September 19, 1949.

Flight 305	*Local day*	*LST*	*GMT*	*Flight Time*
Vancouver	Lv. Mon.	00:05	08:05	
Anchorage	Arr. "	05:05	15:05	7:00
	Lv. "	06:30	16:30	
Shemya	Arr. "	13:00	00:00	7:30
	Lv. "	14:30	01:30	
Tokyo	Arr. Tues.	20:30	11:30	10:00
	Lv. "	22:00	13:00	
Hong Kong	Arr. Wed.	06:00	22:00	9:00
				33:30

Flight 306	*Local day*	*LST*	*GMT*	*Flight Time*
Hong Kong	Lv. Thurs.	14:30	06:30	
Okinawa	Arr. "	20:00	11:00	4:30
	Lv. "	21:00	12:00	
Tokyo	Arr. Fri.	02:00	17:00	5:00
	Lv. "	03:30	18:30*	
Shemya	Arr. Thurs.	17:30	04:30	10:00
	Lv. "	19:00	06:00	
Vancouver	Arr. Fri.	09:00	17:00	11:00
				30:00

**Flights operating via Misawa may depart at 18:00GMT. First flight leaves Vancouver Monday, September 19, and each Monday thereafter. Return flight leaves Hong Kong Thursday, September 22, and each Thursday thereafter.*

A CPA Canadair Four taking off from Vancouver. The C-4 was CPA's first large airliner and the aircraft which made possible its entry into global airline operations. (CPR via R.J. Fox)

company officials and VIPs. The schedule set up for this flight consisted of a 15 hour one-stop flight via San Francisco to Honolulu; a 36 hour layover there then a one-stop to Fiji via Canton Island which again took 15 flying hours; an overnight stop there and then a further 8½-9 hour flight to Sydney to complete the outbound portion of the trip. After two or three days in Sydney the return flight was made using the same layover procedures and times. The entire trip took about 10 days. This would give the crew their normal 75 hours monthly flying time and they wouldn't have to fly again for 18 days.

Only two crews were needed for the southern route as the schedule only allowed for one trip each way every second week. The two senior crews elected to fly this route, relegating the rest of us to the northern route which began in the fall of 1949. The flying on this route proved to be much more interesting but not nearly as pleasant as the South Pacific.

Shemya became our refuelling and crew lay-over point on the North Pacific route. This was a bleak, windswept island. While there, CPA used the facilities of Northwest Airlines. Due to Shemya's location, the weather there left much to be desired. With the cold Bering Sea to the north and the warm Japanese Current to the south, fog was prevalent, and ceilings of 200 feet and half a mile visibility were more the rule than the exception. Coupled with strong winds (usually cross winds) these conditions called for some rather interesting approaches. The US Air Force, however, maintained at Shemya what were probably their best-trained GCA operators.

One of these, a Sergeant Gates, was superb. A southerner, his drawl would come over the air when the aircraft was approaching the minimum altitude and one's knuckles might be getting a little white on the wheel. "You ah slidin' dawn the ole wiah," he'd report, meaning you were right on glideslope and perfectly lined up with the runway centreline. This would nearly always relieve any tension that might be mounting and enable a pilot to relax and do a much better job flying. Often, though, one would make two or three approaches and pull-ups before getting everything buttoned up and completing the landing.

During the five or so years we operated through Shemya CPA had no serious incidents although there were a few damaged tires due mostly to crosswind landings. The runway was excellent as it was 10,000 feet long and 250 feet wide between the lights with 100 feet of paved runway on either side for the entire length. Very unusual for the times but under the existing conditions quite necessary.

One anecdote I remember about McConachie's optimism was a pep talk he gave while we were in the preliminary navigation class, to the effect that Shemya was a piece of cake to operate through due to the runway length and width, plus the adequate GCA unit there. In this he was correct, but his statement that, although the summer cloud ceilings were 200 feet and even sometimes lower, one could almost see Shemya from the north end of Vancouver Island on a clear day, did raise a few smiles.

McConachie was ever the optimist. All of us were grateful that a man of his optimism, faith, and salesmanship was guiding the company during these sometimes difficult but always interesting times. Although we did not always share his unlimited optimism, we always heartily endorsed his vision, drive and courage.

L.C. Stevenson

McConachie had fought hard and long for trans-Pacific rights and his self confidence seemed unshakable. Ron Keith quotes him in reference to this stage in CP history: "Flying CPA should be like getting religion, with every happy customer a convert for spreading the good word." Even though business was slow at first on the new Pacific route, it was to grow into one of the most successful, with the early impetus coming from two unforeseen bonanzas. One was the flying of Chinese immigrants from Hong Kong to Vancouver beginning in 1949. Only those who could afford the $798.60 fare of course! The other was the airlift of American military personnel between Vancouver and Tokyo during the Korean War. Between 1950 and March 1955, 703 trips were made and some 39,000 personnel carried. For both the immigrants and the military, the service was the same — CPA's incomparable first class treatment. In the case of the military, Ron Keith notes in his biography of McConachie, *Bush Pilot with a Briefcase*, that before long the top ranks realized that this service was being wasted on the run-of-the-mill soldier. They quickly remedied things, packing the Canadairs with officers and letting as many of the regular troops as could be bumped find some other, inevitably less enjoyable, means of getting across the Pacific.

CPA advertising of the day reflected the company's goal of establishing itself as the top airline on the Pacific. One ad announced, "Fly Canadian Pacific style to Australia! Flights from Vancouver via Honolulu and Fiji for New Zealand! In the air, Canadian Pacific gives you all the speed of flying plus the superb service and friendly hospitality that have made Canadian Pacific the world's greatest — and the world's most gracious — travel system! You get the extra comfort of sleeping on land at night at Honolulu and Fiji as the guest of Canadian Pacific — with your room and meals at no extra cost to you! When you fly Canadian Pacific style, you fly relaxed, arrive refreshed! Here is the happy balance between air and surface travel that only Canadian Pacific offers."

Misadventure at Tokyo

On the night of February 2, 1952, CF-CPR was approaching Tokyo. The flight, No. 306-9, had earlier departed Manila for Hong Kong at 0141, arriving there 3 hours 39 minutes later. It next left Kai Tak for Tokyo where it arrived after 7 hours 7 minutes. There were ten passengers and seven crew aboard.

The weather was down at Tokyo and it was raining heavily. As the C-4 was about to land with a slight tail wind blowing, the GCA operator ordered a wave-off, as the aircraft was 60 feet above glidepath. The captain, who did not hear the wave-off, continued his landing as he had the airport in view. But he touched down more than half way down the runway. With only 2000-3000 feet of runway left and the brakes not grabbing on the wet concrete, trouble loomed. As a last resort reverse pitch was applied, but seconds later the aircraft left the runway and bounced over a sea wall into the shallows of Tokyo Bay.

The lengthy testimony at the board of inquiry into the Tokyo mishap includes these words from a stewardess. She describes the last few moments

CF-CPR after overrunning at Tokyo on February 9,1950. The sea wall is evident in the second photo. (Public Archives of Canada C117929, C117930)

before the crack-up: "It was only seconds before we hit the wall that I realized it wasn't a usual landing. When we hit, there was no panic among the passengers. All stayed in their seats or returned to their seats." The radio officer reported: "After landing, I felt the brakes applied and the speed of the aircraft was slackened considerably. However, we continued to roll down the runway at a fast rate of speed and after the initial slackening of speed . . . I did not notice the usual pulling action of the brakes . . . As we went down the runway . . . I heard Captain Potter call for propeller reversing switches to be thrown on. I turned to do this, but Captain Sawle was there ahead of me and had the switches on before we reached the end of the runway . . . We left the end of the runway, continued across the unpaved section and finally saw the sea and the retaining wall projecting slightly above the earth. I judged we would strike the wall, which we did, at a speed of approximately 25 mph. There was a rather sharp impact and the next thing, we hopped the wall and were in the water."

Once the aircraft stopped, the crew immediately evacuated the passengers. Help arrived within seconds and as the water was not deep everyone got off quickly and safely. Nine aboard received minor injuries, and CF-CPR was a total loss.

CPA engineer Ferdie Vachon relates that one day a USAF C-54 suffered rudder damage at Tokyo. The wreck of CF-CPR was still on the airport and CPA engineer Red Rose let the Air Force have the rudder from it. The C-54 was soon on its way equipped with a good-sized memento of CPA and Canadair.

Engineer's Views

Fernando "Ferdie" Vachon was CPA's supervisor of maintenance when the Canadair Four was ordered for the Pacific routes. He had begun his aviation career with Canadian Airways in 1937. From 1944 to 1948 he worked for TCA, then he joined CPA, with an early assignment at Canadair while the company's C-4s were being built.

Although he had some reservations about the forthcoming Merlin 620, he was generally pleased with the new aircraft as he watched them taking shape. He had worked closely with the CGTAS Lancasters and was familiar with the Merlin's shortcomings at the time. But once the aircraft were delivered and in service, Vachon was to develop a real liking for them and their engine. He recalls the excellent workmanship produced on the Canadair production line.

Initially there were problems with the aircraft. The Simons-Universal (SU) low-pressure fuel pump was greatly superior to the carburetor in the Lancaster, which had been prone to freezing up. There were initial troubles with the SU pump, but these were overcome. The main coolant pump also presented troubles at first, but it too worked out successfully. Unlike TCA, which switched to the Godfrey blower, CPA stayed with the original AiResearch type and again had success, although the system added to the noise, as crossover exhausts were not adopted. On the advice of Captain Pentland, who did not like its electrical features, CPA inhibited its propeller reverse system so that it could not be used on a routine basis.

After overhaul, Vachon relates, the Merlin's boost control unit had to be calibrated and adjusted to a full power engine run-up. However, such run-ups were to be avoided and under the engine warranty were only allowed once every 15 minutes of engine time. Vachon got around this problem by experimenting with calibration using a large capacity air compressor. This succeeded and no more wide open engine run-ups were needed in the calibration procedure. The DOT approved the technique and it was also adopted by the RCAF.

One day CPA was informed by Rolls-Royce in Montreal that it should immediately ground its C-4s. BOAC was having serious trouble with its Argonauts. The carbon seals in the intercooler coolant pumps were wearing too fast and a spate of engine shut-downs had occurred. But CPA had been getting coolant from its supplier that was 100% glycol, not the usual 60% glycol-40% water mixture. Tanks at the supplier had been improperly labelled and only at this time was the mistake realized. The added lubrication of the 100% coolant had been keeping the seals from failing, and the trouble experienced by BOAC had not hit CPA. Meanwhile a longer lasting seal had been introduced and CPA immediately switched to the 60/40 coolant without having to ground its fleet. CPA's relationship with Rolls-Royce during this period was excellent, with the engine manufacturer providing full back-up for its Merlins. Rolls-Royce Vancouver representative Ernie Gerber was always there when needed and, as Vachon remembers, would not hesitate to roll up his sleeves and go right to work beside the CPA mechanics.

Vachon looks back with favour at the Merlin. As he says, "On the whole it was an angel of an engine. Once the pilot established his constant power setting with the automatic boost control unit, the engines took care of themselves." He adds that on the 12,000-mile inaugural flight in 1949 there were only two minor snags, one with a propeller governor and the other with a carburetor air temperature bob.

Beginning in 1953, the beautiful DC-6B began taking over CPA's international routes. Twenty such aircraft were registered to CPA over the years. With greater speed, range, seating and passenger appeal than the C-4, this Douglas airliner propelled CPA to airline stardom in the 1950s. Seen here is CF-CUQ, which eventually fell victim to an in-flight bomb explosion over northern B.C. on July 8, 1965. (Douglas Aircraft)

C-4s Replaced

The Canadair C-4 put CPA on the map as an important international airline. Statistics show that for 1948 the company flew 4,015,365 revenue miles. This was boosted to 7,280,861 miles in 1951, mainly on account of overseas operations. Passenger revenue figures also show a great increase upon introduction of the C-4: $4,561,803 for 1950, and $10,765,746 for 1951. Yet CPA was finding the C-4 too expensive to operate and decided to phase it out as soon as possible. It made a deal in 1951 to sell its three remaining C-4s to TCA, which badly needed the extra capacity they would offer, and purchased some well-worn DC-4s from Pan American at a very good price. An entry in a CPA report of the day notes that two C-4s sold to TCA plus spare engines and accessories means a $2,182,126 "deductions from property," indicating a very good resale value for the aircraft. The "new" DC-4s were completely refurbished by CPA. Another comment in the 1952 annual report to the directors of CPA says, ". . . the expense per mile flown was reduced from $1.29 to $1.22, the chief factor contributing to this increase in efficiency being the substitution of Douglas DC-4 for Canadair Four aircraft. This change resulted in reductions in per mile costs of engine maintenance, insurance and fuel."

The DC-4s served briefly on CPA's Pacific route, and in January 1953 the airline took delivery of its first DC-6B. This was the aircraft which was to bring CPA genuine long-range capability and the passengers a level of comfort previously unheard of on a Canadian airline.

BOAC: Argonauts around the World

Queen of the airways, BOAC's Argonaut *Astraea* in her element. With the traditional Douglas lines and those of the beautiful Rolls-Royce Merlins, this is surely one of the classic airliners of the propeller era. (BOAC via A.J. Jackson)

BOAC received its first Argonaut in March 1949. Its delivery must have brought a sigh of relief from the corporation, which desperately needed new equipment.

The original BOAC seating allowed for 24 passengers in the forward compartment and 16 aft. A horseshoe-shaped lounge with seating for six was to be provided in the rear of the aircraft, built against the rear pressure bulkhead with two seats having lap straps for use on takeoff and landing by the cabin crew. During service it was found that passengers rarely used the lounge for more than ten minutes at a time, apparently because of aircraft movement. One experienced traveller admitted that even he had become queasy after staying in the lounge more than 10 minutes.

BOAC sent six pilots to Canadair to train in mid-February 1949. Six more pilots arrived on March 18, both groups taking some of the training aboard CF-TEN-X. Also in March, 15 BOAC supervisory engineers arrived at Canadair to train. To assist BOAC in its UK training, Canadair provided a pilot, crew chief and technician for a 2-3 month period during which any use of CF-TEN-X was charged at $400 an hour. The first BOAC pilot to check out on the new aircraft was Captain Tommy Farnsworth.

The first Canadair Four, as BOAC referred to the aircraft at the time, was handed over at the plant on March 21, 1949, this being G-ALHC. It flew on a demonstration flight on March 24 and on March 29 arrived at Heathrow. Aboard were Al Lilly and Smokey Harris of Canadair, Captains Farnsworth and Cane of BOAC, Engineer Officers Jones and Redshaw, and R/O Stewart, also of BOAC.

At a management meeting on March 11, 1949, BOAC had decided to use "Argosy" as the class name for the new aircraft, but a subsequent contest was held among the employees, which was won by Jean Hurst and J. Poole for their suggestion of "Argonaut." This name was officially accepted on May 26, 1949.

Soon after the intensive proving flight of June 14-July 8, 1949, the Argonaut fleet was put to work on the BOAC routes. Inaugurals in 1949 included:

August 23	London-Rome-Cairo-Basra-Karachi-Calcutta-Rangoon-Bangkok-Hong Kong weekly.
August 26	London through to Tokyo weekly.
August 27	Singapore-Hong Kong twice weekly.
November 2	London-Rome-Cairo.
November 7	London-Rome-Damascus-Abadan weekly.
November 8	London-Kuwait weekly.
November 12	London-Rome-Damascus-Kuwait-Bahrein weekly.
November 15	London-Rome-Lydda weekly.
November 16	London-Rome-Damascus-Baghdad weekly.
November 16	London-Rome-Cairo-Bahrein-Bombay-Colombo-Singapore.
November 17	London-Malta-Cairo-Bahrein-Karachi-Delhi-Calcutta twice weekly.

In early 1950 Argonauts began replacing Tudors on the South Atlantic, and on January 12 it was decided that these would be fitted with 36 passenger seats and a cargo bin forward in place of four seats to capitalize on increased cargo along this lucrative route, but the idea did not pay off.

Queen Elizabeth and the Duke of Edinburgh arrive at London from Entebbe following the death of King George VI. *Atalanta*, specially fitted out for the royal tour, brought them home on one of the saddest of Argonaut flights. (*The Aeroplane* via Stephen Piercey)

Another innovation was the experimental fitting out of an Argonaut, less pressurization, to carry 66 troops.

On January 31, 1952, G-ALHK left Heathrow for East Africa with HRH Princess Elizabeth and the Duke of Edinburgh aboard. Within a few days, King George VI passed away and 'HK brought back a queen, not a princess, to England. Later that year a plaque was affixed to 'HK reading, "Their Royal Highnesses the Princess Elizabeth and the Duke of Edinburgh flew from London Airport in this aircraft to East Africa between 31st January and 1st February 1952. Her Majesty the Queen, with the Duke of Edinburgh, returned in it from East Africa between 6th and 7th February 1952."

On August 21, 1952, the Comet replaced the Argonaut on the Colombo service. Comets replaced Argonauts on the Singapore and Tokyo services on October 14, 1952, and April 3, 1953. On October 1, 1953, the long awaited Hermes replaced

Fleet Radio Officer Jimmy Miller and Navigating Officer Cliff Farndell at London Airport before leaving aboard *Atalanta* for Kenya on January 31, 1952, with HRH Princess Elizabeth and the Duke of Edinburgh. Radio Officer Paddy Cussans shared radio duties with Jimmy Miller on this historic trip. (via J. Miller)

Argonauts on all East African routes. Back came the Argonauts on the Comet routes in April 1954 with the grounding of the new jet. On October 5 tourist services were introduced using the newly configured 56-seat Argonauts from London to the Persian Gulf, India, Pakistan, Ceylon, Lagos and Accra.

BOAC continued to investigate an all-freight Argonaut, the final report of November 13, 1952, showing a strengthened floor (130 lb./sq.ft.), provision for a large freight door aft of the rear passenger door (which was to be removed), and all passenger cabin windows to be blanked out. The estimated cost was $200,000 and the scheme was dropped for economy reasons.

In 1952 G-ALHD and G-ALHW were fitted out for royal tour requirements. *Airport News* of April 23 notes in part, "Mr. J. Paine, foreman of hangar five and six, explained that the first stage of the preparation for the Royal flight constituted a Check 4 — a major overhaul of every part of the aircraft including the fitting of new engines.

"Both aircraft were then taken on shakedown flights, How Dog to Bahrein, and How Willie to Nairobi . . . Under the guidance of William Devenney, assistant foreman, the whole of the furnishings and fittings of the aircraft were stripped and replaced with new ones.

"The 32 seats in each plane were the normal type first-class seats, but a special layout of the interior had been evolved and extra partitions and curtains fitted. Attractive personal accommodation is being provided for the Queen and the Duke in the rear of the plane, where the fittings will include two divans and tables. The colour scheme is in dark blue and grey . . . " On April 28, 1952, the royal couple arrived back at London aboard 'HD following their successful Commonwealth tour.

In the February 27, 1953, issue of *The Aeroplane*, R.E. Gillman wrote a pilot report for the BOAC Argonaut detailing some of its finer points.

So rarely nowadays does one hear the purr of a Merlin engine on a civil airfield that one instinctively welcomes the Canadair C-4 whenever it puts in an appearance. As BOAC use the C-4, or Argonaut, on its Eastern and South American routes, the sound of its motors that bring a touch of nostalgia to outstation staff in places as far apart as Rome, Santiago and Tokyo.

In appearance the Argonaut's power plant is deceptive, for the annular radiator and cowlings give it a suspiciously radial look, though the illusion only lasts so long as the engine is silent. The three-bladed airscrews remind one, perhaps, of days gone by, but these are the only features which have an old familiarity about them for the aircraft itself is well up in the ranks of modern air liners.

The flight deck is roomy, with the main controls easy to hand, while most of the electrical switches are set over the pilots' heads on the cockpit roof and laid out neatly in groups according to their associated functions. On the ground, the Argonaut is an extremely handy machine for its size, for the nosewheel is set unusually far back, being positioned almost under the pilots' feet and by virtue of this fact manoeuvres on limited amounts of hard standing are very easy to judge.

The toe-operated brakes rely on the main hydraulic system for pressure and, provided that the gust locks are left "in" for taxiing to prevent the rudder control from moving when pressure on the toe-brakes is applied, it is easy to achieve easy and even braking. The gust-lock lever is held up in the locked position by a "bungee" cord, the other end of which is attached to a rod that actually slides through the throttle quadrant, thus making it impossible for the throttles to be opened up to the takeoff position when the controls are locked. The nose-wheel tiller is in the form of a full wheel and, like most of its kind, rather heavy to operate, but the follow-up action is good, and the very full lock available adds to the handiness of the aircraft on the ground.

The Merlin 626 motors drive Hamilton Standard Hydromatic airscrews which have an overall diameter of 13 ft. 1 in., giving more than adequate ground clearance. These are electrically controlled by toggle switches on the throttle pedestal, or, alternatively, a master lever, when this is switched in, controls all four airscrews at once. The master control is used in the maximum revolutions position for take-off and a row of four amber lights on the throttle box light up when all of them are at the fine-pitch stop.

During the cruise, a synchronizer unit may be switched in, using power units 1 or 2 as the "master," so that the speed of the other three engines will be married electrically to whichever motor is selected as master. However, a built-in limit-band feature restricts the range over which the synchronizer can work to a maximum differential of 90 revolutions, so that, should an engine lose power, or fail, the synchronizer unit will no longer attempt to govern it.

If, during the initial synchronizing of the engines, a motor is more than 90 rpm away from the master, it can be brought into the control range with its toggle switch, but, to obviate such juggling, a resynchronizing button on the control panel widens the control

One of the early Comets graces the clouds. Comets had begun replacing Argonauts on BOAC routes when structural failures forced their grounding and saw the Argonauts taking back the same routes. (de Havilland via Ralph Clint)

Adonis at London Airport on January 18, 1954, looking much in need of a wash. (Peter Keating)

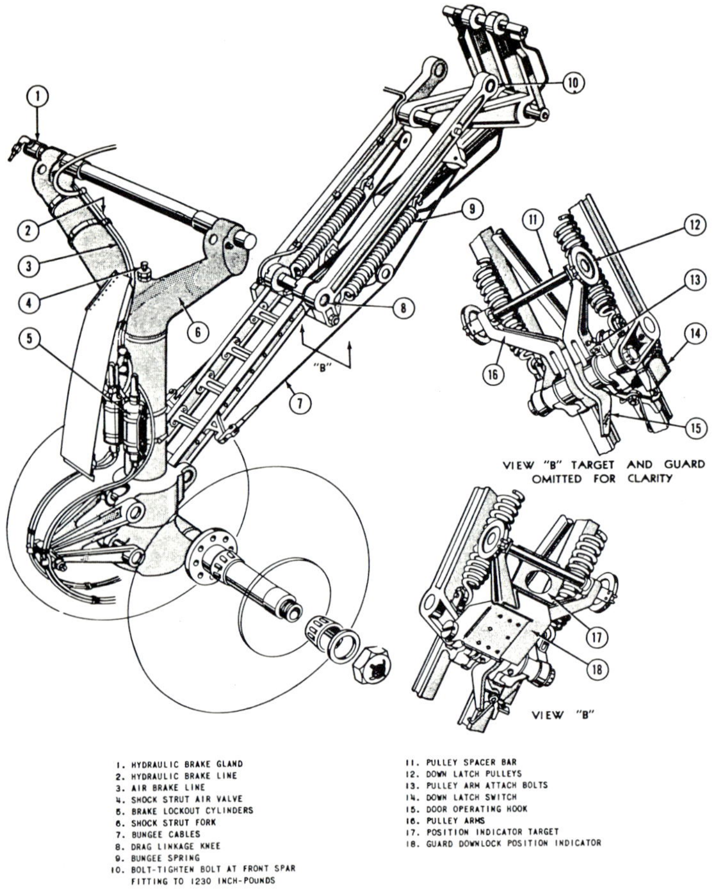

Main landing gear.

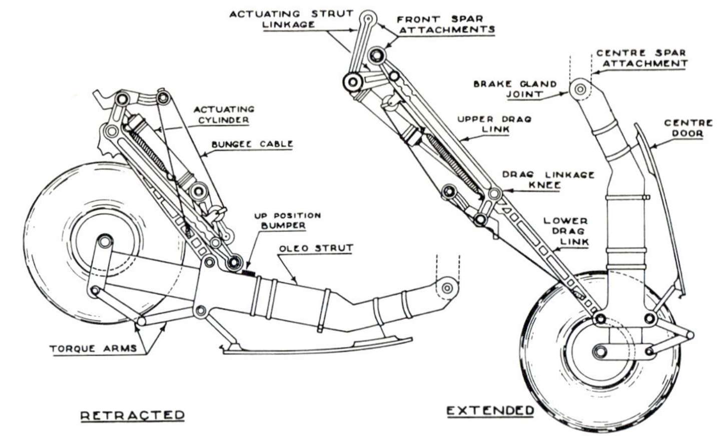

Main landing gear, side view.

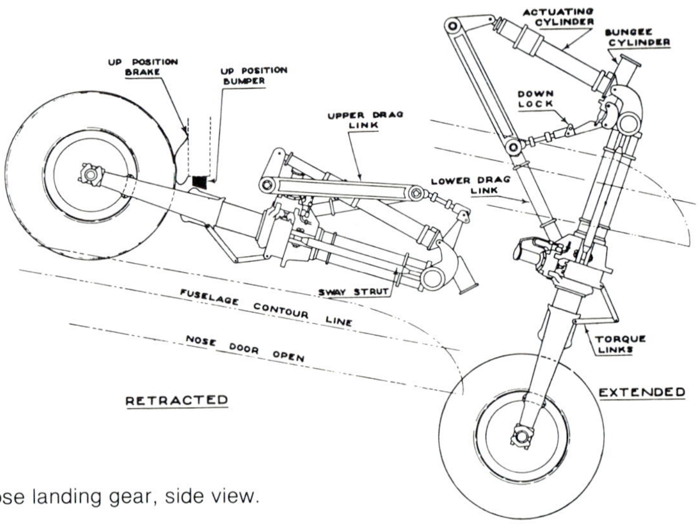

Nose landing gear, side view.

range by a further 90 rpm in order to collect the wandering airscrew. The master control lever over-rides the synchronizer unit while it is moving, so an alteration of power setting can be made with this lever and when the desired revolutions are attained on the master engine, the synchronizer unit will again take over and line up the speeds. This is a vast improvement on the old method of synchronizing airscrews with the aid of their "strobing shadows."

After the normal engine-power checks at the runway holding point, the pre-take-off check runs to 21 items and includes the adjustment of radiator flaps and intercoolers, two features of this type of engine. The auxiliary radiator flaps are set open and the main radiator flaps set to the outside air temperature provided that this does not involve opening them to more than 30 degrees. When the radiator flap switches are set to "auto" during the cruise, the coolant temperature is thermostatically controlled to the desired 104 degrees. The intercoolers are used to control the induction charge temperatures to the optimum 65 degrees. Should this rise during the climb, the injector meters the fuel flow to achieve the correct mixture strength, though this must result in some loss of power.

The Argonaut's take-off is perfectly straightforward, the captain steering with the nosewheel until adequate rudder control is felt, while the power is pushed up to maximum settings of 71.5 inches of boost and 3,000 revolutions. As 110 knots is approached, a long backward movement of the control column is necessary to unstick, the marked displacement of the elevator resulting in rather high stick loads at this point.

As the take-off flap comes up, the change of trim is slight, but the sink is quite noticeable. The optimum climbing speed is 138 knots, but, for the sake of the coolant temperatures, particularly in high ambients, 145 knots is usually held with a power setting of 48 inches and 2,650 rpm. For pilots used to operating with radial engines, the use of such high rpm settings is rather disturbing at first, but the noise level is not seriously increased, since most of the engine noise which finds its way into the cabin comes from the exhaust system. The level of this noise has been considerably reduced by the fitting, by BOAC, of modified exhaust pipes, so that the index is now down to an acceptable level.

The controls are medium weight and crisp for an aircraft of this size and, like the DC-4, which this machine resembles in many ways, the weather-cock stability is good, but a distinct nose-down tendency in turns has to be firmly resisted. The rate of roll is quite adequate and the self-centring of the ailerons is most decisive.

For long-range all-weather cruising, the operating crew could ask for nothing better, for the cabin air-conditioning arrangements extend to the front of the aircraft and no agonies have to be suffered on a frozen flight deck. The cabin air temperature can be maintained between 65 and 85 degrees F. with an outside air temperature range of minus 35 to plus 85 degrees, while the inside temperature can be

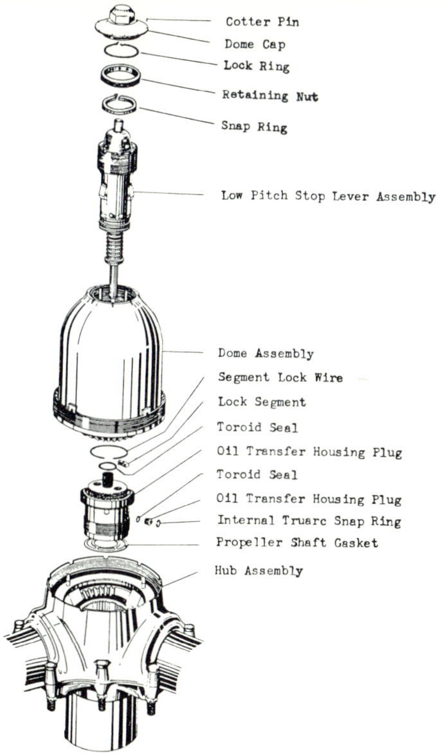

Propeller assembly, North Star M2.

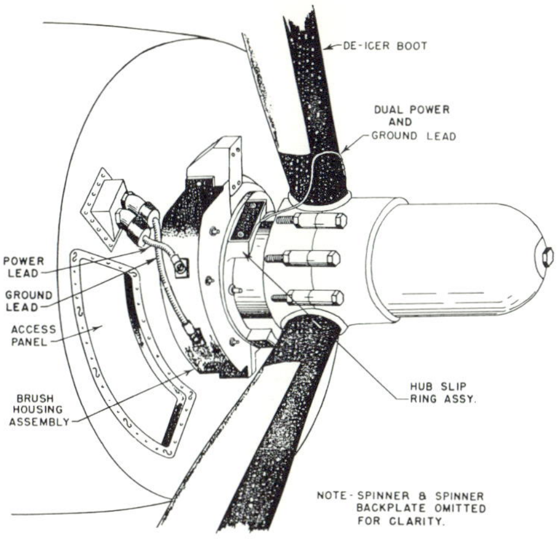

Propeller de-icer installation.

kept at least 10 degrees below ambients of 85-100 degrees. The warmed air is routed under the cabin floor and up the walls for radiant heating before entering the cabin through wall inlets for direct heating. Personal cold vents are situated by each seat row to supply cooler air than that in the cabin and these can be adjusted by the passenger, or shut off altogether.

The temperature rise through the cabin superchargers is usually sufficient to give comfortable warmth, but when the outside air temperature falls below freezing level, a 300,000 BTU combustion heater can be brought into play to keep the cabin cosy. A ground blower is also provided which ventilates the aircraft and provides air to the combustion heater when the engines are not running and ground power is connected, or when engines No. 2 and 3 are running above generator cut-in speed. The hot-air supply can also be directed between the two panes of the sandwich windscreen, the amount of air so directed being regulated to avoid the inside pane misting or, at the other extreme, the outside of the sandwich icing up. With the aircraft pressurized to its maximum differential of 4.16 lb./sq. in. a cabin altitude of 8,000 feet can be maintained while flying at 20,000 feet, and the ventilating rate of the system accomplishes a complete change of air in three minutes, thus, although the passengers can enjoy a comfortable warmth at high cruising altitudes, the air is never stale.

Passenger comfort has become one of the most important aspects of the air line business now that the travelling public has become more discerning in this matter and the international competition is so fierce. Obviously desirable is any facility which assures the customer that his welfare is the company's main concern.

On British air lines the tradition of the captain's rounds is just such a nicety and, provided that this is conducted with care and sincerity, its effect is immediately noticeable. No air line is entirely free from snags and delays occasioned by unserviceability and bad weather, for these are the penalties that fast and complicated aeroplanes bring with them, but it is often difficult for passenger-handling staff on the ground to get this point over to a group of inconvenienced travellers who can know nothing of the factors involved.

However, during the voyage, the captain makes, if possible, a tour of the cabin to inquire after his passengers' comfort and can explain the reason for any en-route delay in non-technical language. The average passenger usually reacts favourably to a reasonable explanation, particularly if he, or she, happens to be an experienced traveller. The captain's round is one of his pleasanter tasks and by his bearing, and carefully chosen explanations, he can do much to instil confidence into his passengers and thus ensure that they make their next air trip on an aircraft carrying the same colours.

Much of the Argonaut route mileage is but sparsely covered with radio aids, so that astro navigation really comes into its own, and the makers of the aircraft have devised an air-lock mounting to take the Hughes periscopic sextant, thus obviating the necessity for an astro-dome. The mounting is housed in the flight deck roof, beside the navigator's position, the dome of the periscope being inserted until it clicks into the first detent. The joint now being air tight, the outer hatch can be opened and the dome pushed up clear of the aircraft. When the machine is pressurized the instrument tends to shoot up at this stage into its mounting, but there is no leakage around it to lower the cabin differential. The sextant is of the pre-setting type and by this method an astro fix can be plotted in less than 12 minutes.

The Argonaut's radio gear is most comprehensive and includes two Marconi AD 107 transmitters and MF/HF receivers, which are high powered and crystal controlled, including ten pre-tuned frequencies which can be used for carrier wave, modulated carrier wave, or R/T, the latter being available also to the pilots once the radio officer has tuned the equipment. Long-range bearings and Consol signals can be received on this set, and two ADFs are available which gives visual indications when used in conjunction with MF beacons, or they can be tuned for aural reception on such aids as radio ranges. A local/remote switch can transfer the control of the ADFs to the pilots, so that they may tune and adjust the volume control of any aid which they are using. One STR 12A VHF/RT set is mounted on the aft face of the control pedestal to give 23 crystal-tuned channels, while an STR 9 four-channel rig is added at Hong Kong to deal with American stations further east.

For short-range navigation and homing, a Mark 2 Rebecca set is mounted on a swinging crate in front of the navigator and thus it is readily available either to him or to the radio officer for homing on a responder beacon, or for use as distance measuring equipment; ILS is fitted as the primary let-down aid.

Three complete compass systems are used: a fluxgate compass, a gyrosyn and the humble P12 magnetic, the first two having their fluxgate and sensing units fitted in opposite wing tips, thus, should one system fail, an

A view of the flight deck of a BOAC Argonaut, including part of the radio panel on the left and the navigator's station in the lower right. (*Flight International*)

ample safety margin is provided.

As a matter of policy, BOAC use a constant-power setting of 950 bhp per engine on this aircraft and on long hauls the rise in true airspeed due to the reduction in fuel weight is quite considerable. A constant-power table is provided for flight-planning purposes; this tabulates the fuel weight used during the climb to the preselected density altitude and the total fuel consumption in lb./hr. at that altitude, while the decreasing all-up weight and the corresponding increase in true airspeed for each hour of flight is also shown.

In computing the total uplift required for the sector, it is then necessary to add the consumption during the descent and the fuel required for flight to the planned alternate field, plus stacking time. The fuel gauges, logically enough, are calibrated in pounds; thus the pilot has an immediate check on the aircraft's weight at any particular time should he find it necessary to refer to the performance tables due to some change of in-flight circumstances. It is only necessary for the makers to produce all fuel gauges marked in kilograms to simplify, still further, the job of flight planning. The flowmeters are marked in lb./hr. and, as no torque meters are fitted to this aircraft, the power is set up on rpm and the flowmeters after referring to the fuel-flow graph, the lowest possible revolutions being chosen for any altitude in order to cut down engine wear and to reduce passenger fatigue.

R.E. Gillman

The Argonaut Crossover

This January 8, 1954, article in *The Aeroplane* by G.W. Boothman describes the BOAC solution to silencing the Argonauts. The article is entitled "BOAC, Keep It Quiet."

Nearly two years ago, the Rolls-Royce "cross-over" exhaust modification was fitted to the engines of a BOAC Argonaut. Passengers and crews had found engine noise in the Argonaut cabin excessive, and the "cross-over" offered a solution to the problem of noise suppression.

Initial trials with the modified exhausts were so successful that the Corporation planned to embody the modification on all four engines of each of the 22 aircraft in their Argonaut fleet. The Merlin power-plants were sent, in turn, to Rolls-Royce for alteration; and by July 1, 1953—the date scheduled for completion of the programme—the last one had the new exhausts fitted and was back in service.

One power-plant, at least, already has over 3,200 hours to its credit since being modified. While it is still too early to assess the full benefit—and cost—of the change, sufficient is now known to make an interim assessment permissible.

The modification is a success. The hoped-for degree of noise suppression has been achieved. And whatever other claims may have been made for the "cross-over," it was as a means of making the Argonaut cabin a more comfortable place to travel in that it was adopted.

To get some idea of the price paid in the interest of quieter travel, it is necessary to know something of the design of the modification, and a brief description may help to refresh readers' memories.

On the un-modified Merlin, exhaust gases from inboard and outboard cylinder blocks escaped through manifolds or stubs on their respective side

Arion being serviced at London sometime in 1960. This Argonaut later served with Overseas, Derby and BMA before being scrapped in 1968. (Peter Keating)

of the power-plant. The "cross-over" exhaust takes the exhaust gases (and noise) from both banks of cylinders and ejects them on the outboard side of the power-plant—the side of the engine remote from the cabin.

To effect the transfer of "cabin side" cylinder gases to the far-side outlet, a "blind" manifold collects exhaust gases from the inboard stubs. Three cross-over pipes, bridging the two cylinder blocks, conduct the gases to the outboard "ejector" manifold, which also serves the outboard bank. All gases and noise are ejected through this manifold's three short stubs.

The temperature increase, caused by exhaust gases being retained for a longer time within the power-plant, called for special cooling arrangements. This is provided by a scoop in the bottom of the front-ring cowling. Two pipes, one on either side of the reduction-gear housing, take air over the front of the engine and direct it on to the cross-over pipes. Trunks taken off the pipes lead a flow of air to the stubs and manifolds.

The cam-covers have been modified to support a "cage," which surrounds the exhaust stubs and directs the air-flow around them. Asbestos lining on the floor of the cage gives additional protection to ignition leads and sparking plugs.

To ensure a controlled flow of air round the cross-over pipes, they have been encased in a "trough." This airflow—after cooling pipes, stubs and manifolds—leaves through a louvred section of the power-plant top cowling, which forms a "lid" to the trough.

Support for the manifolds, cross-over pipes and trough is provided by struts fixed to the main ring of the engine bearer structure. Ball-joints and rubber mountings at all suspension points allow independence of movement between engine and cross-over assembly.

For this purpose, too, the stubs have ball and slide joints mid-way between exhaust-port face and manifold aperture. Cross-over pipes

G-ALHD, *Ajax*, about to touch down at Rome. The Merlins are idling but within moments they will roar back into life as the propellers are thrown into reverse pitch. (British Airways 17089)

are each similarly jointed, half-way between blind and ejector manifolds.

Several undesirable features of the cross-over system are immediately apparent. The weight of the unit, for example, is far from negligible. Transfer pipes, support struts, cooling pipes and trunks, troughs and cages, all mean extra weight to be carried and, consequently, pounds off the pay-load.

The rate of wear and tear is bound to be high. Ball-joints, sliding joints and non-rigid mounts are prone to chafing and rapid deterioration and require frequent inspection and considerable maintenance. Relatively high operating temperatures, too, encourage component depreciation.

The life of each component has not, of course, been finally fixed. Based on their behaviour and condition to date, estimated life figures for the various parts are: Cross-over pipes and ejector manifolds—2,000 hours each; 1,500 hours for the blind manifolds, and only 1,000 hours for the stubs.

It is hoped that most of these provisional limits will eventually be raised. The ejector manifold is now thought to be good for nearer 2,500 hours. Stub life, however, may be substantially less than the 1,000 hours with which it is at present credited.

Stubs serving cylinders one and six come in for the worst hammering. As the three transfer pipes are centrally located in the manifold, these two stubs are positioned farthest from them. Gas and heat dispersal is therefore slower. As a safety measure, stubs one and six have been remodified. The obvious place for stub failure is at the ball and slide joint. On the two extreme stubs, collars or "belts" have been fitted to prevent them sliding into the manifold and leaving a break, with the danger of flame-leakage, at the failed joint.

The time needed for inspection and maintenance of the new assembly has naturally meant an increase in the duration of the servicing period. In addition, that period has been lengthened because the position of the new unit has made servicing of the power plant proper more complicated. The arrangement of "cage" and cross-over pipes, for instance, makes the removal and replacement of sparking plugs more difficult. Reaching the plugs in the "Vee," and beneath the trough, looks like a job for a Houdini!

On the credit side of the balance it has been noted that since the introduction of the modification there has not once been a "route delay" due to exhaust trouble. Difficulties with back-pressure, which might have been expected in a drastic change of exhaust design, have not been experienced—at least, not since they have been fitted to the Argonauts.

In all, the results of the change are pretty much as one would expect. To gain quieter conditions in the cabin, something was bound to be lost. Whether the loss will exceed that originally estimated will not yet awhile be known.

The handicap of extended mainte-

Two views of the BOAC crossover exhaust system. (Rolls-Royce)

▽ An Argonaut with the crossover exhaust system on the inboard engines only. (*Flight International* via Stephen Piercey)

The Merlins of G-ALHD humming along over the Andes. This view shows mods made to the exhaust system. (*Flight International* via Stephen Piercey)

nance time should be reduced as life expectations are more accurately defined and the need for constant supervision lessens. It should be cut, too, as fitters become more accustomed to working on and around the new assembly.

If, as may happen, "beefing up" of some components is necessary, the weight penalty may yet rise, though any great increase is unlikely.

BOAC's bid for noise suppression appears, so far, to have been rather expensive—but, with the down-payment made, instalments from now should be progressively lighter.

G.W. Boothman

G-ALHW at Beirut on August 31, 1957. Compare its colour scheme with that of G-ALHD. (Peter Keating)

The BOAC Argonaut crossover was effective but not perfect, as log books record occasional failures. Captain Peter Cane's log for October 21, 1950, reads, "No. 2 feathered. Fish tail exhaust burnt out." In a lighter way Captain John R. Johnson, recalling the crossover, said in a 1961 interview, "Pilots didn't like it at first. With less noise, they felt the power had been reduced, though of course it hadn't. I remember on our first trip over the Alps, after the modification, my first officer, who was accustomed to shouting at me, leaned over and whispered, 'This silence, skipper . . . it's eerie'."

Life with the Argonaut

BOAC Captain J.R.C. Young recalls some exciting times flying the Argonaut. On November 10, 1953, he was landing G-ALHF at Rome. It was night. The captain was handling the control column and the F/O the throttles. As Captain Young remembers it, "At about 50 feet I called, 'Power slowly off.' Much to my amazement this was followed by a roar of mighty engines and we were in full reverse! I quickly pushed a burst of full ahead into the motors. There was an almighty crunch and the aircraft settled onto the runway. After taxiing clear I went back into the cabin and found that all the rack luggage had fallen on the aisle but that no one had been touched by it.

"On arrival at the tarmac I found that the relieving captain had already called London to report the heavy landing and that he was not taking 'HF on! Indeed, it was clear that damage, confined to the nose oleo, had been done.

"As to the cause of the mishap, it transpired that a Murphy's Law mistake had occurred in reassembling the reverse mechanism in the shop, coupled with the F/O raising the physical obstruction plate in flight to make pulling back into reverse possible whilst still in the air."

On another occasion, writes Captain Young, "I had left London on September 26, 1956, with F/O Goulburn, N/O Williams and R/O Hubbard. All went well and after the usual Idris-Kano-Accra-Kano legs with stopovers in Idris and Kano, we left Kano on October 1 just after sunset for Idris. The weather north of Kano was very bad with severe thunderstorms. We weaved our way northward until at about 1900 hours No. 2 engine started to run rough. After checking to see if it would run on one magneto at reduced power (it would not), it was feathered at 1910. Kano was an hour and a half behind, equipped only with a radio beacon and encircled by storms. Ahead was the edge of the overcast and clear starlight. Accordingly I asked Williams to prepare a three star fix for about 1920 and when we were sufficiently clear of the cirrus overhang he got his fix before we turned around and plunged back into the murk. We landed at Kano at 2140. Of course, our plan had been radioed ahead and beds were ready for us and the

'HS alights in this fine old *Flight International* picture. (via Stephen Piercey)

passengers, and a new engine was being prepared. After minimum rest and a morning cup of tea we took off again at 0613 on a murky, wet morning.

"Goulburn and I took turns on watch and at about 1007 I handed over to him and got up to stretch my legs. I was leaning over Hubbard's desk. Imagine my surprise to read, as he took down a message, 'Priority. Feather No. 2 engine immediately. Internal circlip missing No. 2 propeller.' I turned to Goulburn and said, 'Feather No. 2.' He immediately complied, but afterwards, over a drink, admitted that he wondered at the moment if perhaps the old man had gone completely bonkers!

"We proceeded northward to Idris where we landed at 1553. The flight across the Sahara half on three engines had taken only a little longer than usual. At Idris everything was ready for repairing our propeller assembly. We later determined that the engineer who had changed the engine had gone off to bed about 4:00 a.m. then came back later in the morning to clean up. As he moved some ladders into a corner, there on the hangar floor was the circlip. He rushed to the tower and had the message transmitted to us.

"Of course, all worked out well, but if that engine had given so much as a hearty cough, the propeller would have come off. I greatly appreciated the courage of the engineer in sending the message, and my next trip to Kano, about a month later, we shared a drink together. I paid!"

Captain Young also provides some details in the case of the mysteriously disappearing donkey. He was checking his load sheet at Wadi Siedua airport, Kano, when he overheard a commotion. Another BOAC pilot, Captain Wright, "was expostulating quite vigorously," and as Captain Young relates, "He had seen a donkey on the runway just as he flared for landing. The airport personnel with whom he was debating were saying that that was quite impossible whereas his first officer was saying that there was blood on the flap and that the flap was damaged! As the argument developed it seemed that the authorities were insisting that Wright must have hit the donkey on takeoff at Beirut! Meanwhile, our engineers and station staff were frantically searching for the wounded donkey. It never did turn up, leading us to muse about the world's quickest donkey funeral!"

BOAC Captain R.K.G. Rice has provided some of his Argonaut recollections. Like so many of his fellow aircrew, his memories are generally pleasant ones.

I first met the DC4M, known in BOAC as the Argonaut, on March 17, 1950, in transit at Rio de Janeiro. I was London-bound with York G-AHFG, the last Starliner of British South American Airways Corporation, operating what we knew as the East Coast Route to Buenos Aires. Our merger with British Overseas Airways had just taken place and their first Argonaut was on "our" route from London via Lisbon, Dakar, Natal, Rio Montevideo, Buenos Aires and over the Andes to Santiago.

Captain Bob Rice at the controls of a BOAC Argonaut. (R. Rice)

My next sight of this machine was to be at London Airport when, after some twelve weeks of ground school on the technicalities of the aircraft in preparation for an examination set by the UK Air Registration Board, Base flying training commenced on August 14, 1950, in Argonaut G-ALHN. For those of us who came from BSAAC this was the first custom-built civil aircraft which we had encountered and we were well pleased with our good fortune. The luxury of a steerable tricycle undercarriage, reversible propellers and pressurization — we really thought that we should be able to fly above the weather — an immense leap forward from the converted bombers (Lancastrians) and military transports (Yorks) with which we ex-BSAAC folk and many of our new-found BOAC colleagues were familiar. Some of the latter pilots were ex-flying boat men and to them the transition was even more dramatic — not only wheels, but they moved up and down.

Base conversion flying, almost all at and around London Airport, totalled some 10-12 hours general handling prior to route operating experience, which in the case of the ex-BSAAC men also involved route experience. Two aspects of this conversion stick in my mind — one not without humour, the other deadly serious.

On August 16, 1950, Captain Archie Jackson and I were considered ready, well nearly so, for a "solo" P1 and P2 circuit apiece. We proudly steered our charge to the end of the takeoff runway, there to be greeted by the figure of our guide and mentor, Captain Bill Griffith who was standing beside his car with a large placard inscribed 'SHUT OFF VALVE?' This device, controlled by a red knob on the cockpit floor, in effect rested the engine-driven hydraulic pumps during cruise but, if left at 'OFF' during ground manoeuvring could result in embarrassing problems as hydraulic pressure was used up by the operation of brakes and nose-wheel steering. Bill clearly had his doubts.

In a more serious vein, the problem of minimum control speed (air) which I believe was 110 knots IAS was convincingly demonstrated. At a safe height an outer engine was "failed," speed reduced to around 100 knots IAS, when one was told to apply full power on the live engines *and* to keep her straight." No way — hard rudder — bank — still she turned towards the "dead" engine. The demonstration was usually concluded by the training captain's succinct remark, "See what I mean?" We did — the recovery procedure of throttling back to check the swing and dropping the nose to build up airspeed would not be very practical just after lift-off.

Despite the "leap forward" in comparison with our previous aircraft, the Argonaut was not what we later came to know as a Performance A machine, where at any stage of a takeoff the operation can be safely abandoned, or continued in the event of engine failure. This fact we had well in mind when operating at maximum weight from the often rough and frequently

Date	York	Route	Time en Route
April 19	G-AHFC	London-Lisbon	5:45 hours
		Lisbon-Dakar	9:20
22	G-AHFH	Dakar-Natal	9:05
		Natal-Rio	7:15
23		Rio-Montevideo	6:05
		Montevideo-Buenos Aires	1:15
		Buenos Aires-Montevideo	1:00
27	G-AHFC	Montevideo-Rio	5:50
		Rio-Natal	6:45
28		Natal-Dakar	9:25
May 1	G-AHFD	Dakar-Lisbon	8:00
2		Lisbon-London	5:40

Progress. This South American trip was flown by Captain Gosling in 1948 with an Avro York. Compare the times with those of a similar trip in an Argonaut.

Captain Gosling remarks of the York era, "We cruised the Yorks at around 9000-10,000 feet. They didn't have the ceiling for the Andes, which we crossed at 18,000-20,000 feet with everyone aboard wearing oxygen masks." For the Andean sector of the South Atlantic service Lancastrians were used as they performed better than the Yorks at high altitude. The Andean return trip was 8-10 hours. Overall, the entire London-Santiago-London service via York/Lancastrian was some 10 hours longer than by Argonaut.

Date	Argonaut	Route	Time en Route
April 13	G-ALHM	London-Lisbon	5:05 hours
		Lisbon-Dakar	7:35
16	G-ALHN	Dakar-Recife	8:15
		Recife-Rio	5:15
17	G-ALHK	Rio-Montevideo	5:30
19		Montevideo-Buenos Aires	1:05
		Buenos Aires-Santiago	4:05
		Santiago-Buenos Aires	3:35
21		Buenos Aires-Montevideo	:55
		Montevideo-Rio	4:45
25	G-ALHN	Rio-Recife	5:25
		Recife-Dakar	8:50
30	G-ALHK	Dakar-Lisbon	7:45
May 1		Lisbon-Madrid	1:40
		Madrid-London	3:30

BOAC's Argonaut service to South America. These excerpts are provided by Captain Leslie Gosling from his log for 1952. The trip took some 73:15 hours aloft.

hot (a log book entry for 10.8.57 reads Baghdad — Basra +45°C) airfields of the Persian Gulf, India and the Far East or the "hot and high" fields of East and West Africa.

It was while operating from one such field, Basra, not high but hot and rough, that Ben Taylor, then operating as first officer, and I made a, to me startling, to him painful and startling, discovery. The selection of reverse thrust was achieved by the first officer raising a flap which normally prevented the movement of the throttle levers beyond fully closed. Once the flap was raised, further rearward movement of the throttles gave the desired "reverse thrust," a very useful stopping aid. On 28.6.54 while seeking to depart from Basra in a weighty G-ALHS on a hot and windless night, unusual engine noises and vibration at around 90 knots IAS made a halt advisable. My bellow of "abandon" produced the prescribed response — Ben's left hand sought the flap and raised it some 45° when the closing throttles (my contribution) made contact with his fingers, jamming them against the forward edge of the flap and producing a very effective, if unplanned, locking of the throttle levers, but no reverse thrust. Ben was a forceful character with considerable strength and a fine flow of language. The throttles were forced forward against my rearwards heave, fingers and flaps were freed, and full reverse plus four feet on the brakes brought us to a spectacular stop in clouds of sand and dust as the desert commenced. The cockpit was then regaled with Ben's opinion of engines, reversing devices and ham-fisted captains who sought his mutilation. We perforce night-stopped and two bottles of rum ex Rome restored good pilot relations ere a three-engined ferry to Bahrain was made next day.

This adventure was in the future when on September 3, 1950, my route training/experience commenced under the guidance of Captain Dusty Miller. The voyage took us from London to Rome, on towards Cairo, where fog resulted in a diversion to Nicosia (Cyprus) with a subsequent arrival at Cairo. The voyage continued via Basra and Karachi with my "Eastern Indoctrination" terminating *pro tem* in Calcutta (Dum Dum Airport) and the Great Eastern Hotel. The Karachi slip introduced me to the Speedbird Resthouse, then something of an oasis in a wasteland between the fleshpots of Hong Kong, Singapore and Tokyo at one end of the line and Rome and home at the other. The considerably smaller, but equally hospitable, Speedbird Resthouse at Bahrain filled a similar need in the then wilderness of the Persian Gulf. The latter boasted the only bar on the island — a veritable oasis for the somewhat dehydrated pilgrims from the East. At the former, beer was almost prohibitively expensive, albeit later subsidized for crews on the advice of the medical department of BOAC. Normally crews "laid over" at Karachi east and westbound, giving the opportunity of exchanging news and views of facilities, weather, etc., along the line. My return from Calcutta now in the charge of Captain Stan Gooch was via Karachi, Bahrain, Cairo and Rome, flying a total of some 44 hours route and aircraft familiarization.

A few days "duty crew" at Heathrow involving a few hours of air tests was followed by a second route/aircraft familiarization voyage this time under the eye of Captain Ballantyne. In the course of this trip Rome, Damascus, Basra, Kuwait and Bahrain were visited and some 33 hours flying completed. At the end of this period I was cleared for route operations, having completed some 12 hours of base training and 80 hours of route training which, in the case of ex-BSAAC pilots, also involved route clearance.

Thus we were launched "on the line" with the Argonaut as our tool of trade and an effective and trustworthy tool we found it. Our crew at this time consisted of two pilots, a navigator and a radio officer, the latter being phased out in 1954 with the advent of H/F/R.T. Several took pilot training and stayed with BOAC, one at least to become captain.

The Eastern, East and West African routes of the fifties included airfields which, during the hot season at least, and it was always hot somewhere, posed performance problems but we had been alerted to them and had our "escape routes," or failing one, "soft spots" in mind as we pointed the aircraft down the runway. As with all machines, especially when they fly,

Aethra on the ramp at London Heathrow. (Peter Keating)

just a little bit of luck was on occasion to be desired. Fortunately such occasions were rare on the Argonaut and my engine failure on takeoff at Basra was the only one I experienced in eight years and over 5000 hours of Argonaut operations. In this context there follows a resumé of the incidents which I encountered during my period of Argonaut operations.

In regions where temperature of 30° and more in excess of ICAN were not unusual, liquid-cooled engines clearly had their problems and from the outset coolant loss, with resultant overheating, was a factor as evidenced by the following:

G-ALHI, 16.8.53, between Beirut and Basra No. 3 engine shut down due to coolant loss, and again on 1.5.55 on G-ALHF out of Rome, and yet again on G-ALHM, Bahrain to Beirut, on 9.10.56. Still, only three in eight years and we did perforce "cook 'em" on occasions.

As well as these cooling problems the following incidents show the random scatter which even in the eighties would seem to affect all machinery — especially if it gets airborne.

On 15.2.51, G-AHFD, cleared to descend on a clear night into Cairo, throttles closed. On attempting to trickle on some power, fortunately well above the cleared flight level, the throttles were immovable! After several minutes of unavailing attempts to remedy the situation, hot air was selected and the throttles immediately freed. Presumably this incident was due to carburetor icing, which we had been assured could not happen.

12.4.53, out of Hong Kong for Bangkok in G-ALHK, returned after 2 hours — No. 3 u/s! And on 22.7.53 on the same route returned after 1½ hours No. 4 u/s.

On 23.9.53, Kuwait, tire burst on landing. At that time the runways at Kuwait were officially designated as "hard packed sand," marked over their length by an extra-hard oil-soaked centre strip, which it was advisable to avoid, and at either end by a "lead in" row of up-ended oil barrels. We always took extra care to "lash up and stow" before arrival here. Incidentally this field then boasted no navigation aid of any kind and sand haze was common in the Gulf areas.

29.8.54, G-ALHM out of Rome — No. 4, "crowned piston."

6.12.55, G-ALHW out of Beirut — No. 3 u/s, oil pressure low, temp high.

5.10.56, G-ALHP out of Karachi, returned after two hours — No. 4, supercharger cracked.

19.7.55, G-ALHE, dropping down towards Rome — No. 3, broken connecting rod.

The pièce de résistance was presented by G-ALHP en route from Okinawa to Tokyo on March 30, 1957. Cruising peacefully in daylight in fair weather, a sharp jolt was followed by a banshee howl. No. 4 propeller was over-speeding. Standard drill produced a stationary propeller and a glorious silence — only three Merlins to be heard. However, the erstwhile stationary blades began to rotate gently, then faster, were re-feathered and speed was reduced to its limit. Terra firma seemed desirable so course was set for Nagoya, nothing being nearer. No further trouble, though that propeller was regarded with deep suspicion — then just at dusk the first officer peered hard from his window and volunteered, "I got Lootenut P. Brown — how about you?" His eyes were better than mind for I could not read the name on the bone-dome in the cockpit of the USAF fighter which sat just below my window. So HP came to Nagoya in good company.

With the exception of the last incident none was unduly alarming and all were easily handled by two "just average" pilots, which speaks well for the machines involved. Considering that they were spread over eight years of the fifties, I suggest that the record vis à vis other contemporary types is better than average.

In making this assertion I suggest one should consider that a Far Eastern trip of those days, of which the following is a typical example, involved a great variety of airfields: London (Heathrow) — Rome (Ciampino) — Basra — Karachi — Calcutta — Rangoon — Bangkok — Hong Kong (Kai Tak) — Okinawa — Tokyo — Okinawa — Hong Kong — Saigon — Singapore (Kallang) — Colombo (Ratmalanna) — Bombay — Karachi — Bahrain — Kuwait — Damascus — Rome — London.

One could encounter blizzards, in Europe and Japan, and tropical storms and the Asian Monsoon in the course of the same voyage. Memories of Kallang airfield, Singapore, where the tower on occasion exhorted us to avoid the cottage hospital which loomed just off the normal departure route. Bound for Colombo heavy on a hot, humid, windless day, we devoutly hoped that we should! Or leaving Ratmalanna (Ceylon) where the runway resembled a rutted country lane, we were sometimes asked to avoid the sacred banyan tree which towered slightly off-centre at the runway's end. Again this too was our wish.

Hong Kong (Kai Tak) added its

quota to the joys of the route. Approaching on 13 with a close study of mountain flora, followed by a 90° descending turn, full flap, power off to the end of the runway — not overlong and so often blessed with crosswinds with gusts. The Argonaut was a "natural" for this somewhat specialized manoeuvre as it also was for "motoring and manoeuvring" around the harbour just below a ragged cloud base sitting on the hills with heavy squalls of rain while the local cement works belched smoke across where we trusted the end of the runway would be. On emerging from a heavy rain squall on the approach to 09 — over the flare on the police station roof — one could and did hop over the odd double-decker bus which unkindly drove across the road just beyond the perimeter fence of that rather short runway from which an overshoot would have called for aerobatics or a vertical climb.

East and West Africa posed their peculiar problems, in particular the hot high fields of Nairobi, Entebbe, Kano and Khartoum. On one occasion the skipper handing over to me volunteered, "You have quite a load. I'd back into the bush and take a run at it!"

The Argonaut served all these fields regularly and with little fuss. I think the only equipment we envied other operators was their storm warning radar, especially during the monsoon season over India and the Gulf. We once saw a DC-6 of Sabena on the ground at Basra which had suffered so severely from hail in flight as to be eventually written off.

The "Star" of the preceding saga is without doubt the DC-4M Argonaut which enabled a number of "average" airmen to safely transport very many passengers over long distances through every kind of weather regularly and with little fuss. They served BOAC, their passengers and their crews faithfully and well.

R.K.G. Rice

Captain Leslie Gosling recalls one flight with a new stewardess, Yvonne Van den Hoek, who had just finished her first service with BOAC. Her great ambition was to become a pilot. She had joined the Airways Aero Club a day or two before her BOAC flight, which took her to Santiago. Captain Gosling arranged for her to have her first flying lesson in an Argonaut, which took place at 14,000 feet above the South Atlantic between Recife and Dakar. The captain put Van den Hoek in the left hand seat, disengaged the auto pilot and taught her how to fly the Argonaut straight and level. The 40 passengers were completely unaware that the stewardess was flying the aircraft.

Later on she became a flying instructor, then air traffic controller at Gatwick. In time she became an airline pilot, eventually retiring as captain from BAC 1-11s in 1980. Among her many plaudits over the years are the Brabazon of Tara Trophy (1959); IAOPA Air Traffic Control Award "For the guiding and guarding of air safety" (1965); the Amelia Earhart Scholarship (1967); and the Sir Alan Cobham Award (1969). In 1974 Princess Anne presented her with the Whitney Straight Award "For exceptional courage and determination under difficult circumstances, in pursuit of a career with the highest qualifications in aviation." And it all started aboard an Argonaut!

Two other incidents recalled by Captain Gosling concern "irregularities." The first occurred on his initial trip in command in March 1950. After arrival at Buenos Aires, when the aircraft had been unloaded, he handed it over to the first officer as was normal practice. He was to taxi out to the dis-

Argonaut crew at Santiago on April 19, 1952. Left to right are Navigating Officer Watt, F/O Don Palmer, Stewardess Campbell, Captain Leslie Gosling, Radio Officer Les Evans, First Steward Connelly and Second Steward Bixter. (L. Gosling)

Service personnel put up a major effort to get 'HP back in service in this 1954 photo. (*Flight International* via Stephen Piercey)

Date	Argonaut	Route	Time en Route
May 3	G-ALHR	Khartoum-Cairo	4:35 hours
4	G-ALHS	Cairo-Rome	6:35
		Rome-London	4:40
11	G-ALHE	London-Rome	4:05
		Rome-Cairo	5:40
13	G-ALHX	Cairo-Khartoum	4:40
14		Khartoum-Entebbe	4:55
15	G-ALHV	Entebbe-Livingstone	5:45
		Livingstone-Johannesburg	2:55
16		Johannesburg-Livingstone	2:45
		Livingstone-Entebbe	5:50
17	G-ALHE	Entebbe-Khartoum	4:50

Part of BOAC's African service as flown by Captain Gosling in 1955.

Date	Argonaut	Route	Time en Route
July 27	G-ALHG	Tripoli-Kano	6:25 hours
29	G-ALHW	Kano-Accra	3:25
		Accra-Kano	4:05
31	G-ALHY	Kano-Tripoli	6:45
August 1		Tripoli-Rome	3:55
2	G-ALHF	Rome-London	4:50

Argonaut flying on BOAC's West Africa service, again from the log of Captain Gosling.

Date	Argonaut	Route, Altitude	Time en Route
Feb. 8	G-ALHW	London-Rome, 19,500′	3:38 hours
		Rome-Cairo, 21,500′	6:00
10	G-ALHT	Cairo-Bahrein, 15,500′	4:37
		Bahrein-Karachi, 15,500′	4:13
12	G-ALHX	Karachi-Calcutta, 15,500′	4:55
		Calcutta-Rangoon, 15,500′	2:37
13		Rangoon-Bangkok, 11,000′	1:45
		Bangkok-Hong Kong, 15,000′	5:00
14		Hong Kong-Tokyo, 17,000′	7:20
16		Tokyo-Okinawa, 10,000′	5:16
		Okinawa-Hong Kong, 8000′	5:17
18		Hong Kong-Singapore, 16,000′	6:29
19		Singapore-Hong Kong, 15,000′	7:30
20/21		Hong Kong-Bangkok, 14,000′	5:00
21		Bangkok-Rangoon, 10,500′	1:59
		Rangoon-Calcutta, 14,000′	3:15
22		Calcutta-Karachi, 10,500	6:11
25	G-ALHF	Karachi-Basra, 14,500′	6:04
26		Basra-Cairo, 14,000′	5:23
27	G-ALHU	Cairo-Rome, 20,500′	6:20
		Rome-London, 20,500′	4:19

Details from the log of BOAC senior flight captain Peter Cane showing the long gruelling London-Tokyo-London service of 1950.

persal point with the station engineer occupying the co-pilot's seat. Unfortunately neither remembered to on-load the hydraulic pump after start-up. They had something like 400 yards of taxi way to cover followed by a 90 degree right turn. By the time they came to the turn they had used up the supply in the hydraulic reservoir. The nose wheel steering and brakes were now useless and the aircraft went off the taxi way, down a grass slope and finished up between two trees with a bent wingtip.

The other occurred on a night flight half way from Recife to Dakar. The captain recalls, "I went back into the cabin to talk to the passengers leaving the auto pilot engaged and the first officer on watch. When I returned to the flight deck some 20 minutes later, I noticed that the moon was shining on the port side whereas before I had gone back and it had been to starboard. A glance at the compass revealed that we had made a 180-degree turn and were heading back towards Recife. The auto pilot had become disengaged and the aircraft had made such a gentle turn that it had not been noticed by the first officer, who was busy writing up his hourly report in the technical log. We lost no time, believe me, in turning back on course for Dakar!"

Following withdrawal from BOAC service of the Comet 1 in April 1954, the company terminated its South American routes as the Argonauts were required in taking over from the Comet on other routes. Captain Gosling brought the last Argonaut service out of Rio for London on April 15, 1954.

Tales of a BOAC Navigator

W.G. "Bill" Hunt spent many years as a navigator on the BOAC Argonaut. He has recorded some of his experiences and recollections for use in this book:

FLIGHT CREW

We flew the Argonaut with a crew of seven: two pilots, navigator, radio officer, two stewards and a stewardess. There was a jump seat which could be set up between the two pilots if a check captain was on board, but this was rarely used. We managed fairly well without a flight engineer (*pace*, my one-time fellow trade unionists!), and in the circumstances of the time, on the routes we flew, two pilots and two specialists, one in navigation and one in communications, made a well balanced technical crew.

As for the cabin, a team consisting of a chief steward, a junior, and a stewardess seemed to work out very well. When I joined the Argonaut fleet, aviation offered only a single class of travel, at a standard which subsequently became known as first class. The Argonaut had 40 seats, and three cabin attendants could give very good service. When tourist class was introduced, the number of seats was increased, the crew remaining the same size. To young single men, as many of us were in the early 1950s, the main disadvantage of this arrangement was that there was but one girl among six men. In those days, being a stewardess was a very OK sort of thing which many might aspire to but only a few achieved. Certainly the Argonaut girls were, with few exceptions, a pretty good crowd. Of course there were a few who were deservedly unpopular, the worst type being those we dubbed "Daddy's Daughter," who affected the airs and graces they imagined went with high social positions. The salient characteristics of this affectation were a reluctance to do their share of the work in flight, and an equal reluctance to mix socially with the crew on the ground.

We had on the Argonaut fleet a group of airmen who were for the most part thoroughly competent and a pleasure to fly with. Apart from a few senior men who had been in Imperial Airways before the war, almost all the flight deck crew had been in the RAF and of these most had been on operations. There were a few Fleet Air Arm

One of the finest ever Argonaut portraits. How unfortunate that such a beautiful aircraft should end up in the obituary column as "Disposed of locally for scrap." (British Airways)

men as well. Quite a few had been decorated — we had a number of pilots with DSOs and DFCs, and the latter were not all that uncommon among R/Os and navigators. This background made for a high level of camaraderie and morale.

We also had some excellent stewards, good on the job, and popular with both passengers and crews. Not so many of these had been in the RAF, far more in the army, and a few in the navy. I particularly recall Ron Stone among the latter who, when coming forward to report the cabin secure for landing, would open the flight deck door and bellow, "All secure below, sah!"

THE FLEET

Inevitably, the Argonaut was early dubbed "Juggernaut," and so it remained to the end. But make no mistake, this soubriquet was no mark of derogation, but rather of affection. For me, and perhaps for others, the Argonaut, while not the last aircraft to merit respect, was the last to earn truly affectionate regard. Collectively, the fleet was usually referred to as "Jackie's Juggernauts," after the boss, Captain "Jackie" Harrington.

The individual aircraft names were hardly ever used by BOAC people. Sure, we used to write them along with the registration on the passenger information maps which were passed around the cabin in flight to tell the people how things were going, but in speech the custom was already established, and persists to this day, of referring to the aircraft by the last two letters of the registration. Argonauts were therefore How Charlie, How Dog, How Easy and so on, according to the old phonetic alphabet. I had my own version, derogatory in form but not in spirit: How Calamitous, How Dreadful, How Execrable, How Frightful, How Ghastly . . .

Astra taxis at Rio on the way to Buenos Aires in August 1950. (via Stephen Piercey)

Time off. BOAC stewardess Esme Jones, N/O Bill Hunt and F/O Bob Broadhurst take a spin on the waters at Basra, Iraq, in June 1952. (via W.G. Hunt)

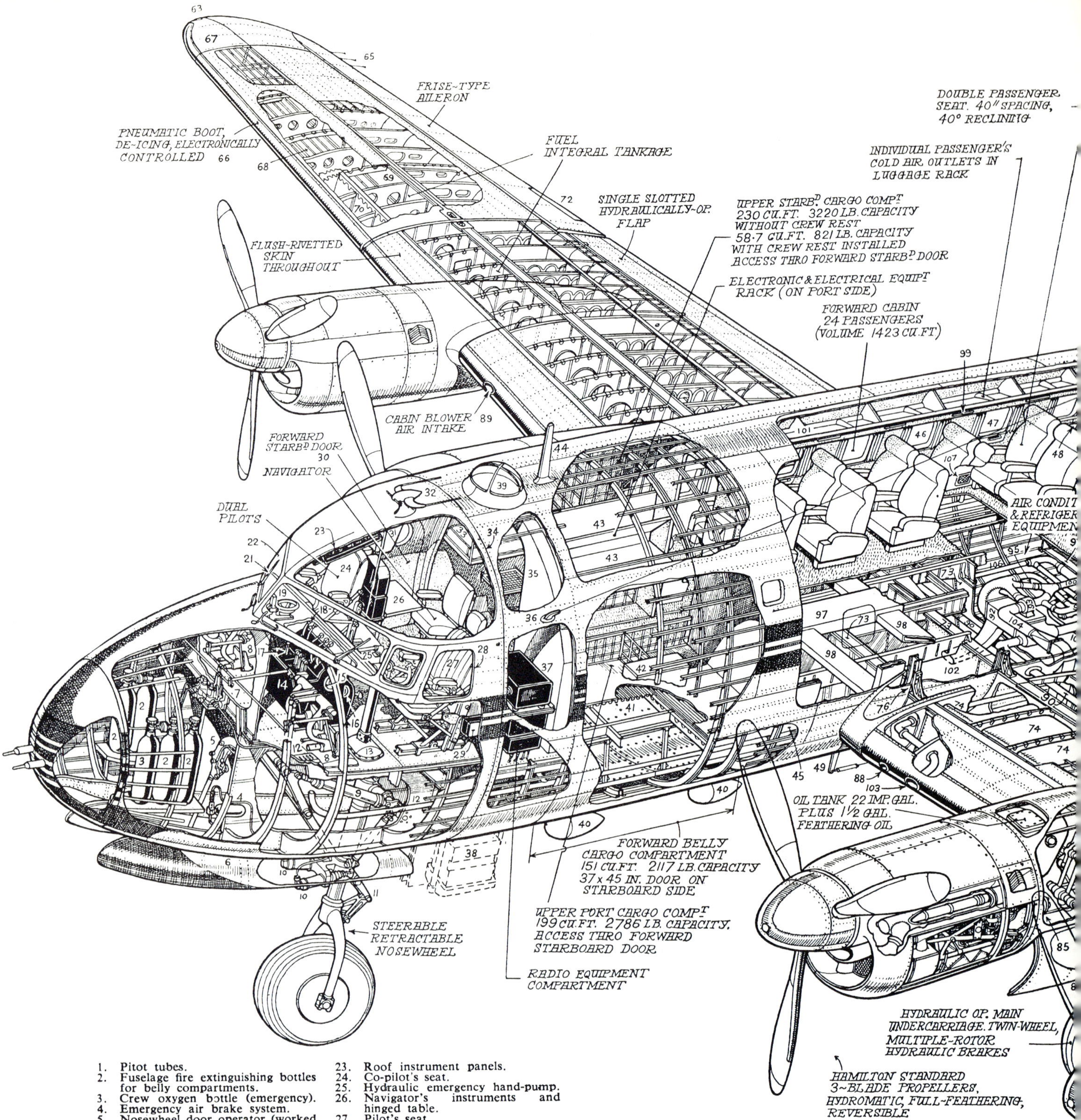

1. Pitot tubes.
2. Fuselage fire extinguishing bottles for belly compartments.
3. Crew oxygen bottle (emergency).
4. Emergency air brake system.
5. Nosewheel door operator (worked by uprising nosewheel leg).
6. Nosewheel doors.
7. Brake control valve.
8. Rudder pedals incorporating hydraulic-brake foot operator.
9. Nosewheel operating jack.
10. Nosewheel steering jacks (67 degrees swivel each way). Nosewheel controlled from handwheel to left of pilot.
11. Nosewheel torque scissors, quickly releasable to permit 360 degrees swivelling for towing or parking.
12. Pilot's footwarmer and hot air trunk.
13. Control column.
14. Control pedestal.
15. Elevator tab control.
16. Aileron tab control.
17. Fuel system controls.
18. Throttle and airscrew controls.
19. Rudder tab control.
20. Fresh water tank (above buffet).
21. Windscreen wipers.
22. Bird-proof heat de-iced windscreen.
23. Roof instrument panels.
24. Co-pilot's seat.
25. Hydraulic emergency hand-pump.
26. Navigator's instruments and hinged table.
27. Pilot's seat.
28. Pilot's side window (can be opened for ground use, or as emergency crew exit).
29. Dual pilots' interphone equipment (repeat on starboard side).
30. Front cabin door, 20 ins. by 60 ins.
31. Main pilot's switch panel.
32. Localizer and glide path antenna.
33. Crew's rest bunk with chairs and table below.
34. Joint between fuselage nose and centre sections.
35. Main electrical and radio junction box.
36. H.F. Radio antenna mast.
37. Radio equipment.
38. Retractable battery elevator.
39. Astrodome, reinforced for additional safety.
40. Automatic direction-finding loop antennæ.
41. Main floor beams and corrugated flooring. Cargo hold below.
42. Invertor and fire detector equipment rack.
43. Electronic and electrical equipment rack.
44. V.H.F. Radio antenna.
45. Hydraulic equipment compartment, containing central control for hydraulic system, main accumulators, pressure relief valves, main regulator, etc.
46. Cabin windows, 16 ins. by 19 ins. clear-vision, double glass. Six windows (47) swing out and forward to provide emergency exits, openable from either inside or outside.
48. Stewardess's handset telephone.
49. Sense antennæ (one each side under fuselage).
50. Cabin attendant's switch panel.
51. Cabin attendant's air conditioning system control panel.
52. Main cabin door, 36 ins. by 72 ins.
53. Joint between fuselage centre and rear sections.
54. Tail bumper hydraulic shock absorber.
55. Wall and mirror cut away to reveal rear dished pressure bulkhead.
56. Rudder control.
57. Tail cone.
58. All-metal tailplane.
59. Metal frame, fabric-covered elevator, complete with servo trim tab
60. All-metal fin.
61. All-metal rudder with spring boost tab (62).
63. Navigation lights.
64. Marker antenna.
65. Static dischargers.
66. De-icing boots, pneumatic, electronically controlled.

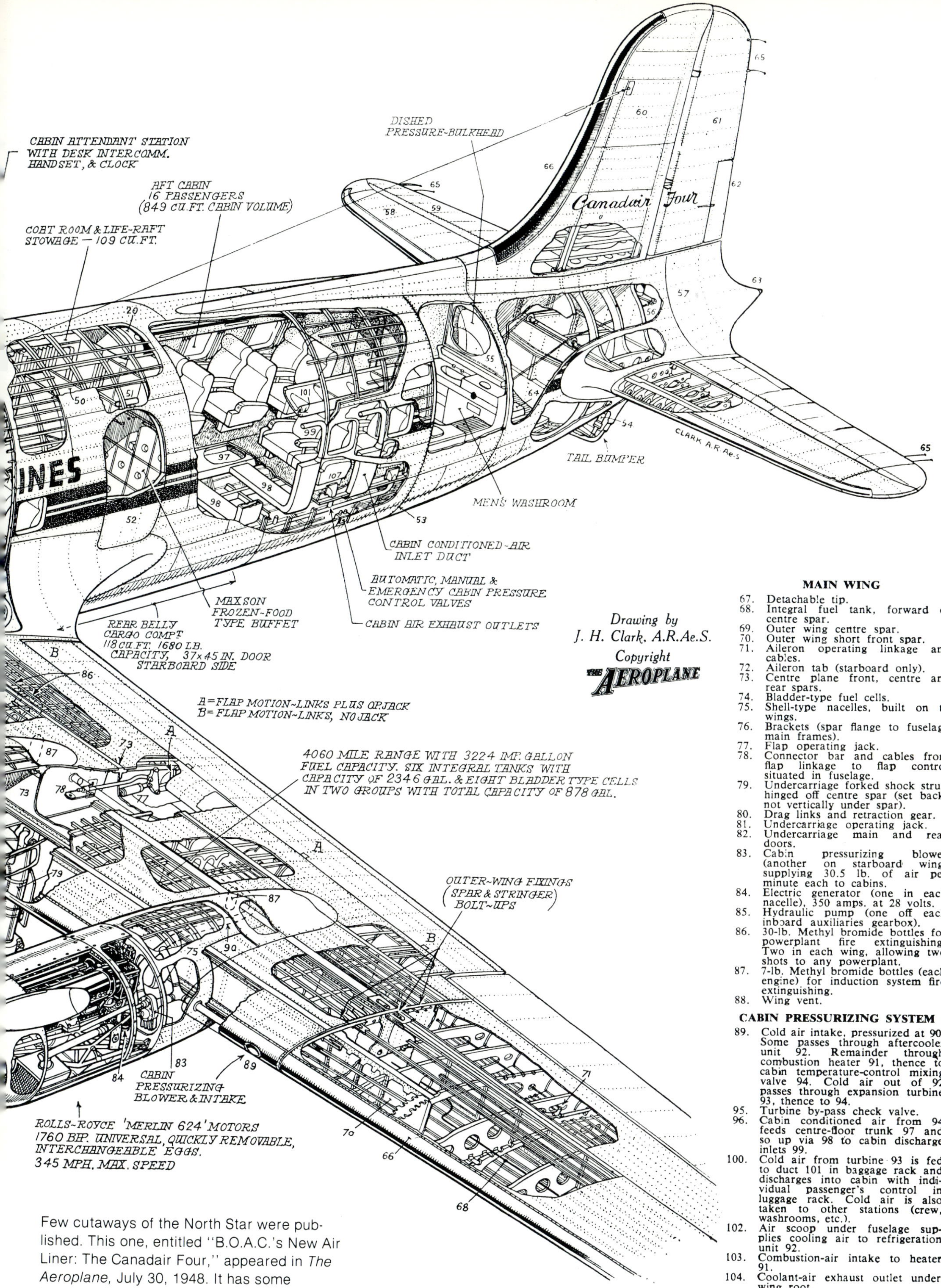

MAIN WING

67. Detachable tip.
68. Integral fuel tank, forward of centre spar.
69. Outer wing centre spar.
70. Outer wing short front spar.
71. Aileron operating linkage and cables.
72. Aileron tab (starboard only).
73. Centre plane front, centre and rear spars.
74. Bladder-type fuel cells.
75. Shell-type nacelles, built on to wings.
76. Brackets (spar flange to fuselage main frames).
77. Flap operating jack.
78. Connector bar and cables from flap linkage to flap control situated in fuselage.
79. Undercarriage forked shock strut, hinged off centre spar (set back, not vertically under spar).
80. Drag links and retraction gear.
81. Undercarriage operating jack.
82. Undercarriage main and rear doors.
83. Cabin pressurizing blower (another on starboard wing) supplying 30.5 lb. of air per minute each to cabins.
84. Electric generator (one in each nacelle), 350 amps. at 28 volts.
85. Hydraulic pump (one off each inboard auxiliaries gearbox).
86. 30-lb. Methyl bromide bottles for powerplant fire extinguishing. Two in each wing, allowing two shots to any powerplant.
87. 7-lb. Methyl bromide bottles (each engine) for induction system fire extinguishing.
88. Wing vent.

CABIN PRESSURIZING SYSTEM

89. Cold air intake, pressurized at 90. Some passes through aftercooler unit 92. Remainder through combustion heater 91, thence to cabin temperature-control mixing valve 94. Cold air out of 92 passes through expansion turbine 93, thence to 94.
95. Turbine by-pass check valve.
96. Cabin conditioned air from 94 feeds centre-floor trunk 97 and so up via 98 to cabin discharge inlets 99.
100. Cold air from turbine 93 is fed to duct 101 in baggage rack and discharges into cabin with individual passenger's control in luggage rack. Cold air is also taken to other stations (crew, washrooms, etc.).
102. Air scoop under fuselage supplies cooling air to refrigeration unit 92.
103. Combustion-air intake to heater 91.
104. Coolant-air exhaust outlet under wing root.
105. Ground blower.
106. Windscreen de-icing hot air.
107. Cabin exhaust air outlets.

Few cutaways of the North Star were published. This one, entitled "B.O.A.C.'s New Air Liner: The Canadair Four," appeared in *The Aeroplane*, July 30, 1948. It has some inaccuracies. Note, for example, the treatment of the engine exhausts. (*Flight International*)

"How Dog" at London Airport, the BOAC Speedbird emblazoned on its tail. After later service with East African, Overseas and Flying Enterprise, this famous Argonaut ended its days on the scrap heap in Copenhagen. (Peter Keating)

Nevertheless, in spite of our own adoption of phonetic letters in place of given names, I was pleased when presented with a signed photograph of the Duke of Edinburgh after I had navigated a royal flight to Mombasa, to find that he had signed the picture and added "Ajax 1956" rather than "G-ALHD."

ROYAL FLIGHTS

Argonauts were used for royal flights on several occasions, the most famous being when the Queen, then Princess Elizabeth, went to Africa in 1952 to return as Queen Elizabeth II after the death of King George VI.

In 1956 a series of three royal flights was carried out. Princess Margaret travelled to Mombasa at the start of an East African tour; the Duke of Edinburgh followed to the same destination about three weeks later, and the princess returned to London from Nairobi some ten days after that. I was chosen as navigator to operate the Kano-East Africa end of the route.

Now as then, I am inclined to cynical speculation about the reason for the fleet management selecting me. On the first flight I discovered that at least three members of our crew, I being one, had been due for leave at that time. This prompted the — I hope unworthy — thought that we had been chosen solely to avoid upsetting the already published crew rosters for the month. Another thought, which may have been nearer the mark, was that in my case the trip was offered as a sweetener to a union representative. Certainly I was among neither the most senior nor the most competent navigators on the fleet, so how else was such a choice to be explained? Be that as it may, it was an honour, and I have never seriously regarded it in any other way. I still have the briefs which were issued before these flights and a couple of menus from one of them.

I can't say much about these flights; they were fairly routine as air operations. Navigation across the heart of Africa was easy with plenty of radio aids in what was still the Belgian Congo, and the French administered territories, and very light winds. The R/O, a small and immensely likeable Scot called Joe Robbie, had the worst job, sending and receiving endless numbers of lengthy diplomatic messages couched in flowery language to and from the administrators of the various territories over which we passed. It seemed to keep him busy most of the trip. On the first flight we arrived a few minutes ahead of schedule, and taxied very slowly for a spot-on arrival on stand.

We had to line up at the foot of the steps to shake the royal hand, arrayed in "best blues" (new uniforms had been issued for the flight) and, we were told, holding our gloves in our left hands. A moment of panic on the flight deck — Derek Villis, the first officer, couldn't find his gloves. There was some offensive individual from Corporation HQ trying to push us around as if we were witless schoolchildren, making humourless hurry-up noises, and generally trying to act as if he and he alone was running the show. His unwelcome appearance on the flight deck at this moment helped not one little bit. Airmen are nothing if not resourceful; I gave Derek one of my gloves, and we lined up each clutching one. Needless to say, no one noticed.

Afterwards we went back on board and mopped up the not inconsiderable residue of HRH's pink champagne, specially provided for the occasion, and some delightful *petit fours.* These goodies, and the new uniforms ("accelerated replacements") were about the sum total of the perks of the job, apart from the photograph from the Duke of Edinburgh.

Layout and seat allocation for royal flight, HRH The Duke of Edinburgh, October 15, 1956. (via W.G. Hunt)

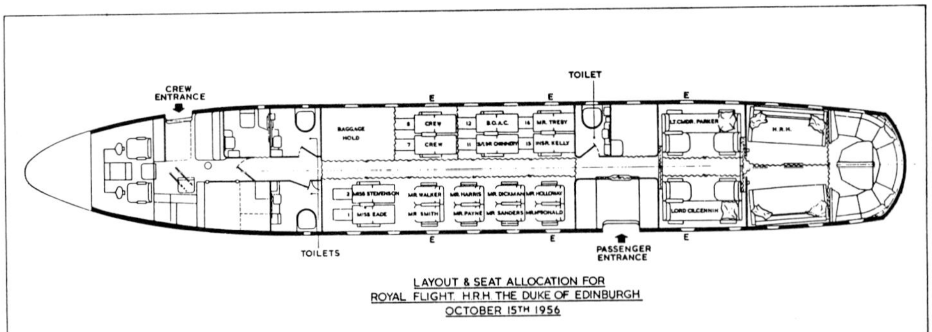

Sept. 27, 1952	London-Singapore	Duchess of Kent	Jan. 27, 1956	London-Lagos	Queen Elizabeth
April 28, 1954	Aden-Entebbe	Queen Elizabeth	Feb. 13, 1956	Lagos-London	Queen Elizabeth
May 1, 1954	Entebbe-El Adam-Libya	Queen Elizabeth	Sept. 21, 1956	London-Mombasa via Gibraltar	Princess Margaret
Oct. 6, 1954	Wadi Halfa-Malta	Haile Selassie	Oct. 15, 1956	London-Mombasa via Kano	Duke of Edinburgh
Feb. 20, 1955	London-Bangkok	Anthony Eden	Oct. 26, 1956	Entebbe-London	Princess Margaret
March 7, 1955	Bangkok-London	Anthony Eden	Nov. 4, 1957	London-Lagos	Princess Royal

Royal tour and VIP Argonaut flights from the log of BOAC engineer Arthur Eagle.

Argonaut operations on the North Atlantic from the log of BOAC engineer Arthur Eagle in 1953.

June 23	G-ALHG	Capt. Holder	London-Keflavik	5:00 hours
24			Keflavik-Goose Bay	8:30
			Goose Bay-New York	6:15
26			New York-London	10:30
July 23	G-ALHM	Capt. Ballantyne	London-Keflavik	5:30
			Keflavik-Gander	9:15
24			Gander-New York	6:50
26	G-ALHX		New York-London	10:45

INCIDENTS

Like all aircraft, the Argonaut had its share of problems. Early in its career a mod to the engine cooling system caused major trouble, culminating in a fleet grounding of several weeks while the problem was sorted out. Even after corrective action, the coolant system was still liable to develop leaks. These were notified by coolant loss warning lights on the panel. The only thing then was to shut down the engine and decide whether to press on, return or divert, depending on the stage of flight where the trouble occurred, the location of engineering facilities, and sometimes, the stabilizing height on three engines in relation to any high ground to be crossed.

In the piston engine era, "mag drops" were one of the commonest causes of delays, and engine run-ups before takeoff were always made to check engine performance on each magneto. Ground engineers always seemed to deal with mag drops in the same ritual way: change the plugs, change the harness, change the magneto, each step being taken only if the previous one failed to correct the defect. There was, I think, no means at hand of analyzing the defect, and this sort of trial and error approach was the best that could be done.

Occasionally one encountered snags of a more subtle nature, and therefore more difficult to deal with. Sooner or later, all airmen become acquainted with the kind of snag which manifests itself in the air but cannot be reproduced on the ground. It is entered in the technical log, and against the entry the ground engineers write

Captain Vic Parker of the BOAC Argonaut fleet. He and Captain Ballantyne shared the captain's duties in flying Queen Elizabeth home upon the death of her father, King George VI, in 1952. (BOAC)

BOAC technical crew at London in 1950 with 'HM in the background. Standing from the left are "Taff," "Philip," George Hazelby, Stan Ellis, Arthur Eagle, Jim Morrison and Paddy Murphy, with Frank Hortop in front. (via Arthur Eagle)

"ground checked and found serviceable." Off you go, and the fault at once recurs. One snag of this kind which I encountered several times on Argonauts concerned the pitch control on the propellers. I am no engineer, and my appreciation of the technical aspects may therefore be inaccurate, but my understanding of the matter is as follows: at some stage of flight, usually soon after reaching cruising altitude, the revs would increase slightly on one engine. Throttle adjustment might ease this temporarily, but not for long, and it soon became apparent that the propeller was slipping into finer pitch. Operation of the pitch control did nothing to correct this, and the engine had to be shut down and the prop feathered before it ran away. Back on the ground, it was unfeathered, and the engine ground run. All was perfect, and the snag could not be reproduced. Off went the aircraft again, and after an hour, or perhaps less, the prop once again started slipping into fine pitch. Back again, with disgruntled crew confronting puzzled and sometimes disbelieving ground engineer, distrustful of pilots' appreciation of technical matters. The first time this happened to me was on December 22, 1952, in G-ALHS, Captain Peter Horn, when we left Rome for Damascus and returned after about an hour. The second time we returned after about an hour and a half, in each case with No. 4 shut down.

The villain of the piece was eventually found to be a leaking hydraulic seal, which allowed pressure to build up in the pitch actuator to the point where the pitch started to slip towards fully fine. The pressure was sufficient to override the contrary pressure in the control mechanism, with the result that control was lost, and the CSU became to some extent ineffective. Since the hydraulic leak was only a very slight seepage, the much greater pressure generated to feather the propeller, acting in the opposite sense, cleared the snag completely, and a ten-minute ground run was insufficient to allow enough leakage to take place to build up pressure and reproduce the snag. In those days, dissemination of information about snags appears not to have been widely practised; experience was less well shared than is the case nowadays, and the second time I encountered this particular snag, neither of the pilots, nor the ground engineer, had heard of it until I recounted my previous experience.

At about the time that economy class was introduced on BOAC routes — I think about 1952-53 — the maximum all-up weight of the Argonaut was increased from 80,200 lb to 82,000, possibly in order to enable the carriage of a greater number of passengers without reducing the range to unacceptable levels. I cannot say there was necessarily any connection between the two things, but after two years of relatively trouble-free flying on Argonauts, I had a year during which I experienced an engine shutdown on almost every round trip. Of course, this could be pure coincidence,

Navigating Officer W.G. Hunt aboard a BOAC Argonaut during 1953. (via W.G. Hunt)

and I would never claim "post hoc ergo propter hoc."

In 1953 I experienced an explosive decompression in South America. The aircraft was G-ALHO, Captain Lincoln "Linc" Lee, and we had left Santiago for Buenos Aires on July 8. We climbed to 25,000 feet to cross the Andes, at which height, with pressure differential at the maximum (I think 4.2), the cabin altitude was about 13,000 feet, and the crew were on oxygen. We had just turned on to heading to cross the mountains, when the F/O's window blew out with a mighty crack, after which there was a sound of rushing air — not, as I first imagined, through the window; the pressure must have equalized within the first second or two; but simply the sound of the slipstream through the open window. The F/O was Russ Ewens, later a captain on the fleet, and his oxygen mask was sucked out when the window went, and lost. Linc at once turned the aircraft round and put the nose down, and we returned to Santiago. The cabin staff told us afterwards that a phenomenon they had been taught to expect in such an incident did in fact occur. After the window went, mist formed at floor level, and rose until about level with the seat arms, then cleared. Russ said he felt his right arm being lifted towards the window, but otherwise no tendency to be pushed bodily out of the aircraft. The perspex window may have had a small crack in it. This sometimes happened, and unless it was unacceptably large, the crack would be stopped by drilling a hole at the end of it, and plugging this, which prevented the crack from spreading.

Most engine shut-downs came into the category of precautionary measures following an adverse indication of one sort or another, coolant loss being typical of these. Sometimes the incident was more spectacular. On May 19, 1953, I was with Captain Vic Parker in G-ALHC, and we had left Cairo for Aden. On the climb we suddenly felt a violent vibration, with loss of revs on No. 4 engine. This was at once shut down, and we returned to Cairo. When the cowlings were removed, the engine was found to have two large holes in the side of the crankcase. A con-rod had broken, and the uneven stresses set up had broken a second one on the same crankshaft throw further back. Prompt action by the pilots on that occasion meant that we suffered nothing worse than an engine write-off.

On January 10, 1958, in G-ALHG, Captain Derry "Boy" Barnes, we were flying a descent pattern into Beirut in fairly rough weather when we were struck by lightning on the nose. The strike made a loud report, heard like a gunshot on the flight deck, and I think that afterwards a discrepancy was noticed between the two pilots' ASI readings. After landing, the aircraft was checked for damage, and one pitot tube was found to be bent down some fifteen or twenty degrees from the horizontal. Worse damage was sustained by the rudder; a piece about six inches square was missing from the trailing edge near the top, together with the static discharge brushes attached at that point. We must have been hit by a very high voltage to sustain damage of this

Date	*Captain*	*Route*	*Time*
April 25	Ballantyne	London-Amsterdam*	1:31
25	,,	Amsterdam-Algiers	4:53
26	,,	Algiers-Accra	9:24
27	,,	Accra-Kano	3:25
27	,,	Kano-Algiers	7:48
28	,,	Algiers-Amsterdam	5:07
28	,,	Amsterdam-London	1:26

**West African service.*

May 11	Houlder	London-Keflavik*	5:20
12	,,	Keflavik-Goose Bay	6:40
12	,,	Goose Bay-Montreal	4:20
13	,,	Montreal-Gander	4:18
14	,,	Gander-Prestwick	9:02
14	,,	Prestwick-London	1:48

**Montreal service.*

June 6	,,	London-Pori, Finland*	4:30
6	,,	Pori-London	5:21

**Charter. Ship's crew. Refuelled from drums.*

Nov. 15	Hartley	London-Paris*	1:31
16	,,	Paris-Athens	5:53
17	,,	Athens-Beirut	3:03
17	,,	Beirut-Kirkut, Iraq	2:28
17	,,	Kirkut-Baghdad	:52
18	,,	Baghdad-Kirkut	:54
18	,,	Kirkut-Beirut	2:35
18	,,	Beirut-Rome	6:16
19	,,	Rome-London	4:07

**Charter. Opening of pipeline at Kirkut by King Faizal of Iraq.*

Argonaut history from the 1952 log of BOAC engineer Ernest A. "Trader" Evans.

order to the airframe. In these conditions, R/O's always earthed the aerials as soon as they could hear static interference building up. I hate to think what would have happened to the radios had that voltage hit them.

One other incident I recall, which I regret to say can only be ascribed to ground engineer's finger trouble. On August 11, 1953, G-ALHG, Captain Pete Mayger, we left Bangkok for Hong Kong. After about three hours we had to shut down No. 1 and the decision was made to return to Bangkok. Examination there showed that an engine change was necessary, and there was no spare engine in Bangkok. We therefore ferried the aircraft on three engines to Rangoon, where an engine was held. After the engine had been changed we ferried the aircraft back to Bangkok to pick up such passengers as had not been put on to other flights, and took off again for Hong Kong. Less than an hour out, No. 1 engine started to surge violently and uncontrollably, and there was nothing for it but to shut it down, dump fuel, and return to Bangkok. There it was discovered that when the replacement engine had been installed, the main fuel feed had been connected correctly, but the nut had not been wire-locked, and had worked loose, eventually allowing the fuel line to become disconnected altogether. This in turn meant that pressure from the fuel pump was causing the pipes to be forced apart until the pressure reduced with petrol spillage, when the pipes would spring together again, allowing some fuel to pass to the engine. This in turn accounted for the surging as fuel passed intermittently to the carburetors. The engine was still dripping petrol when we landed, and we were lucky not to have had it on fire.

A couple of things which were not incidents as such, but which may be of interest. The first I cannot identify in my log book. I do remember that the aircraft concerned had been involved in a royal flight — not the 1956 series on which I operated — and had apparently already been snagged by several crews for poor performance. On the occasion I am thinking of, I remember that it took an hour and twenty minutes to reach 16,000 feet out of Delhi, about double the expected climb time at the loadsheet weight. It was subsequently rumoured that for the royal flight steel reinforcement plates had been added to the floor to give additional protection in case of a forced landing, and that when the aircraft was returned to line flying, it had been forgotten to take the plates out again. If this story was true, the extra weight would not have appeared on the loadsheet, and the aircraft must on a number of occasions have gone off at well over the maximum permissible weight. Certainly this would have accounted for the atrocious performance we experienced that night. The rumour was never, I believe, admitted to be true, but no other explanation was ever forthcoming, and I cannot think that it was a case of the aircraft requiring a rigging check — the deterioration in performance was too great for that to be a likely cause. Shortly after this trip, the aircraft's performance was reported to be back to normal, which was assumed to be the result of removing the steel plates.

Lastly, there was one aircraft — I think it was How Jig — which had a sprung panel in the fuselage. When the aircraft was unpressurized, this was presumably bowed inwards, because at a certain point on the climb, when the pressure differential reached a certain value, the panel would spring out with a loud clunk. After a time, one came to expect it.

WHO'S FLYING THE AIRCRAFT?

Aboard the Argonaut, the navigator's table was fairly wide, and a pilot in the left seat could reach around to his right and place an object on the back of the table without any difficulty. Not all captains were as courteous and considerate as one might have wished, and on one trip the navigator got thoroughly fed up with the captain placing empty plates, glasses and cups on his chart without so much as a by-your-leave, or even taking the trouble to look where he was placing them. Eventually, after yet another empty was pushed unceremoniously onto the table, he tore a message chit off his pad and sent forward an alteration of course 60 degress to port. This was not questioned, and the aircraft was turned on to heading. A few minutes later, the navigator passed up another alteration, 120 degrees to starboard.

"What the hell's going on?" enquired the captain.

"It's all right, skipper," said the navigator. "I'm just doing a dogleg around your coffee cup." The message went home!

Navigator Mick Savill was always

"putting up blacks" of one sort or another. Stories about him were legion. One of his gags, if flying with a skipper who was new to him and whom he sensed did not entirely approve of him, was to prepare five or six message chits, and when after takeoff the captain asked what course Mick wanted, he would go forward, holding the message chits face downwards fanned out like playing cards, and say, "Take a card." When the captain had one, Mick would ask, "Which one have you got?" Shown the chit, he would say, with the air of one who really can't be too bothered, "Oh, that'll do," and return to the table. Of course, all the chits had the same course written on them.

On another flight some crew members were talking about seniority. "Seniority!" said Mick. "There was a time when I owed them seniority but just now I've got three months I haven't even touched!" His view of seniority as a kind of expendable credit akin to an indulgence peddled by a medieval monk appealed to me. Mick's "blacks" were always hilarious, but his humour had no malice. I never heard anyone suggest that he couldn't do his job.

Another navigator story occurred soon after the Argonauts were put on the South American route. For the most part ex-BSAA crews flew this route, but as that airline had not employed specialist navigators, BOAC navigators were used. One was on his first trip there and found himself with a skipper who, though doubtless well intentioned, began to irritate him by the habit of pointing things out to him in a manner which suggested that he thought him a complete novice.

"Look," he said on approach to Rio, "just like Hong Kong, isn't it?"

"Except that it's about four times the size in every direction," thought the navigator, but he kept his counsel.

The trip wore on. At last the Andes were crossed and the Pacific Ocean came into view. The captain called the navigator forward for the umpteenth time. "Look," he said, "Now you can say you have seen the Pacific."

"Yes," replied the navigator, "both sides of it!"

Captain Laurie Arthur, a well-known character on the fleet, was a true raconteur. Laurie was transiting Okinawa, usually a necessary technical stop between Hong Kong and Tokyo. Here he was accosted by an American officer who pointed to the Argonaut and said, "Say, what is that ship? It looks like a DC-4 with Allison motors."

"No," said Laurie, "Rolls-Royce Merlins."

"Oh, you mean Packards," said the officer.

"No, no. Rolls-Royce, Derby, England," answered Laurie patiently.

"Say, they're making them there now?"

There was little point in pursuing the matter!

Nigel Pelly was one of the best liked and most respected captains in the fleet. Unlike some of the older skippers who had been with Imperial Airways or British Airways before the war, Nigel never seemed to feel the generation gap between himself and the wartime airmen. Incidentally, he was the pilot who flew Chamberlain to Munich in 1938. Here is a story about him told to me by F/O Jack "Tosh" Burridge. I don't recall the precise technical details but there was some switch or control that was, shall we say, supposed to be off during takeoff. In the course of pre-flight checks Tosh noticed Nigel reach out and put the switch to on. Nigel was so very senior, and had such a presence and aura of authority, albeit never less than genial, that Tosh was nonplussed as to how he could draw Nigel's attention to the irregularity. As they turned onto the runway, Tosh plucked up courage.

"Excuse me, captain," he said, pointing to the switch, "I don't think they like us taking off with that on."

"My dear boy," replied Nigel, opening the throttles as he spoke, "It's absolutely prohibited!"

In the early days of the Argonaut fleet, in the 40-seat configuration, there were, as well as two toilet cubicles, a ladies' powder room and a men's toilet with urinal and hand basin, on starboard and port sides respectively just behind the flight deck. One night a captain went aft to freshen up, and while he was washing became aware that another man had come in and was standing at the urinal. Only when the captain turned around did he realize that the other was the first officer! The same thought occurred to both at once — Who was flying the aircraft? They hastened forward and found the navigator sitting in the captain's seat with a map, looking for some landmark or other, and blissfully unaware that there was no one in the right hand seat. What had happened was that, with the curtains drawn behind the pilots' seats, and no lights except on the panel forward of the curtain, the F/O had assumed, when the navigator climbed into the seat, that the captain had returned, and that he could therefore go aft himself. It was just as well there was not a runaway prop at that moment.

This happened one night just after I had settled myself in the starboard toilet. It was disconcerting, to say the least, to hear the steady roar of Merlins overlaid by the sound of No. 3 suddenly rising to a fiendish scream. With visions of the prop coming through the fuselage (this had actually happened on a Hermes) I made a hasty and undignified exit, trousers at half mast. Fortunately the pilots reacted quickly and caught the runaway in time.

Chief Steward Paddy Sheeran was a former RAF bomb aimer and off duty sometimes wore a blazer with his squadron badge on the breast pocket. He was in a bar somewhere out east when he was accosted by some young pipsqueak of an RAF officer, postwar vintage, who said, "I suppose you know you're only entitled to wear that if you represented your squadron in some sport." (As it happens, this is not even true.) "What sport did you represent your squadron at?"

Paddy looked at him for a long moment. "Bombing," he said, and turned away.

Sheila Gordon was one of our stewardesses — one of the irrepressible kind, a madcap. She was Scottish and had an identical twin sister, also a BOAC stewardess. It appears that the Corporation would only employ both on the condition that they should always be on different fleets, and the sister was on the North Atlantic. Sheila decided that she wanted to see America and the story goes that she arranged to swap places with her sister for a trip. It was said there were a few awkward moments for both when hailed by "acquaintances" they didn't know from Adam or Eve, but they carried it off successfully.

We sometimes used to play gags on new stewardesses. One was to invite the new girl to look at the sun, moon or whatever through the navigator's periscopic sextant, the eye piece of which had been treated with shoe blacking. On her return to the cabin, the chief steward would wait a while before suggesting that she might go into the toilet and freshen up. This sort of trick depended, of course, on having the right crew, and passengers who would share the joke. So far as the latter was concerned, things were a little different in those days. We were much closer to the era of the flying boat and landplane services where the passengers and crew stayed together for the whole trip, and with such small numbers of passengers as well, there was a much more intimate and friendly relationship between passengers and crew. The sort of people who travelled by air in the 1950s expected a little larking about, or at any rate, were not surprised or upset by it, provided it stayed within reasonable bounds. Somehow I don't see this sort of thing being accepted in quite the same way either by the airlines or their passengers today.

The new girl was treated to a few more legpulls: coming on the flight deck with the basket containing barley sugar, Fox's glacier mints and little packets of cotton wool for the ears, she would be sent back to hunt through the box for the "mag drops" instead! Such practical jokes were more like badges of approval — we only played them on those we believed would take them in good part, and who had thus already been accepted as worthy members of the Argonaut fleet.

A more elaborate gag in which I took part may serve to illustrate the point about passengers enjoying a bit of fun. I think this took place on January 31, 1954, on board How Charlie, Captain Dennis "Bismarck" Briggs, en route from Nicosia to Rome. The passengers had been told that we had picked up a rather well known VIP in Nicosia, who would pass among them after takeoff and give them his blessing. During most of my Argonaut days I sported a beard, and I was therefore regarded by the rest of the crew as a natural choice for the role of Archbishop Makarios, in those days a figure much in the news. BOAC used to issue crew with a separate lining to be worn inside the raincoat in cold weather, and I had mine with me. Worn on its own, this long, dark, sleeveless garment strongly resembled some clerical vestment. Thus arrayed, with the moulded rubber visor from the Rebecca set perched on my head like a bishop's mitre, and the fuel tank dipstick, cruciform handle uppermost, clutched in my left hand, I passed through the cabin, making appropriate gestures to left and right. The passengers appeared to enjoy the joke, and I got a round of applause at the end of it.

We did these things only to show that we did not always take life too seriously. When it came to doing our job we were professional, but as Chesterton pointed out, it is possible to be serious without being solemn.

Fleet modernization being an ever-present process in the airline business, the Argonaut was gradually supplanted on BOAC's far-flung route system. Aircraft like the DC-7C and Britannia, then the Comet IV and 707 took over from the hard-working Argonaut between 1955 and 1960. Here G-AOIF, a DC-7C, touches down at Malton on July 2, 1960. Britannia 102 G-APLL is also seen at Malton, June 15, 1961. (Larry Milberry)

END OF THE ROAD

The Britannia 102 took over bit by bit from the Argonaut. Turbine engines had arrived. Though the Comet I had come and gone, the Comet IV was on the horizon and the Britannia 312 had already started on the North Atlantic. Instead of the full-throated roar of piston engines, the whine and scream of turboprops and jets began to predominate at the world's airports. The aircraft were now larger and faster, and the new engines more reliable. Airports began to reek of paraffin.

The Argonaut continued in service for a while on some routes after I and many others had left for other fleets. There was no particular point at which it all seemed to stop and there was no official farewell party.

For me, at least, the piston engine age went out, if not in a blaze of glory, then certainly with the satisfactory feeling of a job well done.

Subsequently I operated on Britannias with BOAC and after leaving BOAC, on Britannias, DC-6s, VC10s and Boeing 707s. None of these ever replaced the old Juggernaut in my affection and esteem. Terrible cliché, but it was indeed the end of an era. What more can be said?

W.G. Hunt

Captain Frank A. Taylor of BOAC offers the following Argonaut nostalgia:

I have the happiest recollections of the seven-year period during my flying career when I was operating Argonaut aircraft.

When the war ended I found myself, in company with many other former pilots in the Pathfinder Force, flying for British South American Airways whose chief executive was Air Vice Marshall D.C.T. Bennett, who had commanded the Pathfinder Force. When two mysterious disasters struck our fleet of Tudor aircraft, BSAA was dissolved but the operating crews were absorbed *en bloc* into BOAC and were soon undergoing training on Argonaut aircraft in order to restart services to South America.

The pilots were completely familiar with the Rolls-Royce Merlin engine, which powered the Mosquito, the Lancaster, the Lancastrian, the York and the Tudor aircraft. It was without doubt a tried and trusted friend which they were glad to see as the power plant of the Argonaut. In addition, it was a pleasure to operate with an airframe specifically designed for civil use rather than converted from a military counterpart as in the case of the Lancastrian and the York — an airframe, moreover, which did *not* have the built-in unreliability we had found with the Avro Tudor.

In due course, on November 17, 1949, a relatively short time after the collapse of BSAA, we set out with Argonaut G-ALHV on the route proving flight to South America with myself, as flight captain South American route, in command. Our flight was London - Lisbon - Dakar - Recife - Natal-Vitoria-Rio-Curitiba-Montevideo -Buenos Aires-Santiago-Buenos Aires-Montevideo-Sao Paulo-Rio-Natal-Sal (Cape Verde Islands)-Bathurst-Dakar-Lisbon - London, arriving back on December 3. The one thing I distinctly remember, after a lapse of 33 years, is that it was the only occasion in my career when I lost my baggage and recalled somewhat late in the day the aviation adage that "A fool and his bags are soon parted."

An Argonaut training flight to South America in G-ALHF soon followed, leaving London on January 24, 1950 with Captains Walton, Wellwood, Sleight, Johnson, Fieldson, Harrison, Gosling, Holland and myself: then, on February 28, 1950, the first Argonaut service to South America, BA354/001, left London under my command. The aircraft was G-ALHX.

In 1952 BOAC decided to broaden the experience of its Argonaut crews by ending the procedure whereby crews specialized on particular routes. This gave me the opportunity in April 1952 to operate Argonaut aircraft to the Far East via Europe and the Middle East. Later, in April 1953, the Argonaut was taking me to Tripoli, Kano, Lagos and Accra in Africa and later still in July 1953 to Rome, Cairo, Khartoum, Entebbe, Nairobi and Dar-es-Salaam and across the North Atlantic to North America.

Looking back over my Argonaut years, two events stand out in my memory. The first, not surprisingly, was the very great honour of commanding the Argonaut aircraft which brought Her Majesty Queen Elizabeth II and Prince Philip back from their Australian tour in April 1954. The aircraft was G-ALHW, operating Aden to Entebbe on April 28 and Entebbe to El Adem on April 30. The crew was as follows:

Captain	F.A. Taylor
First officer	C.R.W. Taylor
Nav. officer	A.G. Lacy
Radio officer	A.M. Ruthven
Steward	G.S. Plumley
Steward	B.H. Monet
Stewardess	H.E. Tyndall

Shortly before arriving at the final destination of El Adem where the royal party was to embark aboard the Royal Yacht *Britannia*, a message was sent to me indicating that Her Majesty wished to visit the flight deck. I was naturally delighted but my pleasure dimmed somewhat when just before Her Majesty arrived the hydraulic pressure "went for a Burton." I found it anything but easy to produce the appropriate small talk for royal occasions while my mind was busy tracing hydraulic pipelines on a mental plan of the hydraulic system.

A further problem was publicity. The flight had attracted a great deal of public attention worldwide and Richard Dimbleby, the BBC's commentator, was on the ground at El Adem to ensure that the public was kept fully in the picture second by second. If I were to declare an emergency, this undoubtedly would have meant adverse publicity for BOAC and for the Argonaut aircraft. Knowing that with the undercarriage pumped down by hand we could use reverse propeller to act as a brake on landing, and knowing also that all emergency services were at maximum alert anyway, I decided not to publicize our difficulties and am happy to say that we landed without incident. However, I was sorry to have deprived Richard Dimbleby of an opportunity for a most dramatic commentary!

The second event which I recall took place earlier — in May/June 1952. On May 26, 1952 a signal arrived at BOAC HQ in London advising that a Hermes aircraft, G-ALDN, had crash-landed short of fuel at approximately 19°50′ N 12°30′ W, 120 km southeast of Atar (French West Africa). The signal indicated that three military aircraft were searching the area and a Junkers aircraft from Dakar would parachute nurses and radio operators if the Hermes were sighted. As the position of the crashed aircraft was some 1600 statute miles distant from the aircraft's destination of Kano in northern Nigeria, it will be appreciated that the signal came as a real bombshell.

A special Argonaut flight in G-ALHT was organized to fly to the scene of the crash and to locate the wreckage and survivors. The aircraft left the following morning, May 27, 1952, with the following crew/passenger list:

Crew

Captain	F.A. Taylor
First officer	Gaynor
Nav. officer	Lacy
Radio officer	Ruthven
Engineer	Ayres
Steward	Kloter
Steward	A. Godfrey
Stewardess	Fishel

Passengers

Captain J.C. Harrington (deputy operations director)
Mr. Folliard - BOAC Accident Branch
Mr. Nelson - MCA Accident Branch
Mr. Spiegal - Engineer

Supernumerary Captain Houlder (officer i/c training)

'HF refuelling at Khartoum in November 1956. (Peter Keating)

After refuelling at Lisbon en route, we proceeded to Atar to commence our search. The position given in the signal revealed nothing, so we started a square search from this position at a height of 500 feet above the ground with myself flying the machine and Jackie Harrington acting as lookout in the copilot's seat. Before long Jackie shouted, "There she is," and sure enough we came over the Hermes to see crew and passengers running out and waving to us while we tried to make a "count of heads." We made a few passes to convince them that we had definitely located them and then set course for Dakar.

The following week was taken up in trying to get the survivors out of the desert, with various agencies assisting but none producing early results. A motorized detachment had left Atar, which incidentally was the last outpost of the French Camel Corps in this part of the world, confident that it would locate and rescue the survivors without much trouble. The USAF sent a Packet aircraft down from Tripoli with a helicopter stowed on board. Other efforts were made from Dakar, not least by a very gallant Frenchman operating an air-taxi service in a tiny Miles aircraft which carried just one or two passengers at a time from an improvised airstrip made near the scene of the crash.

There is too much detail to recount but with considerable help from Captain Jackie Harrington, who himself went into the desert to join and aid the survivors, all with one exception were eventually brought out safely. The one exception was First Officer Haslam who unfortunately died of heat exhaustion.

One is perhaps permitted to recount an amusing incident to offset the sad and serious aspects of the event. As the representatives of the insurers were making their way across the desert towards the scene of the accident to assess the degree of damage to the aircraft and the salvage value, etc., they passed the Hermes coming the other way — in small sections on the backs of camels led by smiling Arabs. . . .

Finally, may I say "Thank you" to Canadair for what was undoubtedly, in its day, an efficient, safe and reliable aircraft.

Captain Frank A. Taylor, MVO, DFC

Hot-weather Problems and Other Reminiscences

Former BOAC Captains Paul Danc and Vic Hodgkinson recall some of the Argonaut's power plant idiosyncrasies, particularly just after the Hermes were withdrawn and Argonauts were used as substitutes on the Hermes routes: "There wasn't time to train extra crews before the West African route was taken over, so for a short time it was very hard work for the crews already on the Argonaut fleet. Problems arose when ex-Hermes crews started operation in West Africa. Firstly, the Hermes crews were not accustomed to operate without a flight engineer; secondly, in some cases the captains didn't trust the copilots to handle the run-ups and general maintenance state of the aircraft. Thirdly, and perhaps most important, they were not used to operate with engines that were, to say the least, temperature-critical in the high ambient temperature.

"Certain Argonaut operational practices had become normal to keep the temperatures within bounds. From the top of descent to touchdown, rad flaps (manual) open, minimum power possible, use of the electric idle cutoffs after landing instead of using reverse pitch. A quick start-up, taxi as soon as possible, do the run-up and takeoff checks whilst taxiing, so that on reaching the runway, with clearance to go, you went without even stopping. A good crew could be airborne within five minutes of starting up. These methods had kept the temperatures within limits.

"At first, the Hermes crews weren't used to this system. Full run-ups were done at the run-up point, then the before-takeoff checks and so on, with a result that the coolant loss lights were on before takeoff and the engine boiling, so back to the ramp! Flight management, as a result, transferred some of the experienced co-pilots to this route to pass on experience and speed things up. In the meantime, the engineers and Rolls-Royce started a program to increase the takeoff temperature limits and reset the coolant loss warning lights."

Captains Dane and Hodgkinson have a few other Argonaut memories: "The Argonaut route network was very extensive. The 22 Argonauts had to maintain an average utilization of over nine hours per day, at that time a record for BOAC. Even with the problems that beset us, it goes to show how very reliable the combination of a Douglas airframc and the Rolls-Royce Merlin was. The old Argonaut was hot and noisy but generally reliable. It was a stop-gap aircraft that was called upon to stop many more gaps than it was originally bought for. And it stopped them! Yet sadly, the Argonaut was never in the limelight like the Stratocruisers or Connies — it was just a workhorse. In fact, though, it was the backbone of BOAC's postwar fleet.

"Many of our 'web-footed brigade' (ex flying boats) had their first experience with large land planes on the old Argonaut. One can remember landing at night at Cairo with its long and wide runways. One of the web-footed brigade was under supervision as captain and he remarked at 500 feet on final, 'It looks far too short for us to land on!' He was greeted with roars of laughter from the captain and co-pilot."

Long-time BOAC radio officer Jimmy Miller has provided a few details concerning the airline's first-rate radio operation: "The efficient use of modern radio equipment by highly skilled radio officers contributed in no small measure to the economic as well as the operational efficiency of the Argonaut fleet . . . Whenever a fault occurred that could

A BOAC Argonaut is refuelled at Entebbe in Uganda in 1951. (W.G. Hunt)

"How How" was the earliest Argonaut to come to grief. It ran off the runway at Tokyo but was rebuilt, surviving until 1965 when it was scrapped by Air Links. (Air Britain)

not be remedied in flight, the radio officer would signal ahead to the next stop for a fresh unit to be made available. When the station engineer drew the unit from store, the store-keeper would immediately signal London for a replacement. This would sometimes be on its way before the aircraft concerned had landed. On arrival, the station engineer would pass the new unit to the radio officer, who would hand the unserviceable unit out for stores action by returning to London. The new unit would then be installed and tested by the radio officer. This system resulted in nil delays attributable to radio. The enthusiasm of the Argonaut radio officers in keeping the expense of overseas radio maintenance at a zero figure was unbounded!" Miller adds that, "Maintaining communications during a flight on BOAC's extensive route network, under day and night conditions, in climates varying from European winter to tropical storms speaks volumes for the quality of the equipment and the personnel involved."

BOAC steward W.J. "Robbie" Burns spent years flying with the Argonaut fleet and, as he recalls it, "The Argonaut was an easy airplane to work on. The fact of the galley being centre-situated between the 24-seat forward cabin and the 16-seat rear cabin made life much easier. At the time there was no such thing as economy cabins. Everything was first class, with two stewards and a stewardess to take care of the passengers. Economy came later when seating was increased to 56 and the cabin staff reduced to one steward and one stewardess.

"During my six years on the Argonauts I had the opportunity to transfer to other aircraft but was happy working on my old mate. Besides, the Argonaut crews were, with rare exceptions, more of a family affair.

"There were countless interesting happenings during the Argonaut years. One I remember between Damascus and Rome. One engine had to be shut down, followed by loss of coolant. At this point we were over the Mediterranean. Everyone had to don life jackets and hope for the best. Our biggest problem was how to deal with a stretcher case should the worst happen. As it turned out, we landed at Malta where several days of relaxation was the end result!"

One of Burns' VIP flights was aboard *Athena* with Sir Anthony and Lady Eden among the passengers. A letter from the Secretary of State to BOAC following the trip noted: "Now that I have returned from my Asian tour I would like you to know how extraordinarily well we were looked after by the staff of BOAC. We were quite a handful and some changes of plan cannot have made things easier for them. But throughout the journey and at every port of call they were unfailingly helpful and cheerful."

BOAC Mishaps

In over a decade of service with BOAC, the ever-reliable Argonaut established for itself an exceptionally fine safety record. Its first serious incident occurred July 21, 1950. *Argosy* suffered overspeeding in Nos. 3 and 4 engines and a fire in No. 3. The fire was extinguished and an emergency landing was made at Fernando Noronha in Brazil, with the other two engines overheating. A close call!

While landing at Dakar on August 29, 1950, *Adonis* ran into trouble. The weather was poor as the pilot approached for a night landing. The nose gear struck an obstruction just short of the threshold and the aircraft slithered to a stop on its nose half way down the runway. Another landing mishap took place at Tokyo January 19, 1952, when *Attica* ran off the runway, causing substantial damage but no injuries.

Captain Vic Parker remembers the Tokyo accident: "The Argonaut slewed off the runway and hit a large transformer. Initially, it was considered a possible write-off, but in the wake of the first Comet disasters BOAC could not afford to lose any aircraft. The badly damaged machine was loaded on to a Hong Kong-bound ship and sent there for repair. The job took about four months and I went to collect the aircraft. This was before the new runway had been built at Hong Kong and some 1200 m. looked rather short for a test flight on a completely rebuilt aircraft. However, Hong Kong Engineering Co. had done a sound rebuild and there were no defects. The Argonaut was flown back to the UK the following day and put back into normal service."

The first serious accident in the Argonaut fleet happened at Tripoli on January 29, 1955. *Altair* was attempting to land at night in very bad weather. On the fourth attempt to land at Idris Airport the aircraft struck trees and crashed 1200 yards short of the threshold. Fifteen of 47 aboard died. There might have been more casualties had it not been for the quick action of some of the survivors. Outside the blazing wreck, one of the crew, R/O George Berrisford, was told by a woman that her child was trapped inside. The crewman dashed into the Argonaut and emerged with the baby.

On June 24, 1956, G-ALHE, *Argo*, crashed within moments of takeoff from Kano for Tripoli. A takeoff had been made on Runway 25 towards the

G-ALHL, *Altair*, photographed at Santiago in August 1950. Aircraft in the background include two Yorks. *Altair* was lost in an accident at Tripoli on September 21, 1955. (via Stephen Piercey)

west. Things were normal and at 240 feet altitude the vertical speed indicator registered 300 ft./min. at a speed of 125K. Suddenly, though, speed fell off to 103K, just above the stall, and the aircraft began to settle. The pilot used full boost, but to no avail. A crash landing was made and of 45 aboard, only 18 survived.

The crash of *Argo* was weather-related; unknown to the crew or to met people at Kano, a thunderstorm cell was developing right off the end of Runway 25. This had spawned winds the reverse of those forecast, (known today as wind sheer) and likely downdrafts as well. The tragedy was unavoidable.

Active Retirement

Long after all the other BOAC Argonauts had been disposed of, G-ALHJ remained. This BOAC news item describes its ground training role:

The last of the BOAC piston-engined fleet starts up with the bark so few people hear nowadays—the throaty growl of four Rolls-Royce Merlins. The tires are pumped up, the chocks in place, she's ready to go.

But *Arcturus* will never fly again. Once a week the sad little drama of the last of the Argonauts is played out in the Maintenance Area at Heathrow Airport as the engines are run up on G-ALHJ.

Arcturus retired nine years ago after 24,000 hours of active service. But she was not sold, as the other 19 Argonauts were.

Arcturus belongs to the Apprentices Training Unit. It is their own aeroplane and their introduction to engineering and maintenance.

Why should an 18-year-old piston-engined 54-seater be so valuable to an airline which flies only the latest jets? Well a pilot starts learning on a basic single engined trainer. By these standards, the Argonaut is positively advanced.

It's pressurized, has four engines, a fair-sized passenger cabin, and the same basic kind of complicated equipment (up to a point) as the present fleet.

The components in *Arcturus* are systematically taken apart and put together again. As Mr. Arthur Evans, foreman in charge of the training, said:

"You can tell a lad what happens when he presses a button on the flight deck. You can show him drawings, working models and the lot. But if he can actually do it, and see the results, he gets a far better understanding of what it's all about."

Arcturus proved useful in another way recently. When the company filming *High Bright Sun* wanted to get the colour scheme and livery just right for a BOAC Argonaut scene, staff photographed and sketched *Arcturus.*

In the spring of 1970, G-ALHJ was replaced by a Comet 4 in the apprentice training role at Heathrow. The following article, entitled "Operation Hoverlift," published by the British Airports Authority, describes 'HJ's last flight:

Capital investment in aviation is so great, that every hour aircraft and runways are out of commission, the airlines and the Authority can lose thousands of pounds of valuable revenue.

From time to time an aircraft gets stuck on or near a runway. When that happens it is obviously very important to clear the runway quickly. The responsibility of removal lies with the airlines. But, since a speedy operation is naturally in the Authority's interest, we have salvage equipment and operating staff available. These can be hired out to the airline operators, if necessary. At the moment trailers are used. The removal may take anything up to two days, and there is a possibility that the aircraft's fuselage may be damaged.

An interesting experiment was recently carried out at Heathrow. It tested a new technique using the hoverlift principle to transport aircraft through the "soft" on a cushion of air.

The opportunity arose when an old Argonaut, purchased from BOAC for BAA fire practice, had to be moved over the grass from Block 52 (Outer), on the old No. 7 runway, to the BAA Fire Service training ground.

It was the first time that the idea had been tried at any airport, anywhere in the world. The exercise as a whole was BAA sponsored; but facilities and staff were also provided by BOAC. Air Cushion Equipment and Aero Docks; RFD and GLEve Construction. About 90 spectators attended, representing the civil airlines, insurance companies, and firms concerned with the hoverlift itself. They met first in the Control Tower conference room, to hear a welcoming address by Mr. "Tony" Cooke-Smith, Senior Operations Manager. They were later taken in coaches to where it was "all happening."

The hoverlift equipment was divided into three segments. These were "clipped" together, so that each could rise and fall independently as the equipment crawled over uneven ground. The Argonaut itself was supported on airbags which rested on the hover units. The airbags held the aircraft well above the ground, and so prevented possible damage. The Argonaut was moved by two winches—one in front pulling, and one behind steadying the aircraft.

Inevitably, the day had dawned dull

Argo at London on July 22, 1949. This Argonaut met a tragic end at Kano seven years later. (Peter Keating)

and overcast. There was also a cold wind, and a touch of rain for good measure. Not to be daunted, however, the hovercraft's engines were started, and the Argonaut lifted just enough to be pulled along with the hover-platforms' air-filled "skirts" just scraping along the ground. The 250 yard route was over concrete, grass, gravel and uneven ground. The hover-platforms made steady, satisfactory progress over all surfaces. The aircraft was firmly in position on the training ground within an hour.

The exercise was shown on ITV News that night. The great success of the experiment augurs well for the future use of this method of aircraft removal.

Such an Old Dog?

Although there was no shortage of North Star bashing going on early in its career, it is important to note that, regardless of shortcomings, the aircraft really did a satisfactory job.

For BOAC, its Argonauts were a God-send. It got them out of a pickle once when the Tudor project was scuttled and twice later on: when the Comet was grounded and when the Hermes failed to live up to predictions. Other little-known bits of history illustrate just how much operators were depending on the North Star/Argonaut. In July 1951 TCA was almost begging BOAC to sell it some Argonauts. G.R. McGregor wrote to Whitney Straight on July 11, "In my planning I am attempting to walk a long and very thin tightrope, if by so doing I can avoid further purchases of a new piston-engined aircraft . . . If I could be absolutely certain of getting between 6 and 10 Argonauts from you between say December 1, 1952, and February 15, 1953, with further purchases contemplated for the following winter if mutually convenient, I believe we could tide over the critical period, but if I cannot be given a firm commitment covering the first part of such a transaction, I must of necessity order DC-6Bs or Super Constellations forthwith."

After airline service, 'HJ spent 10 years with BOAC's Apprentice Training School at London. Finally, it was towed away for use by the fire services at the airport. Parts of its burned-out hulk were still at Heathrow in 1982. (BOAC via Stephen Piercey)

Replying by telegram, Straight said, "Very sorry we can undertake no commitment as can foresee continuous use of Argonaut fleet beyond dates you suggest." Then in November 1952 Straight was inquiring of McGregor whether he might advise him of the chance of getting hold of some M1s for freighting. This, of course, was out of the question as the RCAF was itself hard pressed for freighting and trooping capacity.

Most interestingly, McGregor wrote this on June 15, 1951, to Sir Miles Thomas, Chairman of BOAC: "Like your Argonauts, our North Stars are doing such a whale of a job under existing conditions of unusually heavy traffic in North America that I live under the threat of having to apologize to you for breaking all the rules of the national ownership game, and declaring a profit in 1951." Sir Miles had earlier written to McGregor that "our own Argonauts are showing up very well from the economic point of view." Later, in 1956, BOAC contacted TCA requesting the lease of several North Stars to augment their strained capacity throughout East and West Africa.

A Gallery of North Star Art

Aviation art has thrived in England and the US, where names like Wootton, Turner and Ferris have become famous. But in Canada, aviation art has remained in the background, though the eminent works of Bob Bradford have received acclaim on numerous postage stamps. Most other aviation art in Canada has been done privately or as commercial art. Much of the collection of aviation art sponsored over the decades by the Department of National Defence is, of necessity, held in storage. The only national publication in Canada that has regularly featured aviation art has been the *Journal* of the Canadian Aviation Historical Society.

So, to provide some fresh exposure to a worthy art form, I decided to present in this book a modest gallery of North Star paintings. Seven artists have produced the work which follows: Robert Bausch (San Francisco), Tom Bjarnason (Toronto), Don Connolly (Aylmer, Quebec), Peter Mossman (Toronto), Les Waller (Toronto), Bill Wheeler (Markham, Ontario) and Graham Wragg (Ottawa). And Des McGill (Pierrefonds, Quebec) has produced the scale drawings in the pocket at the back of the book. In his own distinctive style each artist has depicted an aspect of North Star history. A brief write-up describes the artist and his work.

Leslie Waller began his art career on Fleet Street, London, in 1954. He came to Canada in 1967, working first in Montreal, then in Toronto. Like the other artists represented here, he has had a lifelong affair with aircraft. A member of the CAHS, Leslie's work has regularly appeared in the CAHS Journal *and the CAHS* Flypast. *Besides painting aircraft, Leslie is also a keen glider pilot.*

Leslie Waller's first gallery contribution shows TCA North Star M1 CF-TEK rumbling in for a landing. The painting is a watercolour.

TEK
TCA
CF-TEK
LES WALLER ·82·

Leslie Waller's second contribution shows CPA's Canadair Four Empress of Vancouver *cruising above the Hong Kong coastline in 1949. This painting is in oils.*

Canadian Pacific · Airlines
CF-CPR
CF-CPR
LES WALLER ·82·

Don Connolly left his home town of Kingston, Ontario, to join the RCAF in 1951. After training as a navigator he was posted to 426 Squadron and spent the next three years navigating North Stars around the world. In 1957 he was the first RCAF officer enrolled at the USAF Academy at Colorado Springs. After graduating from there he served in various research capacities at Department of National Defence HQ in Ottawa, then retired from the RCAF in 1966 as a squadron leader. By this time he had had several years of dabbling as an amateur artist, and in following years he continued to develop his painting skills, while running a bookshop and picture framing business in Ottawa. In 1979 Don "went professional" and ever since has produced a wide variety of art, including many aircraft paintings.

Don Connolly's painting shown here depicts in acrylics a scene that he is personally familiar with. He shows a 426 Squadron North Star departing Resolute on Cornwallis Island in the Arctic. The low evening sun casts a long shadow from the cape below out over Barrow Strait as the aircraft heads south for Churchill.

DON CONNOLLY

Since he began painting professionally in 1969, Graham Wragg has become Canada's most productive aviation artist. He has done paintings for such clients as de Havilland, Canadair, McDonnell Douglas Canada, and Air Force *magazine. Several of his paintings, depicting CFB Moose Jaw, are part of the Canadian War Museum collection. As well, Graham has worked on the staff of the* Sentinel, *published the Canadian Skies Calendar for several years, and done many aircraft paintings as private commissions.*

Graham Wragg paints in a realistic style using watercolours. Reproduced here is his painting of BOAC's Atalanta.

B·O·A·C
ATALANTA
G-ALHK

The Rockcliffe Ice Wagon as painted by Graham Wragg. The Ice Wagon was one of three North Stars radically modified for research purposes.

17513
ROCKCLIFFE
ICE WAGON
ROYAL CANADIAN AIR FORCE
EXPERIMENTAL & PROVING ESTABLISHMENT

Peter Mossman has worked in Toronto as a commercial artist since 1955. Ever the aviation buff, he spent a year working in the art department at Avro Canada just before the CF-105 cancellation. Peter was an early member of the Canadian Aviation Historical Society and his illustrations have frequently appeared in the CAHS Journal. *Two of his Spitfire paintings belong to the Canadian War Museum collection, and he has done other aircraft paintings as private commissions for commercial clients. In 1981 he produced the jacket art and the colour profiles for my previous book,* The Avro CF-100.

Peter Mossman renders his aircraft profiles using the air brush technique. Seen here is the gleaming North Star prototype, CF-TEN-X.

TH STAR

Bill Wheeler was born in Port Arthur in northern Ontario. He attended the Ontario College of Art in the early 1950s and spent several years working in Toronto studios before he went freelance, then began teaching art in Scarborough. At present he heads the art department at West Hill Collegiate. Bill has spent many years painting airplanes and his work has appeared in such publications as the Star Weekly *and the CAHS* Journal. *He has also produced covers for aviation books published by Macmillan and McClelland and Stewart, and has had a variety of private commissions. As member No. 5, Bill was in the founding group of the Canadian Aviation Historical Society, and for many years he has functioned as editor of its journal.*

In his painting, Bill Wheeler shows CF-TFC of International Air Freighters over Cuban coastal waters on one of its many Toronto-Havana trips.

International
NORTHSTAR
CF-TFC
W.WHEELER '82

Aviation artist Robert Bausch has spent his life in California's Bay Area and lives today in Belmont. He has been professionally involved in painting aircraft since 1970 and has participated in the USAF Art Program over the years since. Some of his work is displayed in the Pentagon. Other pieces are held by the USAF Museum and Strategic Air Command. His paintings are regularly on show at the Nut Tree, a well known gallery in Vacaville, California. Robert notes of his stylized art, "I don't attempt at all to reproduce photographic realism. I'm aiming more for the feeling of flight, the power or majesty of aircraft.

Robert Bausch's paintings, done in acrylics, reflect a fantasy mood, yet the actual aircraft details are very accurate. Here he shows the much-modified AVCO North Star on one of its night missions over the Kwajalein Missile Range in the South Pacific. Streaking into the atmosphere is its target, a USAF missile nosecone.

Robert Bausch © '82

Tom Bjarnason is one of Canada's best-known names in artistic circles. He has been in the illustrating business since 1955 and worked in England, Germany and Sweden before settling in Toronto. His aviation art has been seen in too many publications to list, including Weekend Magazine, The Canadian *and* Reader's Digest. *Other clients have included de Havilland, Canadian Pratt and Whitney, and Innotech. He has participated in the Canadian Armed Forces Civilian Artists Program and several of his paintings from that program are in the Canadian War Museum. An early member of the Canadian Aviation Historical Society, Tom has produced many covers and other illustrations for that society.*

Using coloured inks, acrylic white and pastels, Tom Bjarnason, in his inimitable style, has produced this dramatic view of a British Midland Airways Argonaut.

IDLAND
BM
G-ALHS

Airlift Champion: North Stars in the RCAF

Early RCAF Operations

On August 1, 1946, 426 Squadron was formed at Dartmouth from elements of 164 Squadron, a wartime operation that had been flying Lodestars and Dakotas. Initially 426 operated Dakotas from Dartmouth, but in March 1947 it moved to Dorval. On September 12 that year the squadron received its first North Star, though training was not begun immediately with the new type. The squadron diary of November 13 notes, "North Star 17502 is at last serviceable and our North Star training has begun. Pilots are confining their activity to circuits and bumps while Wireless Operators are familiarizing themselves with the new equipment and practicing on the Squadron practice frequency. Both Pilots and Wireless Operators seem very satisfied with the aircraft and equipment from crew standpoint. However, passengers might not be so enthusiastic due to the high noise level further back opposite the engines."

Training was slow to pick up due to snags and an entry in the diary for November 14 relating to 17501 and 17504 states, "Still u/s on acceptance check." Nonetheless North Star training for January 1948 included 16 flights for a total of 89:55 hours. Aircraft on strength that month were 17501, 17502, 17504 and 17507; 17510 and 17511 were accepted by the squadron at Canadair on February 11 and 17 respectively. On February 24 the squadron flew its first North Star freight on a trip into Downsview.

Long-range training commenced March 1, 1948, with a trip from Dorval to Edmonton under command of F/O Joe Giles. Noise level checks were made en route and the aircraft returned March 3, making the Winnipeg-Dorval leg on three engines.

On April 1 F/L Sanford took a 426 North Star up to 32,000 feet, and on the same day the squadron operated its first scheduled run: Dorval-

17512 at Watson Lake in the Yukon during cold weather trials in the late 1940s. An American Albatross is parked nearby. (George Jackson)

Newly delivered to 426 Squadron, 17507 is seen at Rockcliffe in 1947 alongside a Vampire. (DND PL38791A)

Moncton-Halifax-Goose Bay. Later in the month several trips were made to Winnipeg carrying sandbags to fight the floods there. Ten North Stars were on strength. Crew training by this time included Link trainer courses provided by TCA at Dorval.

A significant event occurred June 16 with the squadron's first transatlantic training flight, F/O Giles in command. The monthly flying rate was steadily increasing and for September 1948 was 491:55 hours. On May 23, 1949, the first 426 South Atlantic flight left Dorval under command of W/C C.H. Mussells. The route took the North Star in a great circle to Puerto Rico, British Guiana, Natal, Dakar, Gibraltar, Lyneham, Iceland, Goose Bay and Downsview. On June 6 17516 left on the first of six South Atlantic training flights and what was perhaps 426's first Arctic trip began July 5 when 17517 under F/L Dickson's command flew to Resolute via Rockcliffe, Trout Lake and Lansdowne House. The flying rate for July was 720:55 hours. That August, five 426 North Stars showed up in Toronto to fly in the Canadian National Exhibition airshow. On September 1-6 17520 carried Prime Minister Louis St. Laurent to Banff on an early VIP flight. In October six aircraft flew in support of the big continental air defence operation known as "Metropolis." The North Star was steadily working its way into RCAF service. The process was carefully executed and calculated to provide maximum training for and to build confidence in air and ground personnel.

On January 7, 1950, F/L Hammond, captain of 17511, located a downed F-84 and its pilot near Watson Lake in the Yukon. He had survival gear dropped and circled the scene until help arrived on the ground. That month six North Stars supported "Sweet Briar," a joint US-Canada military exercise in the Yukon and Alaska. On January 14-15, 1949, a North Star operated by the Experimental and Proving Establishment at Rockcliffe flew from Vancouver to Halifax in the record time of 8:32 hours. This was Canada's first non-stop coast-to-coast flight. The aircraft was commanded by F/O J. Jolicoeur. On January 16, 1950, 17512 of 426 Squadron, commanded by F/O G. Webb, shaved seven minutes off the previous record.

In May 1950 426 was busy on Operation Red Ramp moving vital supplies and equipment to Winnipeg when the Red River flooded disastrously. Also in May, 426 sent personnel to Canadair on familiarization courses on the C-5 VIP airplane. Medical evacuations were also becoming routine for 426 by this time.

The spring of 1950 saw 426 involved in its first major Arctic activity, Operation Re-supply. This was the annual airlift to remote northern weather stations, in this case Mould Bay, Isachsen, Eureka and Alert. One North Star was provided along with a double crew. Four C-54s, one C-82 and one C-47 of the USAF were the major

A disassembled road grader aboard a North Star about to be flown to Churchill. On the left is Gerald Kennedy, a met officer at Churchill, while LAC St. Louis is on the right. In the background are Cpl Lanteigne and Cpl Johnstone. (DND PL54100)

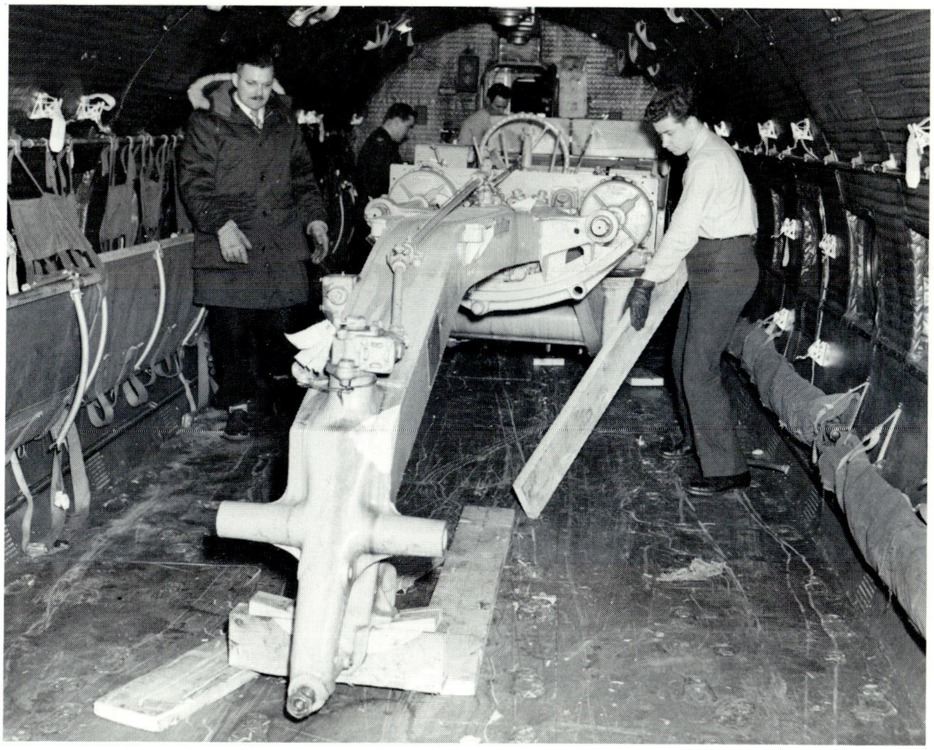

A 426 crew at Mould Bay, April 12, 1952: Sgt G. Howard (in the greasy outfit) and (in the nice clean flying garb) F/L M.D. Broadfoot, F/L O.F. Bradley, F/L W.I. Butchart and F/L L.A. de Quay. (DND PL54043)

△ A pair of North Stars at Isachsen on April 2, 1952, during Operation Re-supply. (DND PL54064)

▽ En route from Churchill to Resolute are weather station personnel and a cargo of survival supplies. Note the passengers earnestly pretending to read, write and relax amidst the din of four Merlins. (DND PL54030)

Unloading Herman-Nelson heaters at Isachsen, April 2, 1952. (DND PL54025) ▽

◁ A Weasel tracked vehicle, somehow squeezed into a North Star, is off-loaded at Isachsen during Operation Re-supply. (DND PL54027)

component of the freighting operation. Supply bases were Resolute for Mould Bay and Isachsen, and Thule for Eureka and Alert. Runways at the weather stations were 4000 x 150-foot ice strips. The first North Star trip arrived at Isachsen on March 24.

Airlift figures for this operation indicate a much smaller payload for the North Star compared to the C-54. The exact reason for this has not been determined; but the figures do not indicate the North Star's true pay load capability.

Max. load to:	Mould Bay	Isachsen
North Star	6,930 lb	7,700 lb
C-54	14,900 lb	15,590 lb

Average load carried/total trips:	
North Star	7,116 lb / 48
C-54	13,780 lb / 77

Part of the summary for "Resupply" reads, "RCAF maintenance again proved to be excellent and no delays were caused by unserviceabilities. Two USAF C-54s were damaged . . . At Resolute a USAF C-54 crashed while attempting a three-engine takeoff and a C-82 was completely written off when an engine failed on takeoff."

"The Lift"

On June 25, 1950, Communist troops invaded South Korea and the Korean War began. Before long Canada and other UN members were deeply committed to this conflict and No. 426 Squadron was to play a major role, providing vital transport of troops and materiel between the US and Japan to aid in the rapid build-up of the United Nations fighting force. On July 7 the squadron was alerted to move to McChord AFB, Tacoma, Washington, and official word came down three days later. The orders issued by Chief of the Air Staff A/M W. A. Curtis read as follows:

NO. 426 SQUADRON OPERATIONS IN SUPPORT OF UNITED NATIONS

1. Immediate action is to be taken to integrate No. 426 Transport Squadron operations with those of the US MATS [Military Air Transport Service] in support of the United Nations campaign in Korea.

2. No. 426 Transport Squadron aircraft allocated to this task will be placed under the operational control of the US MATS insofar as operations directly applicable to Korean Air Lift are concerned. RCAF ATC will maintain administrative control throughout and operational control in domestic operations or on those flights not immediately concerned with MATS operations.

3. All domestic operations of No. 426 Transport Squadron are to be curtailed except those given special authorization.

4. No. 426 Transport Squadron is to be brought to war strength (12 UE aircraft) as soon as possible with such additional personnel who may be required in accordance with the extended nature of the operations.

5. RCAF aircraft are not to operate into Korea.

6. It is anticipated that arrangement with other nations for logistic or other support will be on a CAN. PAY basis.

7. The planning and implementation of all financial, logistic, personnel, organizational and other matters in respect to this operation are to be given the highest priority.

426 Squadron personnel load a North Star at McChord AFB. A C-124 sits gaping in the background along with SB-17 rescue aircraft. (DND PL54838)

On July 25 in a ceremony at Dorval, A/V/M James bid farewell to the hastily readied squadron: six North Stars, aircrew, and 185 ground personnel. The aircraft took off, flew to Ottawa for a flypast in tribute to the late Primc Minister Mackenzie King, who was lying in state there, then made their way separately to Tacoma.

The arrival at McChord AFB was, to say the least, a confusing experience. The base was a madhouse, with aircraft from all over the United States arriving as the Korean airlift picked up steam. The 426 diary describes the situation: ". . . the squadron was confronted with conditions that could best be described as chaotic."

Two of the nose hangars built at McChord to accommodate the 426 North Stars. These facilities were stop-gap and made necessary by the crowded conditions at McChord during the Korean airlift. (DND PL50474)

The RCAF contingent wasted no time in settling in to their makeshift surroundings and within hours their aircraft were put to work. On July 27 three North Stars headed for Japan. The first departure was made by 17502 with crew consisting of F/L Byford, F/O Kuhn, F/O Boland, F/O Bohn, Sgt Barber and LAC St. Laurent. It flew to Haneda via Elmendorf, Shemya and Misawa, returning via Shemya and Elmendorf. Hereafter there was no stopping the Thunderbirds. Day after day they were to launch aircraft across the Pacific operating on the same kind of war footing that they had known just a few years earlier when their Halifaxes had ranged over enemy skies in Europe. Whatever equipment and supplies were needed by the UN forces in Korea and could fit inside an aircraft were likely to be found aboard the 426 North Stars. As well, combat troops were carried outbound, and hundreds of medical evacuation cases on the homebound route.

In "Milk Run to Korea," published

Nov. 1950	Evaluation of McChord-Travis-Hickam-Wake-Haneda-Adak-McChord route to Japan. Conclusion that winds would be generally favourable so North Pacific remained the standard route.
Dec. 2	San Francisco-Honolulu leg in 14:10 hours. Captain, F/L Webb.
Dec. 3	100th trip by 426 on Korean airlift.
Dec. 6	Squadron authorized to "move to Japan and to operate into Korea should its services be required."
Dec. 10	Shemya-Haneda leg in 12:50 hours. Captain, F/O Broadfoot.
Dec. 12	426 did not move to Japan after all. "Much to our chagrin . . . no change in mission or location of 426 Squadron at this time."
Apr. 19, 1951	"Due to faulty let-down clearance issued by Ashiya AFB tower, North Star 17509 was involved in a minor flying accident. The tower operator gave clearance to let-down from 4000 to 3000 feet but at 3400 feet the aircraft collided with tree tops on top of a hill not indicated on the aerodrome area map."
June	189 air evacuees Haneda-Hickam, 173 US combat troops McChord-Haneda via North Star.
July 1	426 Squadron relocated back to Dorval as increased MATS space requirements forced the Canadians out of McChord.
Aug. 20	F/L Jim Morrison flew a North Star from McChord to Japan on the 1000th flight of the Korean airlift.
Nov. 30	Total hours flown by 426 on Korean airlift to date: 5960.

Miscellaneous events noted in the 426 diary, 1950-51.

in *Maclean's* in May 1951, Pierre Berton described the history-making 426 Squadron airlift across the Pacific:

MILK RUN TO KOREA

A little more than a month ago a young RCAF Flying Officer named Dean Broadfoot checked into the Marunouchi Hotel in Tokyo. Broadfoot, who had just arrived from the Aleutian Islands, spotted an acquaintance in the lobby.

"Where you been?" he asked. "I haven't seen you around lately."

"I was in Montreal on leave," his friend said, and promptly departed for Hawaii that afternoon.

One week later the two of them had a drink together in Tacoma, Washington.

The only remarkable thing about this episode is that it has become commonplace. Broadfoot and his buddy are members of No. 426 Squadron, which is the most-travelled unit in the RCAF. At present it is operating out of McChord Air Force Base, Washington, on the great Korean airlift. To the members of this squadron the world has shrivelled to the approximate size of a city block and they have all got into the habit of talking about such unlikely places as Shemya, Alaska; Wake Island, South Pacific; and Tisdale, Saskatchewan, as if they were just around the corner.

It is less than four years since Dean Broadfoot himself was taking geography at Tisdale High. Now he could tell the geography teacher a thing or two, for he has since seen more of the world than most men will see in a lifetime.

He has swum on the beach at Waikiki, lain in the sun on the French Riviera and been shaken by an earthquake in Tokyo. He has lived in an igloo in Labrador and been beaten up in Montmartre. He has drunk *saki* in Japan and *kashasha* in Brazil. He has won money at *le crappe* in Monte Carlo and at Ace-away in Whitehorse, Y.T. He once crossed both the equator and the Arctic circle in the space of nine days. He has flown over the north magnetic pole and the international date line, carried penicillin to Iceland, gunny sacks to flood-stricken Winnipeg, rockets to Korea and wounded soldiers to Honolulu. He has flown more than half a million miles in his time and he has seen London, Paris, Amsterdam, Nice, Gibraltar, Tripoli, Malta, Cairo, Dakar, Natal, Trinidad, Puerto Rico, Bermuda, Fort Nelson, Goose Bay, Anchorage, Itima, Taejon and Alert Bay in the Arctic, which is the farthest civilized point north in the western hemisphere.

When veteran members of the squadron look at Broadfoot and others of his generation they are apt to wince and remark that they are growing old. Broadfoot is just twenty-two. With his thick curly hair, his pink good health and his ability to sleep like a top and work around the

Busy ramp scene at Haneda on February 26, 1951. A North Star shares the scene with transports of the USAF, TWA and PAA. (DND PL51375)

clock, he is a constant reminder that flying is a young man's game. He is a postwar pilot. When the war broke out he was a kid of ten.

It was Broadfoot and his crew who took me recently from McChord Field to Haneda airport in Japan on the gruelling northern leg of the Korean airlift in a North Star transport. The oldest officer aboard, Gus Fielman the radio operator, had just turned 25. Co-pilot Jack Henry, fresh out of University of Toronto, is 23. Navigator Bud Austin, 22, graduated from high school in London, Ont., just two years ago.

Their job is to fly the airlift to Japan, north by way of the Aleutians and back by way of Wake Island and Hawaii, carrying everything from paper cups to helicopters and bringing back litters of wounded men. The airlift has been in operation since the Korean War started last summer and has already surpassed the famous Berlin lift in size and scope. Its planes —US, RCAF and commercial—have covered more than thirteen million miles without a fatal accident and poured ten million tons of freight in and out of Japan and Korea.

The RCAF's No. 426 Squadron joined the lift in July. Before that it was handling regular transport stints to the Canadian North, helping out in flood emergencies and taking training flights south of the equator.

When the squadron got orders to join the airlift it was based in Montreal. Thirty-six hours after arriving at McChord three of the planes were off for Tokyo. And they were back again in eighty-four hours. Broadfoot, who was in Trinidad when the order came, found himself on the other side of the world a short time later. Now he is captain of one of the sixteen RCAF crews who fly the milk run around the giant 11,000-mile circle from the US to Japan and back again.

It was a drizzly grey morning when we climbed aboard Military Air Transport Service North Star No. 7504 on the wet McChord airfield. Because of the changes in temperature they were to encounter in the next week, the men were carrying a good deal of luggage. They had parkas for the Aleutians, dress uniforms for Tokyo, summer uniforms for Wake Island and bathing trunks for Honolulu.

Broadfoot climbed aboard with a brief case stuffed with manuals and documents. He grinned at me. "We got a saying that the weight of the aircraft is equal to the weight of the paper carried," he said.

He had filed a flight plan to Anchorage, Alaska, which is the regular refueling stop on the northern leg—nine and a half hours away. The rain was driving in sheets across the McChord runway as he hoisted himself into the pilot's seat.

"I hope these wipers work," he said.

"If they don't you can run along outside and clean them off," Bud Austin, the navigator, told him.

The radio began to crackle and Gus Fielman, the operator, put away the copy of *The Young Lions* he'd been reading and began to fiddle with knobs.

"Hello, Seven-Five-O-Four. Tower. MATS requires visibility one half, ceiling four hundred before take-off," the radio said. "Visibility is now one quarter, ceiling two hundred."

Broadfoot swore under his breath and all the crew made elaborate gestures of distaste.

A 426 crew at Ashiya: Cpl O'Ray, LAC Rogers, F/L Watt, F/O Spratt, F/O Moberg, F/O Jessup, 1st Lt Lewis. (DND PL51396)

A North Star is serviced at Haneda. This aircraft was wrecked at Shemya on December 27, 1953. (via Jack McNulty)

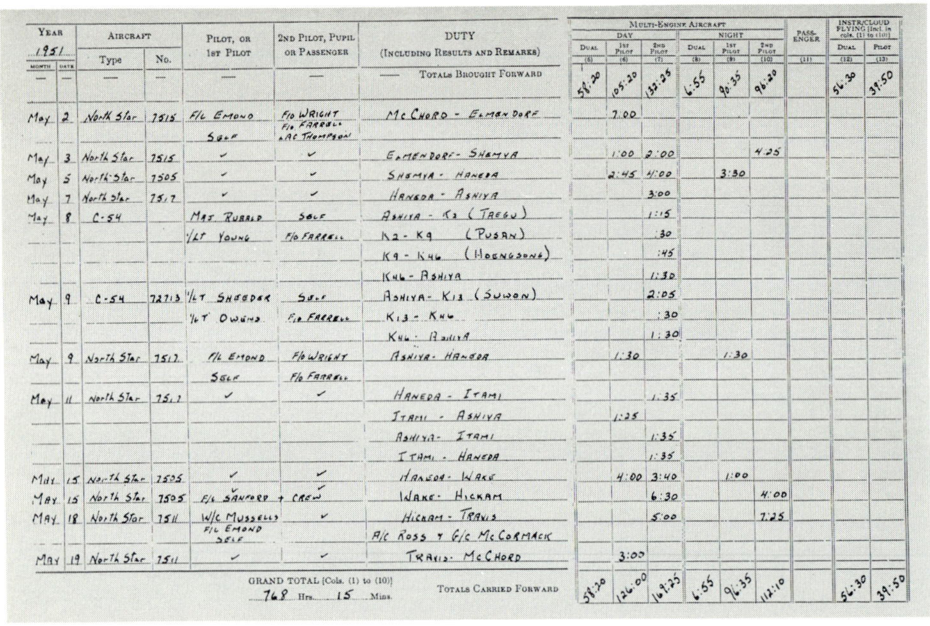

Year 1951 Month	Date	Aircraft Type	No.	Pilot, or 1st Pilot	2nd Pilot, Pupil or Passenger	Duty (Including Results and Remarks)	Multi-Engine Aircraft Day Dual (6)	Day 1st Pilot (6)	Day 2nd Pilot (7)	Night Dual (8)	Night 1st Pilot (9)	Night 2nd Pilot (10)	Passenger (11)	Instr/Cloud Flying [Incl. in cols. (1) to (10)] Dual (12)	Instr/Cloud Pilot (13)
—	—	—	—	—	—	— Totals Brought Forward	58:20	105:20	132:25	6:55	90:35	96:20		56:30	39:50
May	2	North Star	7515	F/L Emond	F/O Wright	McChord - Elmendorf		7:00							
				Self	F/O Farrell + AC Thompson										
May	3	North Star	7515	✓	✓	Elmendorf - Shemya		1:00	2:00			4:25			
May	5	North Star	7505	✓	✓	Shemya - Haneda		2:45	4:00		3:30				
May	7	North Star	7517	✓	✓	Haneda - Ashiya			3:00						
May	8	C-54		Maj. Rubald	Self	Ashiya - K2 (Taegu)			1:15						
				1/Lt Young	F/O Farrell	K2 - K9 (Pusan)			:30						
						K9 - K46 (Hoengsong)			:45						
						K46 - Ashiya			1:30						
May	9	C-54	72713	1/Lt Sheeder	Self	Ashiya - K13 (Suwon)			2:05						
				1/Lt Owens	F/O Farrell	K13 - K46			:30						
						K46 - Ashiya			1:30						
May	9	North Star	7517	F/L Emond	F/O Wright	Ashiya - Haneda		1:30			1:30				
				Self	F/O Farrell										
May	11	North Star	7517	✓	✓	Haneda - Itami			1:35						
						Itami - Ashiya		1:25							
						Ashiya - Itami			1:35						
						Itami - Haneda			1:35						
May	15	North Star	7505	✓	✓	Haneda - Wake		4:00	3:40		1:00				
May	15	North Star	7505	F/L Sanford + Crew	✓	Wake - Hickam			6:30			4:00			
May	18	North Star	7511	W/C Mussells	✓	Hickam - Travis			5:00			7:25			
				F/L Emond Self		A/C Ross + G/C McCormack									
May	19	North Star	7511	✓	✓	Travis - McChord		3:00							
				Grand Total [Cols. (1) to (10)] 768 Hrs. 15 Mins.		Totals Carried Forward	58:20	126:00	169:25	6:55	96:35	112:10		56:30	39:50

Excerpt from the log of F/O G.G. Waddell of 426 Squadron during the Korean Airlift. Shown is a typical McChord-Haneda-McChord trip using the southern route for the return leg. Included is a side trip by C-54 to Korea. These were familiarization trips flown by many 426 crews in anticipation of the RCAF extending their operations straight into Korea. Such trips were officially logged as combat missions!

F/O Waddell's log also shows trip returning by the northern route. This is one from 1951.

July 17	17515	McChord-Elmendorf	7:20 hours
18		Elmendorf-Shemya	8:00
21	17506	Shemya-Haneda	10:00
24	17515	Haneda-Misawa	2:15
		Misawa-Shemya	7:45
28	17512	Shemya-Elmendorf	6:45
		Elmendorf-McChord	7:25

"Hey," said Broadfoot into his microphone, "has anybody up there *looked* out the window lately?"

The voice on the radio said they had.

"Well, the ceiling and visibility are okay out *here*," said Broadfoot, looking optimistically into the rain.

But it was fifteen minutes before they cleared him for take-off into the low wet clouds hanging heavy over the Pacific Northwest. The big North Star, loaded with about six tons of tires, machinery, generators and spare parts destined for vehicles bogged in Korean mud, climbed to 8,000 feet and headed north over the clouds.

For nine monotonous hours we flew north and west with only the sea showing grey and choppy through the ragged clouds. Occasionally the far-off vapour trail of another aircraft, a chalk scrawl in the blue, served to emphasize the loneliness of the empty world of sky around us.

The crew worked at its various tasks. Bud Austin, the navigator, did things with a slide rule. Cpl. Heddie, the air traffic assistant, served bacon, eggs and beans on pink Russell Wright plates. Presently Jack Henry, the co-pilot, pulled a copy of *The Naked and the Dead* from his jacket, climbed out of his seat and stretched out on a bunk in the cabin behind. Gus Fielman tapped away on his telegraph key, sending regular weather reports back to McChord. Fielman is a farm boy from Rosenfeld, Manitoba. "I saw a sign saying, 'Join the Navy and See the World,' " he told me. "So I joined the air force and here I am."

Just before dusk a long horny spine of mountains crept out of the horizon, trailing off into the sea under the flat tiers of sheepskin clouds. It was Alaska.

"There are volcanoes down there," Broadfoot said, pointing. "There's one that has no crater at all—just holes in the side with the stuff pouring out in different places."

An hour later we were over Anchorage, with dusk falling and the red neon lung of the main street blinking at us out of the haze.

"What a town!" Broadfoot said. "No matter where you stand on that main street you're always standing in front of a bar."

He brought the plane quietly down on the landing strip of Elmendorf airport. In the snack bar the crew wolfed minced steak and coffee, then went into the weather office to get a forecast for the next leg. There on the big wall map of the Aleutians you could see our objective: Shemya, a tiny dot at the very tip of the long chain of volcanic islands that swings for a thousand miles out into the North Pacific like a long curved finger beckoning at Asia.

"See that big island next door to Shemya?" Broadfoot said. "That's Attu—the last of the Aleutians the Japs had during the war. There's still a small detachment of Yanks there. I hear they had a hundred-knot gale—blew all the buildings down last month.

"A while back one of these islands blew right up. You see this Great Sitkin volcano near Adak? We pass right close to it. They expect it to blow up, too, in about five years. They keep going up and measuring it with instruments."

The quick northern night had fallen by the time we taxied out onto the runway and headed west along the Aleutian chain for Shemya. We expected to get in about 3:30 a.m. I dozed off and woke about 3:15 to hear Bud Austin say: "We can't get into Shemya. The winds are sixty-five knots on the ground. We're going into Adak instead."

Adak is the warmest of the Aleutian Islands—a US Navy base that lies about two thirds of the way along the chain. In about two hours with the night below us, we were over the island and a voice in Broadfoot's earphones began to shepherd the big plane down onto an unseen landing strip, using the Ground Control Approach system without which the Korean airlift would be a virtual impossibility.

GCA, which Bob Hope after an Aleutian tour claimed stood for "Greetings—Congratulations—Alive!" is a radar system by which men on the ground can talk an aircraft down under conditions which would otherwise make landing impossible. Somewhere below us lay Adak, sheathed in clouds which masked the 4,000-foot mountains and rocky cliff approaches which hang over the narrow airstrip. And down there a man in a yellow truck was staring at the quivering white "blurp" made by the image of

Keepin' 'em fit. Action on the flight line at McChord, September 24, 1950. (DND PL50136)

our plane on the gridded radar screen, and telling Broadfoot exactly what to do and when to do it.

His voice came over the pilot's earphones: ". . . maintain 180 degrees 7504. Coming in nicely now. You are ten feet below the glide path—bring her up a little. You are on the glide path. Turn left to 172 degrees. Left to 172. You are fifteen feet above the glide path. Twenty feet. Come down a little, 7504. Maintaining 172 nicely. Twenty-five feet above the glide path now. Bring her down. Twenty feet. . . fifteen feet . . . ten. You are on the glide path one half mile from the airstrip and coming in nicely. . ."

Ahead and directly below, two parallel chains of lights loomed suddenly, framed on either side by the black bulks of Adak's cliffs. A moment later we had landed. It was 5 a.m. Aleutian time, 8 a.m. McChord time. "Twenty-five hours," said Broadfoot, who had been on the go steadily since 7 a.m. "Not a bad day's work."

On the airstrip the unending Aleutian winds were howling. The temperature stood at one degree below freezing. In the grey light of dawn Adak looked like a lost land, its low black headlands unbroken by any tree or bush, its only vegetation a coarse brown grass ceaselessly blown flat by winds that never die. Far inland a chain of snow-tipped mountains rose out of the bald hills.

We threw our gear onto a truck and drove five miles through the rolling land to our billets. On the way we suddenly passed through an oasis of civilization: huge brick-and-concrete barrack blocks and mess halls (panelled and tiled within); flat asphalt roads with concrete sidewalks and wrought-iron lamp standards; a church and a movie house, and all the paraphernalia of civilization — every blessed board and brick and bolt of it freighted in by sea and air.

Broadfoot and his crew slept twelve hours while the winds of Adak whined around the Nissen hut, rattling the shutters and shaking the windows and straining the walls. By 5 p.m. Weather reported the winds dying down at Shemya. We took off at 7 as dusk was falling, and a little more than three hours later we reached the tip of the Aleutian chain.

If any place merits the designation, "end of the earth," it is this tiny island of Shemya, a pinprick on the map, a piece of volcanic rock erupting from the cold northern ocean, treeless and cheerless as the grave and chill with clinging mist and sifting snows.

We hit the tarmac with a forty-knot crosswind which threatened to blow us off the runway. Down at the other end lay the smashed wreckage of a four-engine B-17, blown right over the bank by a similar crosswind. Even as we landed a sudden blizzard sprang up, obliterating everything.

"Just think," said Bud Austin as he climbed down, "if we hadn't joined the air force we'd have missed all this."

F/O Jack Egan, commanding the sixteen-man RCAF detachment on the station, came out of the swirling snow with a parka tight around him.

A USAF SB-17 rescue aircraft. Known as "duck butts," the SB-17s served as airborne life stations during the Korean airlift. On occasion they shepherded North Stars coming in on three engines. North Star pilots recall how they had to throttle back to give the old Flying Fortresses a chance to keep up with their "crippled" transports! (DND PL54960)

"Welcome," he said, "to Shemya-By-The-Sea."

"Hey Jack," someone shouted. "How'd you get this job?"

"Darned if I know. Just lucky, I guess."

Egan was here for the regular RCAF six-week tour of duty. The Americans, who are here by the hundreds, often stay longer. Some have spent eighteen months in Shemya. Some even volunteer for duty here because it is not difficult to save money. There is a saying that there is a woman behind every tree on the island. But there aren't any trees.

Almost everybody lives in an incredible labyrinth of a building which is actually a series of huts all joined together under a single roof, dug deep into the ground and banked with earth as protection against the hundred-knot gusts which sweep the island. This building, which has all the twists and turns of the Pentagon, also has hallways that run for a thousand feet without a corner. Under this one roof is everything an airman needs: billets, Post Exchange, movie theatre, post office, mess hall, hospital. It's said that some men have never been outside in all their time on the island.

There isn't much to do outside — as we found in our thirty-six hours on the island. Within ten minutes I saw the weather change from brilliant sunshine to blinding blizzard to gale, to sun again and then to hailstorm.

A wartime base, Shemya was doomed to destruction and abandonment, but the Korean Airlift gave it a reprieve. The island is five miles long and two miles wide but there are fifty-nine miles of asphalt roads. It is jammed with Nissen huts, hangars, dumps and installations, most of them in an advanced state of wreckage, for the whole island was in the process of being put to the torch when the reprieve came. This carnage adds to the sense of desolation and decadence which, along with the creeping fog, hangs over the island like a pall.

There is nothing much to do in Shemya except sleep, talk and read. Broadfoot and his crew sat on their bunks and talked mainly of how much better it was going to be on the southern leg of the journey. The first stop out of Japan is Wake Island, where the crews rest for forty-eight hours, swim in the warm green ocean or drink at a bar called The Drifter's Reef.

Wake is sort of a rich man's Shemya and is known around the RCAF for its rats, its skulls and its high prices. The prices are high because Wake is a non-military establishment run by a civilian airline. Airmen who get an allowance of \$9.50 a day while travelling the airlift pay upward of \$10 a night at Wake. The rats were brought by the Japs, who also left the skulls. GIs have been known to dig up skulls and prospect them for the gold in the teeth.

"Used to be one of the darn things by the name of Yorick hanging over the transient quarters," Broadfoot told me. "Some guy swiped it."

The load back on the southern trip usually consists of about thirty air evacuees from the Korean war. The RCAF carries more wounded than US planes because the North Star is faster than the DC-4s and cuts two or three hours off the Pacific trip.

"Last trip we had a flock of those guys with body casts right up to their necks," Bud Austin told me. "Those guys were really high."

The aircraft go right through, stopping only to refuel, but the crews have to rest forty-eight hours at Wake and Hawaii, so that as one crew arrives with a plane the rested crew boards it

North Star 17502 loading mail for Japan at McChord. This was one of the last North Stars flying. It served until 1976 with Air Caicos of Sarasota, Florida. (DND PL54841)

and flies on. Thus a crew takes about eleven days to complete the circuit but an aircraft is back at McChord base in about eighty hours.

At two o'clock Thursday morning a new crew came into Shemya and Broadfoot and his crew rolled out of bed and prepared for the next leg of the journey, to Tokyo.

"How's that aircraft?" Broadfoot asked the incoming captain.

"George isn't working — so the pilot's got to work this trip."

"I guess George is tired."

"Everybody's got to get tired once in a while."

George is the automatic pilot.

soft grey ocean of clouds, headed southwest to Japan.

Just out of Shemya we crossed the International Date Line and it was tomorrow. Broadfoot told me that in the early days of the airlift the planes used to fly the northern route coming and going and it was just about at the Date Line that they passed.

"The guys from Japan would always tell us what tomorrow was going to be like," Broadfoot said. "They knew all about it because they'd just been in it."

At 11 a.m. a lone voice crackled out of nowhere on the radio. Below us somewhere under the clouds was

Russian coast," Broadfoot said. "That's probably old Joe himself."

The voice stopped and a string quartet began to play something that sounded as if it might be Shostakovich. Broadfoot told me that for a while the Russians were beaming 500 voice choirs over the GCA frequencies, apparently trying to jam air radios.

The Kurile Islands were just over the horizon to our right.

"Some of the boys have got off their course and flown over them," Broadfoot said. "Nobody's too fussy about doing *that.*"

He wiggled the dial some more and a girls' choir came on singing Beauti-

North Star 17505, later written off at Shemya, parked serenely at Haneda during Operation Hawk. (DND PL53223)

The crew hopped down to the Weather Office where on the weather maps the whole Aleutian route lay garlanded with isobars curling like snakes around the islands, showing the high- and low-pressure areas, the wind speeds, cloud formations, the cold fronts marked in blue and the warm fronts marked in red.

"Head winds all the way," said Broadfoot, squinting at the map. "It's going to be a long slow trip. Eleven and a half hours anyway."

With the wind still whipping across the runway we took off into the grey Aleutian fog, climbed to 8,000 feet into the pink of the early morning and, with the sun just touching the

United Nations weathership *Sugar*, wheeling in a small unending circle in the empty sea. The voice from the weathership gave Broadfoot his position by radar and a complete weather summary and forecast.

Another voice came out of nowhere: Northwest Airlines No. 404 heading for the Aleutians from Japan.

"Which way you heading?" the voice asked us.

"West."

"You're goin' to have a *sl-ow* trip."

"What are the winds there?"

"Sixty-eight knots at 250."

"That's just ten degrees off our nose," Broadfoot said to me. "Slow is right."

He wiggled the radio dial and a throaty Russian voice came in on the earphones.

"We're only sixty miles off the

ful Dreamer in Japanese. That was Radio Tokyo.

Broadfoot went back and lay down on the bunk and Jack Henry flew the plane. Ahead of us loomed a long straight line of cumulus clouds.

"There's that cold front we saw on the weather map," Henry said.

I remembered the blue line just off the Japanese coast. The plane hit the front a few minutes later and began to bump and toss in the turbulent air. The thermometer dropped five degrees. Then we were out of it again. The clouds parted and down below us lay the wrinkled old hills and tinted mountains of Japan.

We flew south along the sea coast of this neat brown land, with its brown mountains and brown hills, its olive brown rivers and tan-colored cities and cocoa-brown forests — all

A Thunderbird mount on the ramp at Haneda during the Korean airlift. (DND PL51259)

roasting to a turn in the hot noon sunlight.

By 2 p.m. Tokyo and Yokohama sprawled out before us in the haze and a few minutes later we were skipping across the tarmac of Haneda airport in the ruined shadow of a blasted Zero factory.

The crew lost no time checking into the Maranouchi Hotel, which is run for Commonwealth troops by the Australians.

In the grillroom that night they ordered two-and-a-half-pound steaks, buttered toast and Burgundy. They went to the Piccadilly burlesque house and watched some comely Japanese girls with very little on dance about to Brahms' Lullaby and other strains.

The next day they shopped in the great Ginza market where you can buy Japanese imitations of Parker '51 pens and Ronson lighters for a song. Bud Austin got some silk for his girl friend and Jack Henry some satin brocade for his wife. Gus Fielman got some flash bulbs so he could take colour pictures from the Piccadilly's front row.

In the afternoon they walked through Frank Lloyd Wright's famous Imperial Hotel and in the outer gardens of the Emperor's palace and through the huge Tokyo Post Exchange, which is as big as Eaton's College Street store in Toronto. Then they went back to the hotel and ordered everything on the dinner menu, including all three desserts.

And the next day, which was Sunday, they were off again on the southern leg of the long milk run, out over the same Pacific Ocean which once took Magellan the better part of a year to cross.

Pierre Berton

Canadian troops arrive at Haneda aboard a 426 North Star during the Korean airlift. (DND PL51410)

Airlift Mishaps

The Korean airlift, though notable for its safety record as far as 426 Squadron went, was not without its share of close calls. Weather was a constant threat and figured in two North Star Category A accidents. It tried its best to figure in others. F/O G.G. Waddell recalls being almost dragged out of the sky by ice and having to let down near Sandspit in the Queen Charlotte Islands for a possible emergency landing. Lower down, the ice shed, allowing the flight to carry on. On another trip he was approaching to land at Elmendorf and apparently passed perilously close to a DC-3 in cloud. Shemya, of course, was always a challenge. On the flight plan weather and visibility were minimal and were usually worse on arrival. The cross winds were so bad that, on final, the pilot was likely to catch sight of the runway looking straight ahead out of the side window. Fortunately the GCA personnel were skilled and the runway wide, but this didn't prevent one North Star from taking out marker lights first on one then on the other side of the runway when landing one night.

The first really harrowing incident on the Korean operation occurred on April 19, 1951. North Star 17509 was approaching Ashiya, Japan, after a three-hour flight from Tokyo. Aboard were 23 including W/C Mussells, the OC, and S/L Morrison, squadron ops officer.

As the aircraft descended there was a sudden crash-banging and a few moments of confusion, followed by the realization that a collision had just occurred. After a quick assessment of the situation, the OC decided to change the pilots, as there were spare aircrew aboard. F/L J.A. Watt and F/L F. Haley, no doubt deserving some relief, took a rest and F/O R.M. Edwards and W/C J.K.F. Macdonald took over.

On climb-out No. 3 engine had to be shut down due to coolant loss and No. 2 had to be run on reduced power for the same reason, so 17509 was brought down at Ashiya rather than return the 600-odd miles to Haneda.

A walk-around at Ashiya showed just how fortunate those aboard 17509 had been. The antennae were torn off the belly, the pitot-static system was knocked out, there was propeller damage, and the oil and coolant radiators on both inboard engines were punctured. As well, the nose cone was flattened. All from a very close encounter with a hilltop that GCA had forgotten to tell F/L Watt about!

Of the event F/O Edwards wrote that "there was no further damage done except perhaps to those of us who partook too liberally of the hospitality of the USAF, who, sensing a need for special handling in this case, produced several bottles of Bourbon during the operation's debriefing. With the help of the USAF, our Flight Engineers subsequently performed emergency repairs and the aircraft was flown back to Tokyo four days later where more extensive repairs could be carried out."

Cause of the Ashiya incident was a foul-up by GCA. The RCAF investigation remarks: "In flight, descending from 10,500 feet by instrument letdown, aircraft was permitted to proceed unlimited distance southbound prior to procedure turn to base. Aircraft struck trees at indicated altitude of 3400 feet . . . Aircraft lost No. 3 engine due to coolant loss and then carried out an emergency landing . . . Procedure at Ashiya discontinued until proper procedure can be effected."

Though there was no doubt in the minds of those who flew through the trees in 17509 at Ashiya as to the gravity of the situation, the 426 Squadron diary includes a quaint statement describing it as "a minor flying accident."

The next serious incident on Operation Hawk occurred December 27, 1953, when 17505 was written off while landing at Shemya at night in a ferocious 50-knot cross wind and blowing snow. The runway was slippery, reducing braking action. Pilots S/L Hare and F/O Charchuk had been holding for some time over Shemya due to power interruptions to the terminal navigation facilities there, having made the trip on a marginal flight plan in the first place. Once cleared to land, 17505 got down onto the runway but was literally blown off it by gusts, nosing down into a gulley.

North Star 17509 parked at Ashiya soon after its close encounter with a Japanese mountain top. Damage to the nose, nose gear doors, engine nacelles and intakes, and propellers is evident. (DND)

Two views of 17505 after it was blown off the runway at Shemya on December 27, 1953. (Glen H. Fenby)

Moments after its spectacular arrival at Vancouver, 17503 lies spread all over the field as men dash to help in the rescue operation. This is a view looking southwest with the detached starboard wing in the foreground. (DND)

What a mess! Few airplanes in the history of crack-ups have ended up in such a mangled state yet allowed everyone aboard to walk away virtually unscathed. Only one minor cut was reported among 54 passengers and crew flying aboard 17503 that incredible December 30, 1953. (DND)

Fortunately none of the eight aboard was hurt.

Just three days later, 17503 was climbing out of Vancouver when No. 4 engine overheated. The flight turned back but was in such heavy icing that by the time the aircraft was on its final approach its stall speed was greatly increased. Visibility at Vancouver was marginal as 17503 came in. It appeared to approach steeply. First the nose wheel then the starboard wing dug into the runway. The wing broke off between No. 3 engine and the fuselage and the rest of the aircraft flipped over.

Gordon Peters, at the time a volunteer on the airport crash crew, witnessed the mishap. As he recalls it: "We saw the aircraft break out of the low cloud in the west but it didn't seem to be properly lined up. It banked to correct this, but the right wing appeared to hit the ground. With that, the wing, right engines and undercarriage came off and, as I remember too well, came toward us. The aircraft then flipped over and skidded across the runway upside down, producing a great shower of sparks. It continued to slide and came to rest just a few feet from a ditch. By this time we had the crash truck on the way. It seemed we would never get there although it was really a very short distance. As soon as we arrived we got our fire fighting gear out, such as it was. The most incredible thing was that there was gasoline all over the ground, in puddles over my shoes. The miracle was that it didn't all go up as the passengers were lighting cigarettes as they got out. We had a terrible time getting them to put them out!"

Another man on the scene was G/C R.W. "Buck" McNair, who was a passenger aboard the North Star. He directed the safe evacuation of the aircraft and for his efforts was awarded the Queen's Commendation for Brave Conduct.

G-ALHG between flights at London Heathrow towards the end of its BOAC career. Known in the fleet as *Aurora*, it served BOAC between 1949-1961 then was sold to Overseas and finally to BMA. While on a return BMA charter into Ringway, it crashed in the worst ever North Star/Argonaut mishap. (Peter Keating)

One of the BMA Argonauts seen at Newcastle on July 29, 1967. Soon after this, BMA withdrew the Argonaut from service. Note the "radar" nose. In fact, only the RCAF North Stars ever carried radar. (Ian MacFarlane via Stephen Piercey)

The National Aeronautical Establishment's North Star at Resolute in April 1976. CF-SVP-X had frequently visited Resolute during the fifties when it flew with 426 Squadron as 17514. (Stephen Piercey)
▽

International Air Freighters' former TCA North Star parked near the Genaire hangar at Malton in October 1965. CF-TFC had carried everything from sheep to lobsters to watermelons to UN ambassadors during its brief period operating between Toronto and Havana. (Larry Milberry)

LEBCA's ex-TCA "Flying Merchant" as seen by Peter Keating at Kingston, Jamaica on October 15, 1965. LEBCA operated this aircraft on regional freight charters.

A superannuated TCA North Star rests forlornly on a back lot at Dorval. So ended most of the great propliners of the 1950s. (Larry Milberry)

▽

1.	Total flying time to 2359 31 January 1952	26,980:00
2.	Total flying for period	1,605:10
3.	Flights departed for Haneda from McChord during period	15
4.	Flights completed during period (Note: one trip completed during this period was dispatched from McChord prior to 01 January)	15
5.	Flight line crews available as of 31 Jan 52	15
6.	Flight line crews in slip positions - McChord	01
	Shemya	02
	Haneda	01
7.	Flight line crews on leave as of 31 Jan 52	NIL
8.	Average duration of trip per crew (Airlift only)	16.8 days
9.	Average duration of trip per aircraft McChord/Haneda and return	4 days, 6 hrs,21 min
10.	Average flying time per aircraft per trip (Airlift only)	65:03
11.	Aircrew available uncrewed as of 31 Jan 52 - Captains	NIL
	First Officers	11
	Navigators	11
	Radio Officers	09
	Engineers	08
	ATA's	06
12.	Late departures from McChord ramp	05

13. Late departures - Breakdown:
Flight No. 317 - 1 hour 20 mins due loading
Flight No. 318 - 30 mins due traffic
Flight No. 326 - 1 hour 20 mins due TCM weather
Flight No. 327 - 35 mins due de-icing wings
Flight No. 329 - 35 mins due U/S radio

14. Summary:

The total flying time of 1605:10 is broken down as follows:

Tokyo Airlift	975:40	
Domestic Operations	428:20	(Approximately 100 hrs of the operational time was utilized for training)
Training	152:30	
Testing	16:50	
Beechcraft	31:50	
Total	1605:10	

Domestic operations consisted of the following completed trips:

Goose Bay	9 trips	(Rations & Backlog specials)
Goose Bay/Chimo	1 trip	(Backlog Special)
Goose Bay/Frobisher Bay	1 trip	(Special)
Tokyo	1 trip	(C5 engine and spares)
Churchill/Resolute Bay	4 trips	(Freight)
Rockcliffe	2 trips	(Air Cadets)
USA and Northern Bases	1 trip	(Defence College)
Toronto/Churchill	1 trip	(Orenda engine)

Approximately 100 hours of the operational time was utilized for training purposes. This is in addition to the training time shown on the flying time breakdown.

From the 426 Squadron diary, the operations record for the period January 1-31, 1952.

F/O G.G. Waddell with his 426 maintenance detachment at Haneda, Japan. (via G.G. Waddell)

F/O Ben Budgeon, US nurse Mary Orichella and F/L R.J. Brown watch as wounded US soldiers are loaded aboard a North Star at Haneda, February 17, 1951. Thousands of evacuees returned from the Korean theatre to the US aboard 426 aircraft. (DND PL51140)

This newspaper report by John Maffre of Montreal, dated June 5, 1952, described the success of Operation Hawk to date: "The squadron's record of service wasn't easily established. Within 36 hours of their arrival two years ago at McChord Field near Tacoma, Washington, which was not prepared to receive them, three North Stars were off again for Tokyo. Briefing had been sketchy, and the 4446 miles to Japan via the fog-bound Aleutians were a hazard of moody weather and unfamiliar navigation.

"By mid-August a flight was leaving every 24 hours. The crews had begun the bone-wearying schedule that was to last six months, in which they logged monthly flying times of over 150 hours. The ground crews, working more often than not on double shift, slaved round the clock keeping their North Stars in the air.

"So things are easier now. There is double the number of North Stars on the run, and a smooth-running organization has been built up behind them. Maintenance crews rotate every three months at McChord Field, at Shemya in the Aleutians and at Haneda airport near Tokyo. Relief air crews take each aircraft over the different legs of the trip. Even so, the average flier on the squadron spends 22 days out of every month away from his home.

"These are the statistics for the squadron's 392 trips until the end of May this year. It has flown 4,579,736 miles, carried 9197 persons, transported 1434 air evacuees, lifted 3,725,130 pounds or nearly 1900 tons of freight, and delivered 1,020,077 pounds of mail. . . . Last June the squadron moved back to Dorval but still maintains its regular schedule of one flight every two days from McChord Field. Every four days a North Star arrives here for a 150-hour inspection and also carries out some domestic runs."

This story nicely illustrates the

RCAF Station Lachine in the early 1950s. 17502 in the foreground became the world's last commercially operating North Star in regular service, flying with Turks Air until June 1975. Flanking it to left and right are 17509 and 17517, while another North Star, a Bristol Freighter and two Beech Expeditors of the RCAF are in the background. (Canadair Ltd.)

nature of Operation Hawk and the professionalism of the 426 personnel. A North Star was due to depart Elmendorf for Shemya with a load of Marines. The aircraft commander informed the Marine colonel, "We'll wait till dark, then take off. We have a better chance of finding the runway at Shemya if the lights are on."

The colonel couldn't believe his ears and was clearly on his first trip through Shemya. "You guys don't know what you're doing," he replied to the pilot.

The pilot said, "Well, colonel, I'll tell you . . . I don't know what your orders are but mine are to take this aircraft to Tokyo."

The colonel then repeated his comment to which the Canadian's retort was, "Colonel, that's your opinion, but when this aircraft gets to Tokyo without you . . . We leave in 30 minutes."

The North Star took off, Marines and all, and got safely into Shemya, changed crews then carried on to Haneda. The RCAF pilot who had been in the discussion then went on to Tokyo with the next airplane where he found a message waiting to call the Marine colonel. This he did and the colonel invited him and the whole North Star crew to dinner. Later in the evening the colonel admitted to the North Star pilot, "You've taught me one of the best lessons of my career!"

North Stars into Korea

During the Korean airlift at least four 426 Squadron North Star trips were made to Korea. This has not been widely known, most people assuming that all RCAF crews visiting Korea made the trip as extra crew aboard USAF transports, generally C-54s. From 1952 onward the Korean airlift subsided for 426 Squadron and operations became more diverse.

Other 426 Activities

During the second half of 1951 there were two relief operations involving North Stars. In August shipments of Red Cross aid went to Jamaica following a severe hurricane, and in November two North Stars carried aid supplies to Italy during floods there. On February 5, 1953, the first of several North Stars, 17517, left Dorval on a major flood relief operation to the Netherlands, and in August 426 flew relief to Athens following an earthquake. At home 426 continued its schedule of domestic service flights, ration runs and re-supply to northern sites. The newly formed Air Division depended greatly on the squadron to move supplies, equipment and personnel to the UK, and combat training exercises continued to count on the North Stars for logistical support.

On May 28-June 9, 1954, the 599th and last North Star trip to Japan was flown, the aircraft returning to Dorval under F/O Bruce Ingall. Over the 46½ months of operations to Japan some 13,000 personnel and 3500 tons of freight had been carried by the North Stars. Though operating under difficult circumstances, Operation Hawk had been without serious injury or fatality in the air. Leadership, rigorous training standards and deep commitment from squadron members enabled the successful completion of its Korean mandate.

Those connected with the Korean airlift look back with many a fond

Date	Pilots	Aircraft	Route
May 11-12	F/L Sanford F/O Spratt	17517	Haneda-Pusan-Taegu-Haneda
June 7-8	F/L Brown F/L Haley	17517	Haneda-Seoul (Kimpo)-Haneda
Nov. 16-18	F/L Herbert F/O Harrison	17506	Itazuki-Kimpo-Iwakuni
Dec. 11	S/L Torontow	17512	Iwakuni-Kimpo-Iwakuni

The four North Star trips into Korea as they appear in the 426 Squadron diary for 1951.

July 5	Dorval-Churchill	6:50 hours
	Churchill-Resolute	5:55
6	Resolute-Churchill	5:00
	Churchill-Coral Harbour	2:30
	Coral Harbour-Churchill	2:55
	Churchill-Dorval	5:25

Northern scheduled service as noted in F/O G.G. Waddell's log in 1951.

memory and a great sense of pride to a seemingly impossible job completed to perfection. F/O R.M. "Bob" Edwards flew many a North Star across the Pacific during the airlift: "It was a severe test for the aircraft and for the men who flew it; yet both came through with a record that would be difficult to duplicate today in terms of performance under less than ideal circumstances. The North Star didn't have the payload capability of the C-54 and the Honolulu-Travis leg was slighly beyond its safe range, but because of the dependability of the aircraft itself and the willingness on the part of both the air and ground crews to regularly work long days, the relatively small fleet of North Stars operating out of McChord AFB established and maintained a record that often brought high praise from the USAF Military Air Transport Service. Without exception, we were all proud to have played a part in that aspect of the Korean War."

Navigating the North Star

Navigator Don Connolly joined 426 Squadron in March 1952 along with many newly trained aircrew. These were augmenting the generally experienced 426 personnel, many of whom were World War 2 "retreads." Here Connolly gives a detailed look at the navigator's contribution on major RCAF long-range operations such as the Korean airlift:

As I studied basic navigation at Summerside, P.E.I., in 1950, I aspired—as did most of my classmates—to be assigned on graduation to 426 Transport Squadron. We had heard much from our instructors of the challenging routes and exotic destinations which that outfit was experiencing and of the highly tuned efficiency of its crews. It seemed the perfect post for honing our uncertain skills and simultaneously becoming world travellers while drawing generous pay and allowances. Besides, they were flying North Stars, which were in those days, from the navigator's point of view, the most glamorous mounts available.

Alas! Not one from our class was called to 426 Squadron on graduation, and we all packed off to Trenton, Ontario, to take a mandatory two-month course in bomb aiming. After that we were to be sent to search and rescue, maritime, and other miscellaneous units far from the Shangrila of Tacoma, Washington, where 426 was based. But while I was home on leave after the bombing course, a signal came despatching me to 426 Squadron, and within several days I was shipping out to Tacoma as a passenger, my ears ringing to the roar of unmuffled Merlins — something

Famous air portrait of 17514 with a trial radar installation in the belly. In 1956 the RCAF equipped its North Star fleet with APS-42 radar in the nose. These were the only radar equipped North Stars ever. (DND PL55533)

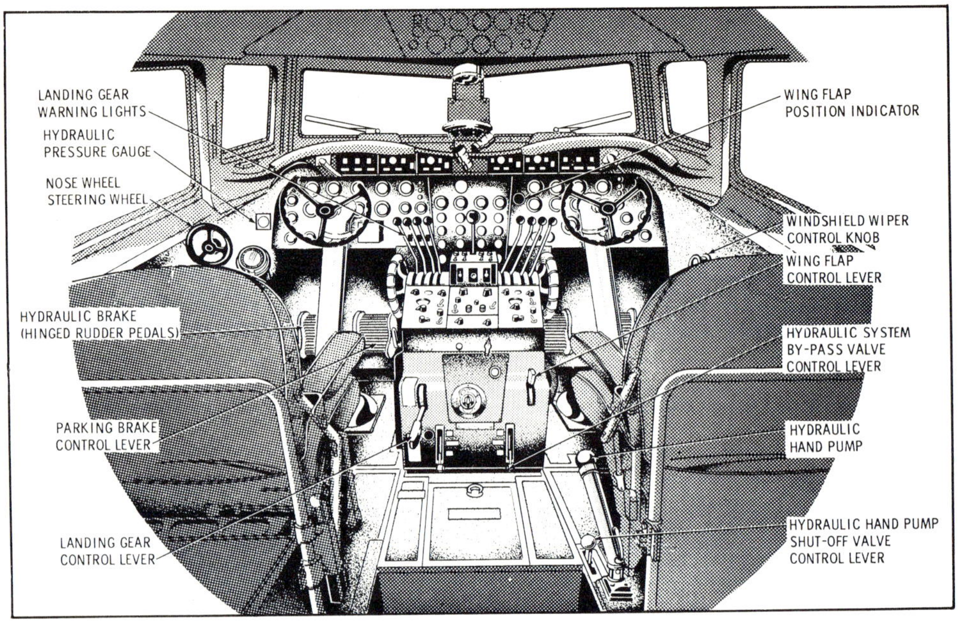

Hydraulic system controls, RCAF North Star.

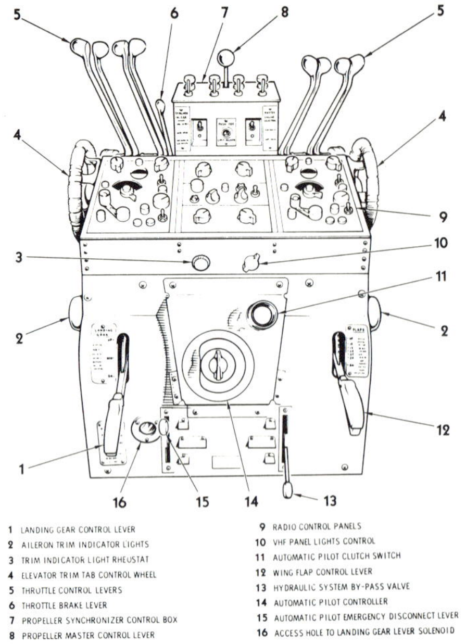

Control pedestal, RCAF North Star.

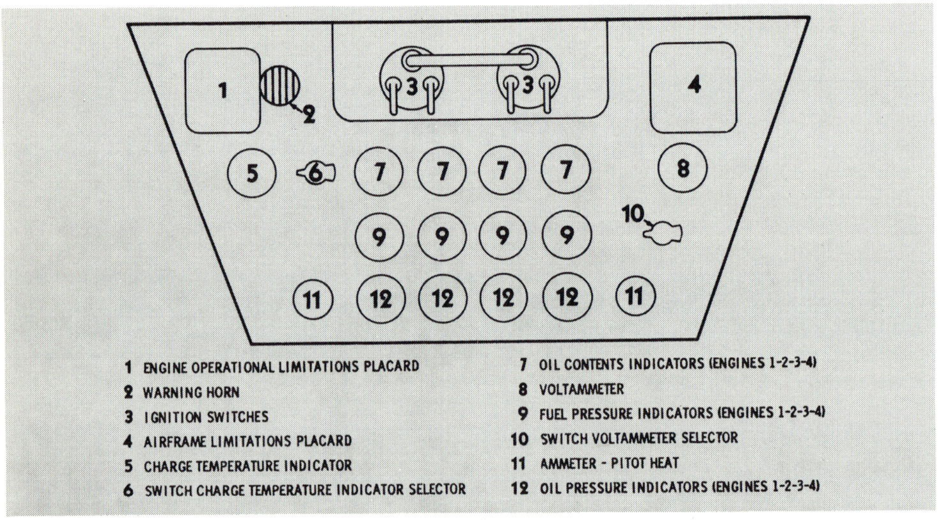

Upper instrument panel RCAF North Star.

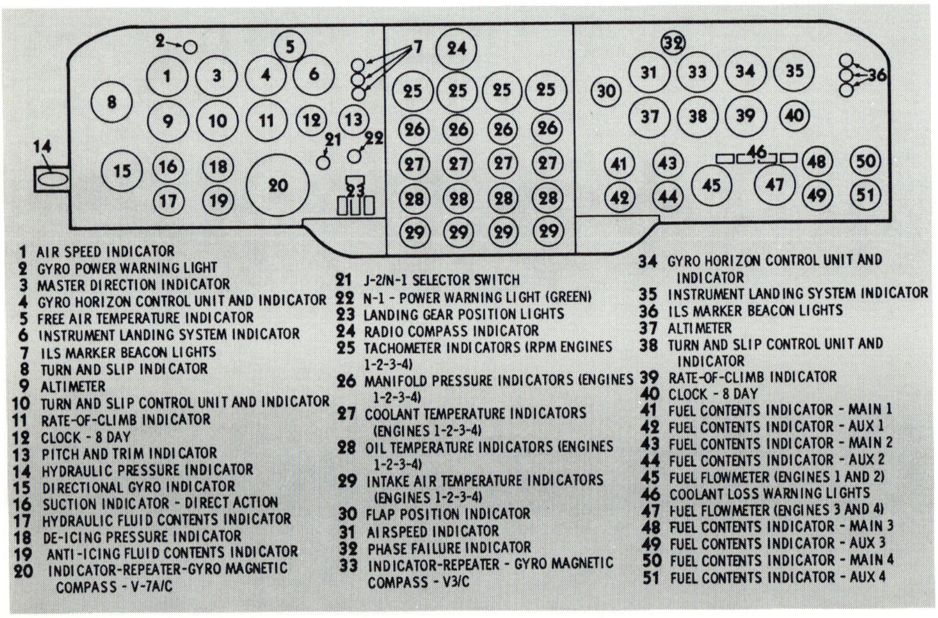

Main instrument panel RCAF North Star.

they still do, I swear, thirty years later. Noisy she was, but I soon learned that there was little else to criticize about the lady. Others will dwell elsewhere in this book on her special virtues of power, reliability and tractability, but I will focus only on the special traits that endeared her to navigators.

First of all, we felt welcome. You see, navigators are considered to be a somewhat temperamental breed amongst aircrew and the poor design and placement of navigation compartments in most aircraft only serves to confirm their poorly-concealed belief that they don't get the respect they deserve. This was definitely not so in the Star. The layout of the navigation "office" was near to perfect. We felt needed.

The flight crew was positioned compactly; an effective arrangement. The navigator sat back to back with, and a step lower than, the co-pilot, facing aft. To the nav's right and directly behind the pilot sat the radio officer, facing to port. A removable jump seat positioned the flight engineer roughly in the centre of this quartet, slightly aft of the two pilots and facing forward. The high noise level made conversation difficult, so important information was usually transmitted by interphone or on a chit, although a well-directed shout, poke or wad of paper could attract attention if necessary.

Undoubtedly this tight arrangement did much to engender good teamwork. In effect, there were always several

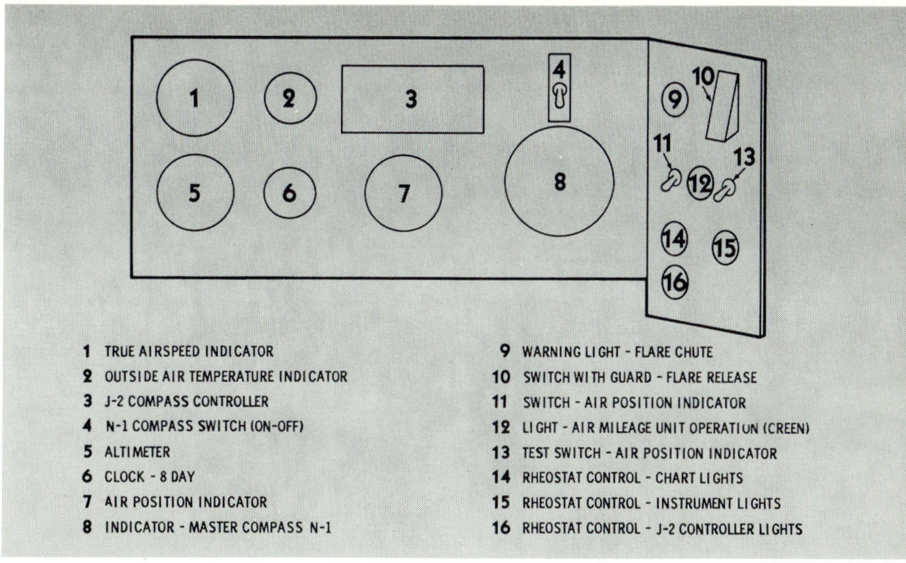

Navigator's instrument panel and sub-panel RCAF North Star.

"officers of the watch" on duty, and we each stayed alert to the hazard of all the others dozing off on the long and grinding over-water passages. This closeness also seemed to stimulate an above-average interest in the other fellows' trades. For example, radio officers and navs would often play at being co-pilots, flying her straight and level and occasionally doing gentle turns and radioing position reports to ground stations as we passed. Similarly, it was not uncommon to see pilots taking sextant sights. In fact, several with whom I flew regularly were capable of shooting, reducing and plotting astro position lines with very little coaching. This spirit of cooperation must have contributed in part to the remarkable safety record posted during Operation Hawk. Certainly the odds favoured accidents, as the weather was frequently lousy and the crews were often pushed to fatigue by disrupted schedules and the stress of the long and noisy flights.

If the nav station was well positioned, it was also, by the standards of the time, quite adequately equipped. Firstly, the table was a generous size and the seat comfortably padded. Although the nav's window was not much larger than a large watch crystal, there was always a good view forward over the pilots' shoulders. Also, in a pinch a glimpse rearward could also be had by looking through the driftmeter which was located just beneath the nav's right elbow, on the aisle. The astrodome was conveniently close by, just one step right and one step aft. The sextant used was a British Mk. 9, a rugged item; but, just in case, a spare was always carried. When in use, the sextant was suspended from a hook at the apex of the dome where there was also a mount for the astrocompass. By the mid 1950s both of these instruments had been replaced by a periscopic sextant. Astrodomes were replaced by flush aluminum plates which carried the ring mount and hatch for the new Kollsman sextants. This system had been developed basically to eliminate the drag and pressure-blowout problems created by astrodomes in faster, pressurized aircraft, but although these conditions did not obtain in the Star, there was the advantage of having two instruments in one. Also, it was now possible to install a polaroid twilight compass, which was valuable in Arctic flying when caught in the extended half-light of the North, when neither sun, moon nor stars were visible.

Nevertheless the change to "periscopics" caused a mild claustrophobically-induced panic amongst some navigators. Not being able to see more of the sky than the narrow 15-degree field of view now allowed by the sextant, they had to rely almost entirely on mathematics to be sure they were pointing at the correct star. Conventional star recognition methods were no longer possible. To add further to the confusion, the 2X magnifying optics of the new instruments also made objects seem brighter. It took some getting used to.

The navigator's instrument panel displayed the usual flight instruments: airspeed indicator, pressure altimeter, thermometer and compass. Arrayed around these were the electronic aids: Loran, a radio compass and a radar altimeter. The radio compass got lots of exercise on the overland and coastal routes, and the other two over the oceans. Ironically the simple little radar altimeter was often the star of the show, as it was the key to the use of pressure pattern techniques which we often fell back on when all else failed in bad weather. Many of the aircraft also carried Rebecca/Eureka, a radar homing device, but because the ground stations were few and rarely operative (they were mainly in the Arctic), it was used very infrequently.

The aircraft had three steering systems: a gyro-damped Gyrosyn magnetic compass, a conventional Pioneer liquid-floated unit such as you see in most light aircraft, and a rather erratic directional gyro, basically a pilotage device, to be used with great caution.

It was not an overly imposing inventory of navigation equipment by today's standards, but it proved completely adequate, the best proof being that we never wrote off an aircraft because of a navigational error. There were close calls, but in an aircraft that cruised at a pleasant 200 knots and could stay aloft for 15 hours, there was admittedly a comfortable tolerance for the errant or confused navigator.

The foundations of the navigation "art" had been established in the 1920s and 30s by the generation of adventurers that included the likes of Brown, Byrd, Chichester and Ellsworth. To this was added the experience of World War 2, most significantly a multitude of new electronic inventions. We were the postwar inheritors of these two legacies. Soon to come, however, was the next major crop of inventions: the costly, sophisticated, self-contained systems that converted the art into a craft, albeit a most refined one, and made the classic air navigator an obsolete species. Today, for example, the Boeing 747 crossing the oceans of the world is guided by an inanimate triumvirate of inertial navigation machines, their disparate deductions of position being adjudicated by a computer. A second pilot usually monitors this sytem's vital signs, interceding only when another more accurate system such as radio distance measuring (VOR-DME) advises that the primary inertial navigator is slightly in error. Ho-hum.

The vintage-1950 North Star navigator chaired a committee of instruments that were often unreliable or simply unusable because of weather or distance from ground stations. The challenge and art was to select the best at any one time, analyze the outputs or readings, and weigh the quality of the findings, much in the manner of his predecessors. Experience and common sense counted highly in rendering the right judgements. Fortunately, the route structure was so varied that experience came quickly — and common sense usually accompanied it. A trip in imagination around the Pacific routes of Operation Hawk will serve to illustrate this diversity and some of the typical problems encountered.

On the first leg, from home base Tacoma to Anchorage, Alaska, the aircraft was first pilot-navigated via radio range beacons to the northern tip of Vancouver Island. During this period the navigator only monitored the procedure sufficiently to compare the actual winds with those predicted, so that he could make adjustments, by "guesstimation," to the winds forecast for the over-water legs. After clearing the Island, course was set for the Queen Charlottes, and once by there the principal over-water leg was begun. The aiming point was Middleton Island, a small piece of sod lying just south of the mountainous Alaskan coast. From there on in, radio range navigation was resumed and the navigator could relax and stare at the intimidating parade of sharp, snow-encrusted ridges that are topped on the way to the river basin that harbours Anchorage.

The over-water leg called for a variety of navigational techniques. Loran coverage in this area was reasonably good, but it varied considerably in quality according to the time of day and the weather. To explain, Loran was (and still is) an electronic aid that evolved from fixing systems developed in World War 2. By simultaneously receiving signals issued by synchronized ground stations operating in pairs, an airborne monitor can measure the time difference between receipt of the two separate signals and can thus establish position as being somewhere on a line between the two stations. The line is, in fact, a section of a hyperbola. Needless to say, the "time-difference" is very small — the speed of radio transmission being what it is — and a special receiver and measuring technique are needed to make the necessary readings. Signals are displayed on a cathode ray tube in the form of sinuous, shifting waveforms. In fringe areas, where signals were weak, these waves were often hidden in a smear or hash of radio static and accurate measurements became very much a matter of skill. Results, therefore, were frequently suspect and had to be treated with caution. At such times more emphasis would be placed on astro, bearings from coastal radio stations, and pressure pattern techniques.

This menu of disparate information could be further augmented by drift and groundspeed measurements taken through the driftmeter. Although the last technique was in the early days of air navigation a mainstay of dead reckoning and the virtual bedrock of position estimating, my own experience was that spot winds so obtained could often be misleading. If there was a unique problem on this leg it was that caused by the mountains to the east. They not only entrap cloud (as west coast dwellers may have observed), but they also occasionally produce freak wind effects felt well out to sea. A strong westerly funneling through a mountain pass can accelerate due to the venturi effect and produce a dangerous sucking stream that can blow an unwary aviator to an uncalculated and sudden "landfall." Several rather mysterious crashes in the coastal range in 1950 were attributed to this phenomenon.

I experienced this effect one night while flying from north to south, in a reversal of the usual counterclockwise routing. We were solidly in cloud and the Loran was useless. I was relying entirely on dead reckoning, pressure pattern methods, and weak bearings from coastal radio stations a hundred or so miles away. When these signals also weakened, I resorted to the radio range technique of listening for beam passage, a method wherein the aural detection of a solid or steady tone indicates that the aircraft is passing through the so-called "beam" which separates one coded sector from another. At one point in this flight I could hear the signal swell and blend as we approached one beam that pointed almost straight out to sea, intercepting our flight path at right angles. But as the strength of the signal mounted higher and higher and the solid tone persisted, my initial bemusement turned to apprehension. We seemed to be sliding along the beam, virtually drifting sideways towards the coast. It defied reason! Had a 40-knot wind suddenly more than tripled in speed? Finally, as the growing signal volume began to assume the role of an alarm, I ordered a 30-degree turn to seaward. By now my ears were straining to detect a change in signal tone and strength, and my eyes were stuck to the unwavering needle of the radio compass which still seemed to indicate a remarkable 50 degrees or more of drift. I called for another radical alteration of heading — 30 degrees more — and finally we started to break through the beam and away from inhalation by the mountains. An hour later we made a belated landfall at the Queen Charlottes, approaching from an angle 70 degrees askew from our planned track. I bought the beers that night — gladly.

I won't dwell on the second leg from Anchorage to Shemya, at the extreme tip of the Aleutians, because it was not very notable from a navigational point of view. Loran coverage was quite good and the radio beacons were favourably placed. The biggest problems were fatigue and the chronically poor terminal weather at Shemya where minimum ceilings and visibility and strong crosswinds seemed more common than not. The collision of Arctic and Pacific currents really set things a-boiling. Fortunately the well exercised GCA crews were second to none.

The third leg, from Shemya to Japan, was as good a test for the navigator as was a Shemya landing for the pilots. Radio aids were sparse, cloud cover was often thick, and the folks living to the west were reportedly hostile, so we didn't want to stray too far to the right, ending up inside the Kurile Island chain. Radio jamming, although sporadic, was also a fact of life. There were radio beacons at both ends of this long route, but they were basically only effective for final homing, during the last 30 minutes or so. The navigational mainstays on this leg were usually astro and pressure pattern.

The basis of this apparently arcane system of navigation was the fact that an aircraft flying at a constant pressure height traces an undulating vertical path across the sky. As it approaches a high pressure area it gradually climbs, and similarly it slides downhill towards low pressure zones. During all of this, the pressure altimeter reading remains constant, but measurements taken with a radar altimeter will show that the true or absolute height is varying, often at a rate of several hundred feet an hour. Now, because the winds in the northern hemisphere blow clockwise around high pressure areas, and the reverse around lows, aircraft approaching highs will be blown (i.e. drift) to the left. Entering lows, they are forced right. The stronger or deeper the pressure zone, the steeper will be the slope the aircraft traces, indicating stronger winds and greater drift. By periodically comparing pressure height and absolute height, a navigator can peg the steepness of the slope quite accurately. Then, by applying some simple math, he can calculate the degree of drift or displacement right or left.

Unlike the traditional system of drift measurement through an optical sight, this pressure pattern method works in cloud; and it averages out the drift over a period of time, thus eliminating the pitfalls of using erratic "spot" drifts. There is one catch: it works only over water, as the elevation of land varies too much to permit an accurate measure of absolute height with respect to the sea level datum that is required.

In another practical application of this elegantly simple principle, an accurate forecast of the absolute height of a given pressure surface (i.e. your planned cruising altitude) above departure and destination points enables a quick calculation of a "single heading." If carefully followed, this will in theory guide the aircraft along the fastest flight path — the path of least resistance, so to speak.

I was always fascinated with the zen-like qualities of pressure pattern navigation and used it with confidence at every opportunity. There were some of my contemporaries who did not share this zeal, and treated it with some suspicion.

As a platform for astro, the North Star was in the *ne plus ultra* category — considering that it usually slogged along at altitudes near 10,000 feet, where the air is not the smoothest. Without a stable mount beneath him, the navigator can find astro to be an extremely frustrating and time-wasting experience. In a steady aircraft, on the other hand, it is perhaps the classic expression of the art of navigation. When I came out of basic training I was, like almost all my contemporaries, uneasy about relying on astro to find the way with any confidence or precision. After that the North Star was a revelation. Given a chance, the bloody system really worked! A succession of cleanly etched three-star fixes parading across the chart at regular 40-minute intervals was not uncommon, and it was always a delight to find the way with this elegant technique that is so much a part of the history of Man's venturing.

Astro was used most on the three long return legs across the mid-Pacific: Tokyo to Wake Island, Wake to Honolulu, and Honolulu to San Francisco. Cloud cover was rarely a problem and on-top flights were the rule. Loran coverage was reasonably good, but pressure pattern, which thrives on strong pressure gradients, was not quite as useful in the more gentle weather patterns of the south.

Terminal homing was more critical, especially at Wake; and with good reason it had one of the most far-reaching homing beacons in the world (we could sometimes receive a strong signal from as far away as Shemya). Despite the usually good weather, coral atolls are very difficult to see. Cloud shadows are extremely deceptive, and one could easily empathize with Earhart and her navigator Noonan when in 1937 they searched in vain for a similar destination, Howland Island.

While the Pacific routes provided the best training grounds possible for newly graduated navigators, we soon learned that there were equally challenging territories closer to home. And the big one was the Canadian Arctic. Flying from home base at Dorval, 426 Squadron North Stars regularly plied the milk run to Churchill and Resolute Bay. From Resolute we staged outwards to the smaller posts such as Alert, Mould Bay, Isachsen and Eureka. These were the more or less regular runs, but side trips all over the region were quite common. We flew with special respect for this

Head on view of 17510. The built-in drag of the "circular" Merlin is evident. (DND PL132808)

unforgiving environment, and with perhaps a little more concern that the navigation be exact.

Aside from weak radio beacons at a few points, there was at that time a paucity of radio aids in the north. Although a casual study of an atlas would seem to suggest that map reading should be fairly easy, that was usually not the case. Much of the terrain — the Arctic Islands, for example — is relatively featureless and disappears under snow and ice, especially in the favoured flying period, the spring, when the sun is up but all is still frozen below. Later, when the water emerges, it breeeds cloud, and the map-reading navigator is often shut off from the view. In the long Arctic night clear skies abound, but few lights twinkle in the black void to provide reference.

At many times, then, the navigation techniques used were those that served best on over-water flights, with astro the preferred choice of most. While fixing was often a problem, the single biggest concern was steering. Because our operations were more or less centred on the north magnetic pole, the standard compasses were virtually useless and gyroscopes were used for steering. However, until the mid-1950s, when a new generation of instruments became available (developed primarily to meet the requirements of Strategic Air Command) we had to make do with fairly primitive devices, somewhat less than adequate for precise navigation. So bad were these gyros that it was standard practice to check the heading with the astrocompass every 20 minutes. Needless to say, the loss of visual contact with sun, moon or stars for any protracted period was cause for anxiety. With average drift rates of 10-20 degrees an hour and "rogue" sessions when these drift rates might increase twofold or more, it was not entirely inconceivable that you could unknowingly be carving large circles inside the clouds.

The problem of steering or direction was compounded by the need to abandon the basic navigational convention of using the north pole as a direction reference and substituting an imaginary "Grid North" which hovered somewhere way out in space. This was a practical necessity because the rapid convergence of meridians near the pole made steering by that reference a decided nuisance. Besides, the gyros preferred space navigation, so were completely compatible with the grid system. The trouble was we humans found it more difficult to assimilate this abstract concept, and the occasional confusion resulted. Many pilots, for example, would have a strong innate sense of "true" direction and would challenge any doubtful steering orders given by a navigator. This was a good form of flight safety insurance. However, when the grid system was in use, this instinctive sense of orientation was lost and the hazard of mis-direction was correspondingly greater.

A good example of this phenomenon happened on a flight leaving Resolute destined for Churchill. Now Churchill is, in conventional terms of direction, almost directly south from Resolute. But in the parlance of grid navigation it is west. When the navigator set his gyros prior to takeoff, he incorrectly set them to read "true" as opposed to "grid" values. As the aircraft took off on the south-pointing runway, the gyros therefore read 180 degrees — "true" south. Unfortunately the navigator then directed the pilot to take up a "grid" heading of 270 degrees — "grid" west. If the reader is now somewhat confused, he can better understand why the pilot obediently turned 90 degrees right and headed towards Alaska!

The remainder of this ill-starred flight was in marginal weather, and the totally disoriented crew continued to discard or rationalize all navigational clues that disagreed with their faulty preconception that they were heading "true" south. After many hours of this flying in a meteorological and mental fog they had an incredible stroke of luck. The pilot had tuned his radio compass to the frequency of the beacon at Churchill, and the signal came in strong. However, the call letters were wrong; instead of "YQ" — the Churchill designator — he heard "PB." They were approaching Point Barrow, Alaska! Some felt that they must have been led in by the ghosts of Ellsworth, Amundsen and Wilkins, who had used the same place on their pioneering trans-polar flights. They landed safely there, some 1500 miles distant from their original target.

As the Korean War wound down, the focus of RCAF North Star operations swung eastward to the Atlantic sector. The transition was made quite smoothly as the new region presented no unique problems and the crews had acquired a depth of experience on Pacific and Arctic missions. Air traffic control procedures were more intense and critical, as traffic density was higher; but this was primarily a problem of pilotage and the training syllabus of the newly formed operational training unit (in 1952) was tailored to acquaint new crews with the intricacies of European air space.

From the navigator's point of view the Atlantic was perhaps a bit easier to passage than the Pacific. There was a more even distribution of nav aids and the selection was usually better in the terminal zones. At the eastern end he could enjoy the novelty of CONSOL, a system originally installed to assist the navigators of U-boats in World War 2. CONSOL was similar to radio range except that the "beam" swung slowly back and forth in a regular 60-second cycle. By counting dots (or dashes) as the beam approached, and the reverse after it had passed, the navigator could position himself as being somewhere on a specific equivalently-coded line on a CONSOL plotting chart. When two or more stations could be received, a series of converging lines would show his position. It was the lazy navigator's delight.

If there was any special problem, it would be the notorious North Atlantic weather. Astro was available on a catch-as-catch-can basis and the steady work rhythms of trans-Pacific and Arctic flights were usually denied us. Atlantic navigation was more the realm of the opportunist who could adapt easily to changing conditions.

As the Star passed into middle age in the late 1950s, the quality of the navigation gear was gradually upgraded. A modest forward-looking search radar was added, allowing a form of all-weather map reading. A highly effective directional gyro was also installed, greatly easing the burdens of steering a steady course in the Arctic. Better Loran receivers and radio compasses also improved matters a bit. But that was enough. Even with the original inventory we felt that we could find the way to any destination. Besides, as I said before, we only cruised at 200 knots true airspeed and we had 15 hours worth of fuel, so there was plenty of margin for error!

Don Connolly

The original Rockcliffe Ice Wagon, seen at Rockcliffe. (DND RE3577-3)

The Ice Wagon in flight. J.T. Bain records, "The Ice Wagon was promoted by TCA through C.D. Howe. Up to that time we were using grooved rubber shoes on the prop blades to carry alcohol out from 'slinger rings.' We started using them on Lockheed 14s. They were a TCA invention. Without those prop shoes our operations would have been "caput" for long periods each year." (DND PL53288)

Rockcliffe Ice Wagon

In the early postwar years the National Research Council conducted extensive research into the phenomenon of icing. At first, an RY-3 was operated on icing research by the Experimental and Proving Establishment at Rockcliffe on behalf of the NRC. But, as noted by Carl Vincent in his book, *The Liberator and Fortress,* "despite its sensitive equipment and the excellent work it managed to accomplish, it cannot be denied that the aircraft was a disappointment. Apart from inherent problems, spares were an enormous headache . . . No spares were obtained in 1946, and the Ice Wagon was kept going by using the few remaining ex-RCAF Liberator spares . . . Maintenance problems were exacerbated because the RCAF Experimental and Proving Establishment, which was responsible for the aircraft, was unable to give it permanent hangar space, and it spent most of its time outside." Carl Vincent goes on to note that in the 1946-47 icing season the Ice Wagon was unserviceable for 108 out of 180 days. Thus it was with much relief that EPE and the NRC learned that the RCAF would be supplying them with a replacement for the Liberator and North Star 17513 was delivered to the RCAF in March 1948 and converted for ice research purposes. The Liberator was sold as scrap.

F/L J.W. Wilkins, while serving with EPE, wrote the following article in *The Roundel* describing 17513 and its work in ice research:

Ice may occur on the fuselage, wing and tail surfaces, and propeller(s) of an aircraft in flight. Icing on the airframe breaks down the aerodynamic airflow over the aircraft's lifting surfaces and adds extra weight that may be, in certain circumstances, dangerous or even disastrous. Propeller icing, which is similar in nature, makes itself felt by a loss of thrust that necessitates the use of more and more power to keep the aircraft aloft.

The weather conditions most conducive to airframe and propeller icing are temperatures from 0°C to -18°C, the presence of supercooled water droplets, and turbulence. The more severe these factors become, the greater will be the amount of ice. In rare instances, ice can form at temperatures as low as -40°C.

Active de-icing research development was begun in 1939, leading to the electro-thermal method of propeller de-icing. After initial flight tests in 1942, over a hundred sets of propeller thermal de-icing pads were installed (in 1945) for service trials on aircraft of the RCAF, RAF and TCA. The method is now coming into general use, and aircraft de-icing remains the subject of extensive investigations in

The Rockcliffe Ice Wagon at Rockcliffe in January 1952. (DND PL53287)

the laboratories of the National Research Council of Canada at Ottawa.

More recently, the electro-thermal method of de-icing has been extended to the wings with very promising results; and this is one of the major projects upon which the "Ice Wagon" (as North Star 17513 is affectionately known) is currently engaged.

The Ice Wagon is a Canadair North Star, built in Montreal and powered by four Rolls-Royce Merlins. It is a normal RCAF transport North Star except for the following major modifications:

- A dorsal fin mounted on top of the fuselage and fitted with heating pads.
- Two observation posts in "blisters" mounted on either side of the fin.
- Propeller-heater shoes fed with power via slip rings.
- Two observation posts in the forward cabin, from which, by use of a strobo-viewer, the propellers can be observed in the "stopped" condition.
- Two special 60 kw. electrical alternators driven by the outboard engines, to supply power for heating de-icing equipment.
- Various experimental installations and instrumentation in the fuselage.
- Reversible-pitch propellers, which enable the aircraft to make emergency landings in small fields or on icy runways.

The wings and tail-unit are at present de-iced by the standard pneumatic overshoes, whereas the engines have the normal warm air intake (to guard against icing in the carburetor, etc.) and also, as an additional precaution, alcohol de-icing equipment.

Complete Arctic emergency equipment is carried during test flights. An auxiliary power-unit in the tail makes the aircraft independent of ground batteries for starting, and airline seats afford some comfort for the ground-crew who are carried on all flights. The aircraft is completely self-sustaining when operating away from its home base, the Experimental and Proving Establishment at Rockcliffe.

The dorsal fin (which is the only one of the aircraft's abnormalities that really hits the uninitiated between the eyes) is used for wing de-icing experiments during the initial phase of the project. It is eight feet high and ten feet from leading to trailing edge, and represents a high-speed symmetrical-section wing. In addition to facilitating observation and wiring, the use of the dorsal fin precludes interference with normal aircraft operation in the event of heating pad failure. Its leading-edge is easily removed, and, since it has two interchangeable leading-edges, de-icing pads can be changed quickly and conveniently. The dorsal fin has no noticeable effect on the handling or performance of the aeroplane.

A closer view of the Ice Wagon showing the ice-gathering fin and observation blister. (DND PL53285)

The Ice Wagon is used for the proving (in flight under natural icing conditions) of the experimental de-icing and anti-icing equipment developed by the Low Temperature Laboratory of the NRC. These tests are supplemented by laboratory tests under simulated icing conditions in the icing wind tunnel.

The three main projects on which the aircraft is engaged are:

- Electro-thermal wing de-icing.
- Electro-thermal propeller de-icing.
- Measurement of cloud properties.

There are three general methods of getting rid of ice:

1. Mechanically, breaking it off—using, for example, pulsating rubber overshoes such as are fitted to the wings;
2. chemically, by adding, say, alcohol to the ice in order to lower its freezing point; or
3. thermally, by melting the ice with heated wings or propellers.

Of the above three methods, the last seems to be the most effective; and the most convenient, reliable, and easily controlled means of supplying the necessary heat appears to be electricity.

To melt all the ice as it forms, however, would require so much heat that the electrical equipment needed to supply it would be both unwieldy and expensive to operate. Therefore, in the system employed on the Ice Wagon, the ice is allowed to form for a short time, and then only sufficient heat is applied to produce a thin film of water between the ice and the propeller blade or wing. This destroys the adhesion of the ice, which is then thrown off by the mechanical or the aerodynamic forces acting upon it.

Ice is allowed to form on the blades of the propellers, over the rubber pads, for a period of one minute or longer. Electrical power is then supplied to the pads for a few seconds, heating them and so melting the ice. Centrifugal forces acting on the ice soon throw it off.

Rubber is used for the de-icing pads on the propellers because of the abrasion caused by dust, stones, hail, etc., which rubber withstands well. The heater element is made up of conducting rubber (developed by the NRC Rubber Laboratory) protected and insulated by non-conducting rubber.

On the fin, a "parting strip," one inch wide on the leading edge, is continuously heated so that ice cannot form on it. This divides the ice so that it forms on either side of the heated strip. As before, the ice is allowed to form for a few minutes, and then the "shedding zone" is heated, thus allowing the ice to be blown off by aerodynamic forces. If the leading edge were not continuously heated, the ice would form a stable cap over it which could not be removed by aerodynamic forces, the result being that melting rather than shedding would ultimately occur.

During the final phase of the project, after the scheme has been satisfactorily proved on the dorsal fin, the fin will be removed and electro-thermal de-icing fitted to all wings and tail surfaces in place of the existing pneumatic overshoes.

Each of the fin's heater elements which have been tested are made up of a fine grid of wires, interwoven with cotton or fibre-glass to hold them in place and insulate them from each other. The element is protected on either side by an insulating layer of plastic or rubber. Rubber, which appears to be more suitable because of its resistance to the eroding action at high speeds, is being used on present pads. The plastic fin pad is approximately .065-inch thick and weighs about 0.5 lb per square foot.

The physical properties of the clouds in which icing is found are measured by a number of instruments. The main items of interest are the temperature, the water content and the size of the water droplets. These are recorded as follows:

Temperature. Measured by a number of thermometers. The great problem is to keep the thermometer elements from becoming wet or iced up, so they have to be protected by housings of special design.

Water Content. Ascertained by a rotating disc-type rate-of-icing meter mounted on top of the fuselage, ahead and slightly to one side of the fin. It can also be measured from a cloud density meter mounted below the aircraft. Other types of instruments are being developed but are not yet fitted.

Water Droplet Size. The cloud droplets are actually photographed in flight as they pass the aircraft by a special under-slung cloud-droplet camera. This camera, designed and made by the NRC Instrument Laboratory, is unique.

Four observers are carried. Their duties are:

Propeller Observer. Seated in the forward compartment, he can observe the propellers in motion by using a stroboviewer. He controls the de-icing of the propellers.

Instrument Observer. Seated in the forward compartment, he has control of the meteorological instruments, and can service them in flight.

Meteorological Observer. Seated in the right-hand blister, he observes weather conditions, besides controlling the meteorological instruments. He can assist in observing the fin as well.

Fin Observer. Seated in the left-hand blister he observes the fin, wing, propellers and tailplane, and can control the fin de-icing.

The Ice Wagon at Hatfield in the UK in August 1950. Paul Hartman recalls: "I had one interesting trip in the Ice Wagon when we encountered ice build-up at the rate of 9 inches per hour over the Gulf of St. Lawrence on January 9, 1952. F/L Doug Biden was with me. We were flying at 4000 feet and were quite pleased because you seldom encountered ice when looking for it." (*The Aeroplane* via Stephen Piercey)

The test flying comes under the direction of the NRC observers, who are responsible for all the experimental equipment, while the operation and maintenance of the aircraft are taken care of by the Experimental and Proving Establishment of the RCAF.

The entire project is directed by Mr. J.L. Orr, Head of the Low Temperature Laboratory of the NRC. The NRC flying crew includes Mr. D. Fraser, Mr. E.H. Bowler, Mr. K.G. Pettit, Mr. R.C. Brown, and (until recently) Mr. J.A. Lynch. The RCAF crew consists of Flt. Lt. J.J. Higgins (captain), Flt. Lt. T.J. Evans (co-pilot), Flying Officer H.A. Smith (navigator), Flying Officer T.A.R. Harris (radio operator), Sgt. A.E. Frieday (who recently replaced Sgt. F.E.J. Dewan as flight engineer), and two airframe technicians, three engine technicians, one electrical technician, one radio technician and one instrument technician.

On August 9, 1950, the Ice Wagon was flown with its complete crew on a liaison visit to the United Kingdom. This visit was arranged in order to demonstrate the aircraft to the Ministry of Supply and other interested parties. Apart from the demonstration, the trip was intended to establish a close liaison with research and development establishments working in the fields of aircraft and jet engine deicing, cloud physics (including induced precipitation — or rain-making), and nephelometric instrumentation (including the use of radar for cloud and storm detection).

The aircraft was based at the Royal Aeronautical Establishment, Farnborough, from August 14th to 18th, where it was available for inspection by government organizations, civil air line operators and manufacturers. Each day lectures were given by the NRC personnel, followed by conducted tours of the Ice Wagon. Later the aircraft made a series of visits to manufacturers' airfields — De Havilland, Rolls-Royce, Napier and Son, A.V. Roe, Bristol Aeroplane — as well as a trip to the Aircraft & Armament Experimental Establishment at Boscombe Down. Discussions were also held at such places as the Meteorological Laboratories at Harrow, the Ministry of Supply Icing Committee, and at the Clarendon Laboratory in Oxford. During their stay in the UK, personnel were afforded an opportunity to attend the SBAC (Society of British Aircraft Constructors) show at Farnborough.

During the short time that the North Star Ice Wagon has been in service — since March 1950 — it has ranged over the length and breadth of Canada in search of icing conditions, from Labrador to British Columbia, and as far north as the Yukon Territory. A total flying time of 150 hours was accumulated in two months of operational flying last year, of which 20 hours, or 13%, was spent in actual icing conditions.

The Roundel

Ice Wagon crewman John H. Milsum of the NRC at work at his station. (DND PL53289)

Besides its work with ice research, 17513 was used on numerous other projects. John Waldie's log book notes some of these. On June 30, 1954, he was aloft 4:45 hours in 17513 flying from Goose Bay on an eclipse observation mission headed by Dr. Millman of the NRC. On August 27-28, 1954, he flew two trips from Rockcliffe noted as "McGill Fence" in his log. The McGill Fence was the Mid-Canada Line defence system. As the "fence" was activated when penetrated by an aircraft, it seems likely that 17513 was being used to evaluate its effectiveness that day. On January 8, 1955, he spent 7:10 hours flying GCI (Ground Control Intercept) calibration in 17513, checking out several GCI ground stations including the one at Edgar, Ontario.

On one radar calibration flight John Waldie recalls getting the Ice Wagon up to 32,000 feet. Glancing back into the cabin, he noticed a pair of feet sticking into the aisle so he sent an airman back to investigate. The airman plugged into the oxygen bottle near the passed-out crewman, then he too blacked out. The bottle was empty! Waldie realized the trouble, cut the power, dropped the gear and soon had the North Star down where there was air to breathe and where the two crewmen quickly recovered. On these high-level flights, cases of the bends were also common.

In March 1956 17513 was hangared at Dorval and had apparently just been converted back to a standard North Star. On March 19 fire raged through the hangar, destroying one of the most unusual North Stars ever flown.

From the Log Books

"Following Officer Air P posted to 4 T OTU Trenton rep 8 Mar firm date TBA by unit for inclusion North Star First Officers Course 39 commencing 9 Mar PD from 2 AOS Winnipeg 224776 FO GM Martin PD on successful completion of OTU course further Trans instructions to a unit in ATC TBA ATCHQ PD Officer to be in possession of valid instrument rating prior to proceeding PD Charge to AFHQ." So read the signal of January 23, 1959, from CANAIRTRAIN to RCAF Station Winnipeg. For F/O Garth Martin it meant the beginning of a new stage in his air force career.

Garth Martin had joined the RCAF in 1954 and commenced flying training at 4FTS, London, Ontario, on June 22. After completing his Harvard training there, it was off to 1AFS, Penhold, Alberta, for multi-engine training. This entailed some 160 hours on Mitchells and 38 on Expeditors.

In September 1955 Martin was posted as an instructor to the Air Navigation School at Winnipeg. His first flight at the ANS was in Expeditor 2301 on September 13; his last was in a Mitchell on January 28, 1959. In between were over 2500 hours in the air, mainly on Mitchells on Airborne Intercept flights training upcoming CF-100 navigators.

His posting to the North Star Operational Training Unit was welcome news for Garth Martin. At the time the North Star was still the RCAF's

Five North Stars and two C-119s parked hurriedly on the ramp at Dorval on March 19, 1956, the day 426 Squadron's hangars were destroyed by fire. (*Montreal Gazette*/Public Archives of Canada PA129378)

17521 sits amidst what little was saved from the fire at Dorval. (*Montreal Gazette*/Public Archives of Canada PA129379)

largest aircraft. It was well liked by Air Force pilots. The new job also had global prospects as opposed to the prairie ones Martin had been confined to for the previous couple of years.

On March 31, 1959, Martin went on his first North Star training flight from Trenton. It was a familiarization ride in 17508 captained by F/L Misener. He first took the left seat on April 7 after six trips, and the course concluded May 1 after 49 trips and 105:55 hours in the North Star.

Preparing to fly to Chile on earthquake relief duty is the 426 crew of F/O Cole, Capt. B.T. Saunders (USAF), F/L J.R. Morrison, F/O J.B. Bussières and Sgt Joe Sopaz. (DND PL118676)

Making the best of it. Air and ground crew of 416 Squadron ride a North Star from Rabat, Morrocco, to home base, Grostenquin. They had just completed a gunnery camp at Rabat. Everyone is studiously trying to ignore the noisy realities of North Star flying. (LCol Syd Burrows)

On May 4, 1959, Martin was posted to 426 Squadron at Dorval (RCAF Station Lachine) and commenced flying there on May 6. May 13-22 he made his first overseas trip aboard 17516 under F/L Fawcett. The aircraft left Dorval and flew to El Arish on a UN support mission. Intermediate stops both ways were at Gander, Lajes, Shannon, Langar, Pisa and Athens.

Garth Martin's hours built up rapidly. When he arrived at the OTU in March, his total time was 2412:25 hours. On June 10 he passed 3000 hours en route Lajes-Langar. This was one of the benefits of long-range flying.

Besides routine flying, there were occasional special trips in the North Star. On June 25, 1959, Martin flew in 17514 to Baltimore, where a cargo of nerve gas was loaded, then flew to Suffield, Alberta, on a 9:55 non-stop. On arrival there, ground crew looking like spacemen in their protective suits marshalled the North Star to a remote area. To the 426 crew this seemed a bit odd as there had been no special treatment or apparel back at Baltimore.

On July 18-20 Garth Martin was F/O aboard 17517, carrying the baggage for the royal tour across Canada. September 1-11 he was aboard 17511 on "Western Weal," a transatlantic ferry operation. The North Star was shepherding twelve T-33s to Turkey. The T-birds were a gift from Canada under Mutual Aid. The ferry route was Trenton-Goose Bay-Søndestrom-Keflavik - Prestwick - Baden - Malta - Eskisehir (Turkey). Flying amounted to 36:05 hours with the longest leg of 5:35 hours being Goose Bay - Søndestrom. By the end of 1959 he had logged 18 Atlantic crossings.

On January 16-17, 1960, Martin made his first trip north with 426, flying Trenton-Churchill in 6:45 hours. A trip on June 2 aboard 17517 operated Key West-Panama in support of relief operations in Chile where an earthquake had struck.

In 1961 Garth Martin made his first trip to the Congo. On October 16 he left Trenton in 17521 and proceeded to Kano by October 18, travelling via Gander, Lajes, Marville, Pisa and Idris. At Kano he took over as slip crew captain on 17516. He flew to Leopoldville (5:50 hours) on October 20, returning to Kano the same day (6:00 hours). He flew the same "milk run" October 24, 27 and 31, and on November 3, 7, 9, 10, 14, 17, 21 and 25. On the 26th he flew back to Marville in 17510 and on the 28th, with F/L R.T. "Dick" Merrick as captain flew to Lajes and was back in Trenton November 30. All told, the operation had kept him busy with 36 trips over 20 flying days for 218:40 hours. Trips averaged 6:00 hours, the longest being 9:25, Lajes-Harmon AFB. This kind of flying was typical for 426 Squadron during its Congo interlude.

Operations from Kano, located in Nigeria's desert north, were a bit dicey for North Stars. Temperatures during the day were generally around 110° F. North Stars taxiing for takeoff generally did so on two engines with the rad flaps open, trying to keep the engines as cool as possible. For takeoff, the rad flaps would be put on auto and part way down the runway would close half way. This cut down on drag but the temperature gauges would shoot up to about 125°, perilously close to the limit. There was little relief in the form of cool air till the aircraft passed through 4000 feet and pilots used to wonder nervously what in the world they would do if they lost an engine on takeoff. To try to remedy things, the Air Force altered its schedule from Kano and evening departures were tried. By then the temperature had relented, but only slightly. The trade-off was in wind. The daytime breezes subsided in late afternoon, so the North Star's takeoff performance was not much helped anyway.

From April 18 to 23, 1962, Garth Martin was first officer aboard 17504 on a northern trip: Trenton-Winnipeg - Churchill - Resolute - Alert -

Thule - Resolute - Churchill - Winnipeg - Trenton. The longest leg was 5:50 hours Churchill to Resolute. The whole trip involved 33:50 hours of flying. On September 17 Martin feathered No. 2 taking off from Nellis AFB in 17516 and ended up flying back to Trenton aboard Hercules 10304. Ironically the one-hour Nellis to Nellis flight was to be his last in a North Star. In 3½ years he had flown 2539:55 hours on North Stars, finishing up with a snag on his last flight, one of but a handful of minor problems he had faced on type. Overall the North Star had provided a pleasant and reliable flying experience.

In November 1962 Garth Martin flew as first officer aboard Hercules 10303 from Trenton to Zweibrücken via Cartierville and Shannon. Time in the air including return to Trenton via Gander was 25:30 hours. A similar trip via North Star from Dorval to Langar via Harmon and Lajes, and back via Shannon, Lajes and Harmon took 41:00 hours. With aircraft like the Hercules and Yukon taking over in the early sixties, it was clear that the North Star's days were numbered.

S/L J.S. "Stew" Parmelee began flying North Stars in June 1956, finished the OTU a month later with 91:30 hours of North Star flying, and joined 426 Squadron on July 9. One of his early trips (July 13) was a medical evacuation, carrying a child from Montreal to the Mayo Clinic in Minnesota. From here on, he was to fly his quadruple Merlins back and forth across the globe on every sort of transport mission. An interesting once-a-year northern trip he remembers was ferrying stevedores from Montreal to Resolute, where they were needed to unload the annual supply ship. The stevedores were a rowdy mob who raised the roof with boozing and brawling while the crew kept themselves prudently locked in the cockpit.

Some overseas flights included shepherding CF-100s and Sabres via Goose Bay and Greenland or Iceland to their respective Air Division bases, and carrying personnel to and from the RCAF gunnery and rocket ranges at Deccimomannu on Sardinia and Rabat, Morocco. Many a flight was made into El Arish in support of the United Nations peacekeeping forces in the Sinai. For night operations at El Arish, an arriving aircraft would call in a few minutes from landing and groundcrew would quickly set out flare pots as runway markers. Despite this shrewd timing, Bedouins had a knack of coming out of nowhere and stealing the pots even before the aircraft got down!

One of the famous Larouche cartoons, this one a parody on North Star flying. It appeared in the March 1961 edition of *The Roundel.*

Aug. 8	Dorval-Gander	4:30 hours
9	Gander Lajes	6:40
10	Lajes-Gibraltar	5:55
	Gibraltar-Malta	5:50
11	Malta-Beirut	5:55
14	Beirut-Bahrein via Damascus	6:15
	Bahrein-Karachi	5:15
15	Karachi-Calcutta	6:45
16	Calcutta-Saigon	7:00
19	Saigon-Hong Kong	4:55
23	Hong Kong-Saigon	5:00
24	Saigon-Calcutta	7:30
25	Calcutta-Karachi	6:45
	Karachi-Bahrein	5:10
26	Bahrein-Beirut	5:35
28	Beirut-Malta	5:50
29	Malta-Gibraltar	5:50
	Gibraltar-Lajes	6:05
30	Lajes-Gander	7:40
	Gander-Dorval	5:15

Captains F/L Longworth and F/O John and First Officer Parmelee made this annual troop rotation trip to Viet Nam for the International Control Commission in 1957. North Star 17504 was their aircraft.

In June 1960 Stew Parmelee participated in the worldwide earthquake relief mission to Chile. The area worst stricken was south of Santiago and all communication/navigation aids had been knocked out. At Puerto Montt there was a short gravel strip and North Stars would head there from Santiago loaded with aid. They would follow the mountainous coast south, using radar, then let down over the sea with aid of their radar altimeters and map-read carefully to a railroad which they would then follow inland to Puerto Montt. Once down, the heavily loaded North Stars would taxi quickly up to the only hardened apron area before they bogged down in the soft gravel.

On one trip to Puerto Montt, the North Star crew discovered a woman sitting under the wing of their aircraft. She was giving birth to her baby. The Canadians immediately flew mother and baby to Santiago.

Crews always faced the unpredictable. On February 15, 1957, Parmelee took off in 17511 from Lajes heading for Gander. Six hours 50 minutes later he and '511 were back on the ground at Lajes, beaten by the Westerlies. To show the fickleness of the winds, Garth Martin's log for August 13, 1960, shows a rare event — a flight straight through from Lajes to Trenton without the usual refuelling stop at Gander. The trip took 13:45 hours.

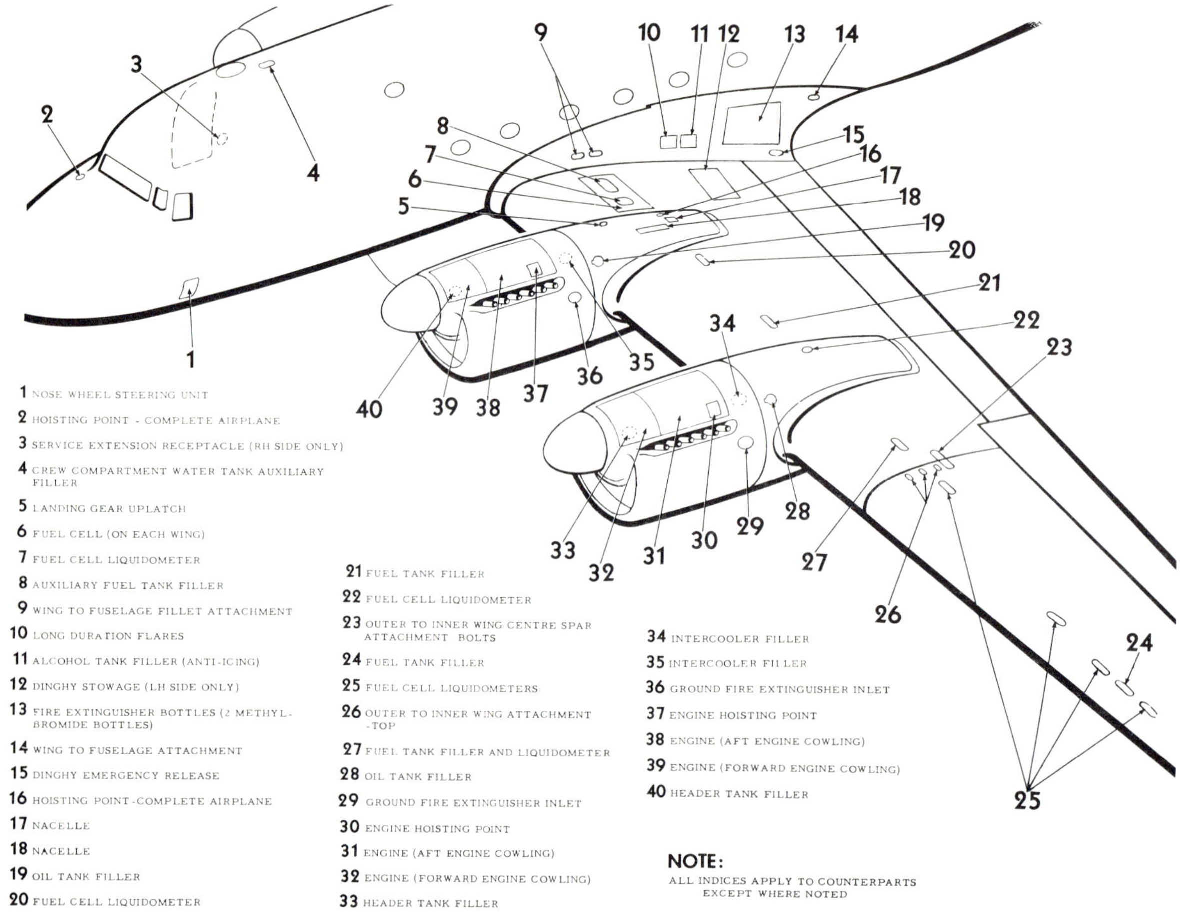

Access doors and panels, North Star

North Star flight line at Trenton, December 27, 1960. (Larry Milberry)

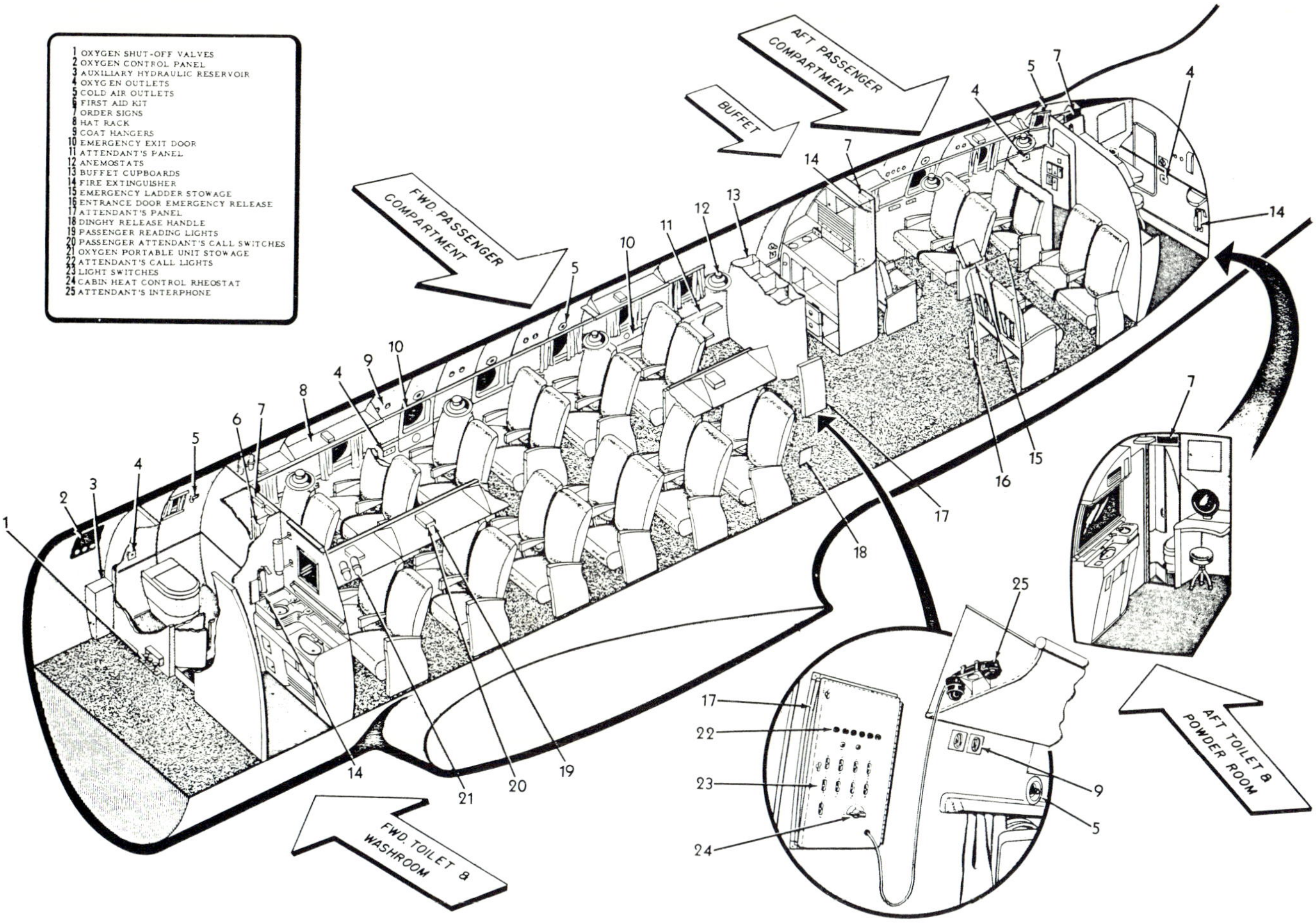

Passenger configuration RCAF North Star.

17508 crosses the threshold at Trenton while in service with No. 4 (T) OTU. As it is winter, the rad flaps are retracted. (DND PL88328)

Date	Aircraft	Captain	First Officer	Route	Time
Mar. 16	17511	S/L Parmelee	F/L Forest	Trenton-Gander	5:15
	,,	,,	,,	Gander-Lajes	7:20
Mar. 18	17512	,,	,,	Lajes-Langar	8:25
Mar. 19	17515	,,	,,	Langar-Pisa	4:30
Mar. 22	17502	,,	,,	Pisa-Idris	4:20
	,,	,,	,,	Idris-Kano	7:00
Mar. 23	,,	,,	,,	Kano-Idris	6:40
Mar. 24	,,	,,	,,	Idris-Pisa	4:10
Mar. 26	17516	,,	,,	Pisa-Langar	5:10
Mar. 30	17510	,,	,,	Langar-Shannon	2:25
	,,	,,	,,	Shannon-Lajes	6:45
April 1	17521	,,	,,	Lajes-Gander	7:55
	,,	,,	,,	Gander-Trenton	6:00

North Star flying in 1961 with S/L Parmelee and his 426 crew of F/L Forest, F/L Romaine, F/O Beck, Sgt. Corbeil, LAC Lloyd, plus supervising crew of W02 Howard, Capt. Saunders and F/O Davidson. This excursion took the crew across the Atlantic, over much of Europe and into equatorial Africa. On nine flying days they flew aboard seven different North Stars for an ear-shattering 75:55 hours.

Date	Aircraft	Captain	First Officer	Route	Time
June 1	17517	F/L Hayes	S/L Parmelee	Trenton-Charleston	4:10
	,,	,,	,,	Charleston-Albrook (Panama)	6:15
June 2	,,	F/L Lynch	,,	Albrook-Lima	8:00
June 3	,,	S/L Parmelee	F/L Spence	Lima-Santiago	7:40
June 4	,,	,,	,,	Santiago-Puerto Montt	3:30
	,,	,,	,,	Puerto Montt-Santiago	3:00
June 5	17502	,,	,,	Santiago-Puerto Montt	3:35
	,,	,,	,,	Puerto Montt-Santiago	3:35
June 7	,,	,,	,,	Santiago-Victoria	2:10
	,,	,,	,,	Victoria-Santiago	2:15
June 8	17512	,,	F/L Genge	Santiago-Lima	7:40
	,,	F/L Hayes	S/L Parmelee	Lima-Albrook	7:00
	,,	F/L Lynch	,,	Albrook-Charleston	8:30
June 9	,,	,,	,,	Charleston-Trenton	4:25

In 1960 the 426 crew of S/L Parmelee, F/L Spence, F/O Neihause, F/O Psaila, W02 Kelly, LAC Falardeau et al participated in the relief mission to Chile after the earthquake there.

North Star navigator Larry Motiuk reports on another interesting 426 operation. In December 1960 two North Stars flew to Ascension Island in the South Atlantic in support of Operation Lookout. That operation involved CF-100s flying missile tracking missions for the USAF. The North Stars, one captained by F/O Red Williams, the other by F/O Herb Gawne, rendezvoused at Ramey AFB in Puerto Rico on December 11 then flew the trip to Ascension via Atkinson Field (British Guiana), Belem and Recife. The return trip, December 15-20, was via Recife, Zanderij (Surinam), Ramey AFB, Patrick AFB (Florida) and Quebec City to Trenton. The purpose of the trip was to escort CF-100s 18439 and 18453 back to Canada at the termination of Phase 1 of Operation Lookout.

Thunderbird Squadron Diary

The compiling of squadron diaries has been a dutifully, if not always enthusiastically, conducted activity since the formation of the first military flying units. In the case of 426 Squadron, keeping the diary generally appears to have been taken seriously. For researchers years later, such documents are gold mines and their worth becomes obvious as we read books and articles which could never have been produced without them. This extract from the 426 diary gives a detailed slice of squadron life. The time is May 1958:

May 1. S/F 53/54 departed for El Arish via Pisa and Malta at 1100 hrs local. The crew will go to the UK while the aircraft, with its 13 passengers, goes to El Arish.

An advance warning order received today states that 426 (T) Sqn North Star will given the flying support to 17 CF-100s that will depart Ottawa on May 7. This aircraft will carry all necessary supporting equipment and personnel during the actual ferry.

May 2. S/F 53/54 that departs Montreal today is to adhere to its scheduled itinerary via Malta even if some talk has been made mentioning the replacement of Malta by Athens as fuel stop on route.

A 426 North Star banks gracefully for the photographer and shows its classic lines. (DND PL101530)

May 4. 426 (T) Sqn has been detailed to send an aircraft to Goose Bay to ferry A/C Carpenter and party. The special flight is sent to Goose Bay to carry surplus of passengers.

Special 212 took off this date at 1230Z. Two North Stars carry the RCAF Staff College to tour the USA. This is a yearly commitment for 426 (T) Squadron.

May 6. The Jump Moat III operation starts tomorrow. The North Star 17525 departed for RCAF Stn Uplands and where its captain will get his briefing on the operation. The aircraft will carry the supporting equipment and personnel for the maintenance of 17 CF-100s. This is the delivery of the 3rd group of the 53 Avro CF-100s all weather jet interceptor being provided to the Belgian Air Force under joint Canadian-United States mutual aid arrangement. The route for the flying support as well as the jet interceptors themselves is across the North Atlantic to Beauvechaine in Belgium by way of Goose Bay, Labrador and Keflavik, Iceland. The North Star will return by Langar, Shannon, Lajes, Gander, Uplands and Canairdiv will utilize space available on the westbound route.

May 7. The revised schedule for S/F 53/54 has been approved by ATC. The main changes are the following: an overnight stop at Pisa en route to El Arish with a fuel stop at Athens and an overnight stop at Athens and fuel stop at Pisa on the return. Athens has replaced Malta on the revised itinerary.

May 8. The first aircraft to follow the new itinerary left Montreal at 2359 tonight. The captain F/L Fawcett will hand over the aircraft to F/L Lewis in Lajes. In turn F/L Lewis will be the first captain to study the new route.

The CO's parade was held this morning at Station Lachine. W/C Mackie, OC 426 (T) Sqn, and at the present time acting CO, took the salute at the receiving base. Today's parade is held in preparation for the inspection of the AOC Transport Command, A/C Carpenter, that will be held on 7 Jun 58.

The Operation Starflight requires two North Stars to position Stn Trenton on the 9 May at 1800. The captains will report to the Starflight Commander at Trenton. This ferry operation might terminate in Turkey.

The Goose Bay Ration Run departs Dorval at 0830. The captain F/L Schunk is to route Goose Bay, Ottawa, Dorval if the 24 persons of the Jump Moat III advance party are ready to return to Uplands.

May 9. At 1030 EDT, the two 426 (T) Sqn North Stars departed for RCAF Stn Trenton to get the necessary information to carry their duties to Eskisehir, Turkey. The two North Stars are used as flying support in the delivery of eight T-33 Silver Star Jet Trainer aircraft to Greece and 17 to Turkey as additional Canadian contributions under mutual aid arrangements. The jet trainers, all Mark 3's, are flown across the North Atlantic in one operation to be known as Starflight. Taking off from Mountain View, Ontario, the jet trainers will be ferried to the RCAF's 1 Air Div Fighter Base at Marville, France, for delivery to Greece and Turkey. The aircraft will stop at Goose Bay, Labrador; Søndestrom, Greenland; Keflavik, Iceland; and Prestwick, England; en route to Marville. From Marville they will proceed to Decimomannu, Sardinia; Eleosis, Greece; Eskisehir, Turkey; and back to Trenton. The North Star aircraft carry 10,000 lbs payload Eastbound.

May 10. The Saturday Standby Crew gave a tour over Montreal City to 89 Air Cadets of RCAF Squadron, Verdun. While a

The 426 Squadron Detachment at Langar in 1957. Standing from the left are S/L J.S. Parmelee, LAC Leduc, LAC Thibeault, Cpl. Terrien, LAC Kutzver, LAC Pinay, LAC Fleming and Sgt Campbell. The others are Cpl Dollimier, LAC MacDonald, LAC O'Connor, LAC Morin, Sgt Pelland, LAC Gibbs, Cpl Huber, LAC Murphy and LAC Benson. (DND)

North Star 17504 at Prestwick during Operation Starflight, when it helped shepherd a number of T-33s between Canada and Turkey. (DND PL95790)

Tinker, tinker. Groundcrew of 426 Squadron seen at the interminable job of keeping their Merlins in trim. (DND PL50035)

Heave! LACs A.E. Corbin, Max Firestone, Brian Coen and Eugene Ralph lean into a Merlin power pack at RCAF Station Lachine, Dorval. The date is August 18, 1959. (DND PL115828)

group was in the air the other one visited the hangars and 426 installations at the Dorval site. The cadets were also shown the Safety Room Display of 426.

May 11. After take off from Lajes the North Star 17515 had No. 1 engine feathered due to faulty SU pump. The captain decided to return to the airport to have the flight engineer repair it and determine the cause of the engine failure.

May 12. ATC have requested three North Stars from 426 (T) Sqn for Air Force Day on 14 Jun. The squadron have estimated a maximum contribution of 3 North Stars of which two will be required for passengers and display flying and one for static display at St. Hubert.

May 13. S/F 51/52 has been cancelled this Tuesday due to special operation. Operation Starflight report: The 25 RCAF T-33 jet trainers landed today at 1 RCAF Fighter Wing, Marville, France. RCAF officials said that great credit for the success of the flight was due to the work of 50 ground crew ferried across by two 426 (T) Sqn North Stars. The ground crew carried out in-flight servicing and refuelling in record time. The total Canadian mutual aid contribution, in the form of military equipment and services amounts to more than one billion dollars.

May 14. The S/F 5/6 detailed to go to Resolute Bay has been put back to the 21 May. No aircraft is available.

May 16. ATC has planned a new program for aircrew of various ATC units applying to receive Trans-Oceanic familiarization training with 426 (T) Squadron.

Since local flying practice fulfils a large proportion of the training requirements for pilots on ATC units, units should give priority to Observers when selecting aircrew for this training. In order that the programme is to produce useful results, units should select candidates who are sufficiently qualified to reap some benefit from these flights. Obs/Nav should be categorized to "A" standard and Obs/Rad to "B." Also, Captains of 426 aircraft should permit full participation by supernumerary personnel in crew activities. These should, on the ground, include attendance at Met briefings and de-briefings. In the air, supernumerary crew should be permitted to work consistent with their ability under instruction by the regular crew. Units will evaluate this training and advise ATCHQ by the 30th Sept 58 if continuation beyond that date is merited.

May 17. A suitable flying program was arranged by 426 (T) Sqn for the 555 RCAC Squadron (Westmount). Approximately 90 cadets were given a tour in a North Star. Cadets not flying were given a tour through aircraft on the ground and through the Engine Bay, Safety Equipment sections and other points of interest.

May 19. Today statutory holiday - Long weekend.

May 20. S/F 51/52 scheduled for today has been delayed 24 hours on the basis of special operation.

Because of impending summer load at 2 AOS and a hold over of basic Long Range Graduates at Winnipeg, three officers were sent to 426 (T) Sqn for useful employment till the 14 July. 426 has to maintain the trade proficiency of these officers at as high level as possible without jeopardizing operational capability or unit training.

May 21. The delayed S/F 51 is leaving at 0800 EDT to permit the crew a twelve hour crew rest in Lajes. In turn they will take the S/F 53 leaving Montreal at 2000 EST.

426 Sqn personnel formed No. 2 squadron at the Acting CO's parade this morning. The parade was held in preparation for the AOC's inspection on the 6 Jun 58.

May 24. 426 (T) Sqn aircraft flew 45 Air Cadets of 325 RCAC Squadron Cornwall today. The stand-by crew flew the cadets for a period of one hour. They were also given a tour through an aircraft on the ground and points of interest at the Dorval Site.

May 26. W/C Mackie has ended his duties of Acting CO of RCAF Station Lachine due to the return of W/C Turner.

May 28. Two trips to Resolute Bay departed today. One Special 1443 and the scheduled S/F 5/6.

The special 1443 called Special "N" trip carries two North Star Crews plus 14 CNS officers. The aircraft will be used on 14

North Star 17511 flies formation with a Lancaster and a C-119 over Toronto around 1960. (Larry Milberry)

seven hour flights from Resolute Bay. The return departs from RB not later than 1200Z 8 Jun. This is a yearly commitment. The flying time involved is 128 hours.

May 29. Lajes field is below minimum. S/F 53/54 is then delayed 24 hours in Gander.

F/L Longworth was the captain of the training North Star 17521 when it had the nose wheel yoke sheared at the left hand attachment point while turning onto runway for take off. The accident caused a short delay at Dorval Airport but the aircraft was removed in the shortest time possible.

May 30. The Standby crew was called to take the available aircraft in order to carry spares to the Special "N" aircraft that left on the 28 May.

May 31. The Saturday Standby crew flew

Bearing the UN Flag, 17517 sits at RCAF Station Lachine (Dorval), July 26, 1959. (Larry Milberry)

Another view of '517, this time wheeling over the Canadian National Exhibition during the early 1960s. Note the minor changes in markings. The soot streaks on the fuselage are from the Janitrol heater exhaust. (DND via G.M. Martin)

Moving day. Members of 426 Squadron dismantling facilities at Dorval on August 18, 1959. They are headed for their new facilities at RCAF Station Trenton. Seen left to right above are LAC C.Y.G. Dumont, LAC Dan Spivak, and LAC H.F. Branch. Below are LAC C.P. Barrett and F/Sgt J.J. McWilliams. (DND PL115825)

55 Air Cadets of 648 RCAC squadron Ayersville. They took the cadets in the air in two groups and each had an hour of flying time over Montreal area.

A/V/M Kerr, Air Member for Personnel is a passenger on S/F 51 departing this date. He will arrive Langar 1 June 58.

African Interlude

Large-scale Air Transport Command involvement in the United Nations Congo operation brought back to 426 Squadron memories of the Korean airlift. For 426 the Congo undertaking began in July 1960 when, on short notice, the squadron's North Stars were dispatched on the first of hundreds of flights to Leopoldville — a far cry from the howling blizzards of Shemya, but completely in keeping with the versatility and go-anywhere-anytime commitment of 426. An article by S/L R.M.L. Bowdery in *The Roundel*, September 1961, outlines the first year of Canada's Congo involvement:

The Congo, a country relatively unknown to most Canadians, now conjures up vivid memories for RCAF personnel who have been involved there during recent months. Cargo for the Congo is a phrase which describes an operation, now entering its second year, that has developed into one of the largest airlifts ever provided by Air Transport Command.

When the UN called upon Canada to provide assistance in the weeks-old Congo Republic last year, Air Transport Command reacted immediately. But even before the first supply-laden North Star left Canada for that troubled land, ATC's Air Officer Commanding, Air Commodore F.S. Carpenter, AFC, was already there surveying the UN's airlift requirement. Because of his work the UN specifically requested Canada to provide an air operations staff to set up and operate air movements within the Congo, in addition to requesting the Canadian Army Signals personnel and equipment to provide an internal communications network.

The first aircraft despatched to the Congo left Trenton on 18 July 1960. Within the hour it was followed by three more North Stars of No. 426 Squadron. All four carried loads of powdered milk and canned pork for distribution by UN authorities in the central African nation. A subsequent airlift, entitled "Operation Mallard," commenced 9 August. For this commitment eight special North Star flights were made to the Congo, seven from Canada carrying personnel and radio equipment and one flight originating in Pisa, Italy, where a North Star on an RCAF scheduled run to the Middle East was diverted to carry World Health Organization officials and UN observers to Leopoldville.

Initially, one North Star left Trenton every 12 hours carrying army signals personnel and their equipment. From 17 August 24-hour departures were scheduled. By 26 August the army airlift commitment had been met for the time being, so two scheduled flights per week were established on a continuing basis between the UN staging unit at Pisa and Leopoldville.

In order to administer the UN's air transport operations within the Congo G/C W.K. Carr, DFC, and his 10-member staff directed flying activities over an area one quarter the size of Canada with 13 different types of aircraft and with aircrews of eight nationalities: Swedish, Norwegian, Ethiopian, Yugoslav, Brazilian, Indian, and Argentinian. While some of the aircrew arrived fully trained to fly the various UN aircraft, others were unfamiliar with the transport planes, thus it was necessary to set up a training program for them.

Many serious handicaps, in addition to the country's unstable political situation, confronted the air operations: shortages of supply and maintenance, language barriers and inadequate aircrew training being most obvious at the outset. During the initial phase of the interior airlift, UN crews were often flying over unexplored territory with virtually no navigational aids. For several weeks RCAF personnel worked around the clock to keep troops, official UN passengers and vital freight moving from the main staging point in Leopoldville. Crews took off for airfields where security was unknown and whose facilities were marginal.

Despite these obstacles, regular air transportation with all major cities in the Congo was quickly established, providing the only reliable means of transport and supply in a country where the basic methods of road, rail and river travel had almost completely

A sample trip on the United Nations Congo operation in 1960, again from S/L Parmelee's log. The crew included S/L Parmelee, F/L Black, F/L Thomson, F/L Lashley, F/Sgt. Couillard, and LAC Gorow.

Date	*Aircraft*	*Captain*	*First Officer*	*Route*	*Time*
Aug. 12	17510	S/L Parmelee	F/L Black	Trenton-Gander	5:10
	"	"	"	Gander-Lajes	6:45
13	17511	"	"	Lajes-Dakar	8:30
14	17521	"	"	Dakar-Accra	7:00
15	"	"	"	Accra-Leopoldville	6:25
16	17512	"	"	Leopoldville-Dakar	12:25
18	17507	"	"	Dakar-Lajes	8:55
20	17510	"	"	Lajes-Gander	8:45
	"	"	"	Gander-Trenton	5:20

broken down. The aircraft were 5 C-119s, 10 Dakotas, 4 Otters, 4 Beavers, 2 H-19 helicopters, 6 H-13 helicopters, 6 Piper Cubs, an SAS Convair and a Swissair DC-6.

It was soon apparent that, with such a heavy program before them, the ATC operation was in need of a direct radio circuit between command headquarters at Trenton and the RCAF component at Leopoldville. Because of the urgency of the requirement, temporary antennae were erected as an interim measure. The first two-way conversation on the new circuit was carried out on 11 August 1960. A telephone link has since been added between Trenton and Ottawa so that AFHQ can communicate directly with the RCAF detachment at Leopoldville more than 6600 nautical miles away.

On completing his tour of duty the original senior RCAF officer in the Congo, G/C Carr, returned to Canada and became Commanding Officer, RCAF Station Namao. His replacement was A/C C.G. Chapman, DSO, who in turn was replaced by A/C H.A. Morrison, DSO, DFC. The air commodore and his staff occupy a part of the fifth floor in a modern apartment building in downtown Leopoldville. They will serve a six-month tour of duty in the Congo working on behalf of the UN. The staff is composed of representatives from 10 of the 30-odd nations serving with Organisation Nations Unies du Congo (ONUC). The air commodore's senior air staff officer is an Italian, Colonel F. Peirotti. His chief of air operations is another Canadian, W/C A.J. Mackie. His chief technical officer is Norwegian Lt.-Col. Hans Teien. Other names such as Major Sasso of Italy, Capt. Machado of Brazil, Major Hereid of Norway, F/L Bhada of India and Sgt. Lelgi of Sweden testify to the

Garth Martin and his crew at Kano. From the ground up are Garth Martin, Gil Jutras, Keith Scott, Don Wynne, Paul LeClair, LAC McLean and LAC Tebbutt. (DND via G.M. Martin)

Under the Rock. North Star 17506 photographed while passing through Gibraltar. (DND PL107404)

A North Star is readied for its next trip from Trenton. The rope dangling from the cargo entrance was standard equipment on all RCAF North Stars and was installed for emergency escape purposes. (Harry Tate)

Idyllic scene with 17517 refuelling at Idris, Libya. (DND PL 146104)

cosmopolitan group that directs the air transport operation.

N'Djili is Leopoldville's major airport. It has one of the longest runways in the world, a modern terminal building and high pressure hydrant refuelling. The airport at Leopoldville also has UN troops armed with Sten guns on guard at various vantage points. The Congolese, too, have armed guards on the airfield, a constant reminder that the airfield is under "joint" operation. The terminal building shows scars, reminiscent of more hectic moments when tempers ran high and bombs and bullets were readily exchanged. The same sense of uneasiness prevails at ONUC headquarters where barbed wire entanglements lie close by, ready to be pulled into position at a moment's notice. Armed troops also guard the gateways to this building and its parking lot. Trim, handsome and impeccably neat Nigerian policemen serve as doormen at the entrances from where overworked elevators shuttle up and down on their never-ending journeys between the ground floor and the offices above.

The present air transport force has a fleet of 78 aircraft. Forty of these

The first North Star built, 17525, formerly CF-TEN-X. It was photographed at London Airport on March 23, 1957 while being serviced. On June 21, 1959, it was burned out at Athens after a hard landing. (Peter Keating)

are military; the remainder are ONUC registry or on charter to the UN. The charter aircraft, 17 DC-4s and 6 C-46s, fly in UN livery as do all of the aeroplanes operating directly under ONUC control. Dakotas, C-119s, Otters, Beavers and helicopters complete the line-up. The C-119s are maintained and operated by Italians; DC-3 aircraft by eight nationalities; the Otters, Beavers and helicopters are the responsibility of Swedes and Norwegians.

RCAF aircraft are not involved in the internal airlift in the Congo but North Stars of No. 426 Sqn. fly to Leopoldville twice a week on scheduled UN flights from Pisa, with stops at Idris, Tripoli and Kano, Nigeria, en route. In addition to these twice weekly scheduled flights, termed trips or maintenance flights, Air Transport Command puts a further two flights a month into Leopoldville, purely for the rotation of Canadian troops serving in the Congo.

Even before independence, the aircraft was a major method of transport in the Congo. Today, the air is an essential link between Leopoldville and ONUC bases throughout the country. Men of many nations are working diligently to keep this important transport system in effective and constant operations.

R.M.L. Bowdery

North Star 17508 burns furiously at Watson Lake in the Yukon on December 19, 1949, after a refuelling fire. Though badly damaged outboard of the engines, the aircraft was later repaired and returned to service. (K.M. Molson Collection)

Thrills and Spills

With a fleet of aircraft flying 1500 hours a month, the occasional scare was bound to happen. Ashiya, Shemya and Vancouver are cases in point for the RCAF North Stars, but there were others. Quite a few others. To illustrate: One day a North Star was taking off from Heathrow under the able guidance of pilots Adamson and Herbert. The aircraft was at all-up weight and carrying 44 passengers for the Azores. At 95 knots going down the runway No. 1 engine began acting up and was immediately feathered. The day was very hot, making takeoff an extra chore, but the North Star staggered into the air, skimming rooftops and heading straight for a looming hill with Windsor Castle perched atop it.

The aircraft was right on the stall and someone kept pleading, "Don't turn, Herbie! Don't turn!" People remember TV aerials whizzing by as the groaning North Star tried to make up its mind whether to keep flying or not. Then there was enough forward speed to manoeuvre away from the hill and an emergency was declared. Fortunately conditions were VFR and a cool tower operator helped the North Star down. After everyone collected his thoughts, Dave Adamson called the 426 Detachment at Langar where WO George Jackson answered. As Adamson puts it, when he heard Jackson's voice, "it was like God answering!" He and his crew had the North Star back in service within a few hours (as usual).

On January 28, 1962, VIP North Star 17520 left Ottawa on a DEW Line inspection tour. Twenty-three passengers and crew were aboard, senior among these being A/C M.E. Pollard of Air Defence Command. Overnight stops were made at Cape Perry and Cambridge Bay, then at Hall Beach (Fox Main) on February 1.

Next morning 17520 was scheduled to return to Ottawa. The local weather was below takeoff limits with ice-fog conditions, but suitable alternates such as Cambridge Bay and Churchill were clear. The passengers were boarded in the hangar, then preflight procedures were conducted outside where

What was left of 17525, the original North Star, after its poorly executed landing of June 21, 1959. An excerpt from the accident report comments, "On approach to land at Athens airport, the aircraft was landed short of the runway, tearing off the starboard tires. The impact shock to the undercarriage was sufficient to cause a substantial fuel leak from the starboard main plane immediately aft of the No. 3 engine. The aircraft ended up on fire approximately one third of the way down the runway. Result was a complete write off." A sad sight indeed, for the once gleaming prototype. (DND)

North Star 17520 about to leave Montreal on its final trip, carrying personnel of the Canadair Variety Show. (Canadair Ltd. 30068)

17520 come to grief at Hall Beach. (R.T. Merrick)

Approach chart for Hall Beach used by F/L R.H. Whelan to show the fateful flight of 17520, February 2, 1962.

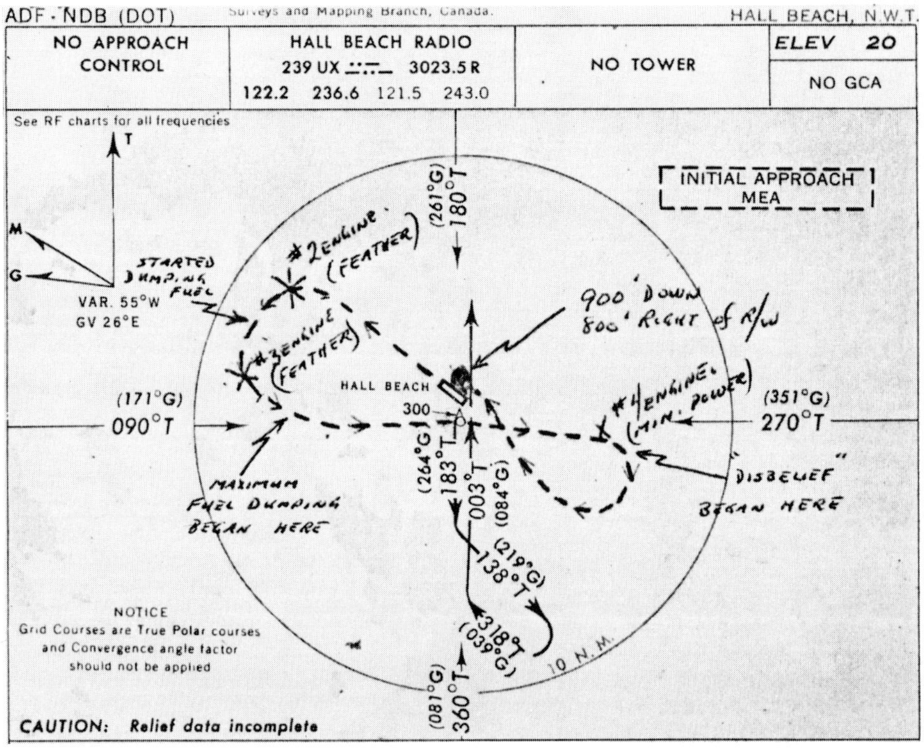

it was about -40°. Pilots F/L R.T. "Dick" Merrick and F/L R.H. Whelan then taxied out and took off. But no sooner was 17520 airborne than the inboard engines suddenly lost oil pressure, ran rough, vented oil and quit.

As soon as trouble arose the pilots began organizing to get back to the strip. F/L Whelan began a left turn for Hall Beach. Before the turn was finished, No. 2 had to be feathered. Within seconds, No. 3 failed. Back in the cabin two of the passengers looking out windows on opposite sides commented to each other that there was an engine out. "Not there," one fellow said, "Over here." They were both right.

The aircraft proceeded downwind dumping fuel and just before turning to land No. 4 began to fail. Things were now very critical and to make them worse the ground was obscured. Only the tops of a few structures could be seen. By now the North Star was almost uncontrollable and F/L Whelan was using all his strength to keep it flying straight as he aimed in the general direction of the strip. Then the aircraft touched down, bumped along with a slight washboard effect and stopped. As the door was opened, one of the passengers was surprised to see the ground so close. He hadn't realized it had been a wheels-up landing!

The aircraft was quickly evacuated and within a few minutes snowmobiles from the base arrived. All had ended well.

Next day a Hercules flew the passengers and crew back to Ottawa. The investigation which followed concluded in part, "The aircraft was left outside for 90 minutes or more in a temperature of -37° with a 15 mph wind." A more recent comment notes, "The aircraft sat outside in the sub-zero weather for too long a period prior to oil dilution. As a result, structural damage was done to the engines when the engineer ran up for oil dilution."

As a result of the accident several recommendations were made, among them that the Engineering Orders and Pilot Operating Instructions be amended clarifying cold weather operations, and that the runways at places like Hall Beach be equipped with bright markers to aid pilots making approaches in white-out conditions. When 17520 went down, there were no runway markers in use. Oddly enough, fuel storage tanks near the approach were painted white. F/L Whelan could not see these and just cleared them through good fortune. The airplane was left at Hall Beach.

Another RCAF North Star was written off at North Luffenham on May 8, 1954. The incident is described here by navigator Don Connolly who was aboard the flight:

Outwardly there did not seem to be much wrong with North Star 17522 as she stood abandoned in the middle of a large puddle halfway down the runway at North Luffenham, England. When you moved closer you could see that the tires of the starboard undercarriage looked flat, but aside from that she looked sound. The first real clue to her problems came with the first whiff of high octane gas. The puddle was aviation fuel. 17522 and her crew had come within a whisker of being scorched.

On May 5, 1954, 17522 left her home base at Trenton where she was on the roster of No. 4 Transport OTU. Aboard were instructors and students on a standard transatlantic training mission to Paris and return. First stop was Montreal where an overnight stay was made and a mixed payload of cargo and passengers was taken aboard. The next leg was to Goose Bay where another layover was made in preparation for an early morning departure for Keflavik. After an 8.4-hour flight and another crew rest we left Iceland early next day, May 8, and after a short 3.8-hour leg, landed at Prestwick, Scotland, to refuel. At 1545 GMT we took off for the short 1.6-hour flight to the RCAF base at North Luffenham. As lead or instructor navigator I was able to relax as we followed radio range stations down the airway, with instructor and student pilots and one of the trainee navs all monitoring the procedure. One of the student pilots took the left-hand seat to execute a standard VFR approach and landing.

The trouble developed only yards from the runway threshold. The landing approach was too low. Not sufficiently so to be dangerous under normal circumstances, but low enough in this case that the undercarriage swiped the top of a mound of earth that bisected the terrain just beyond the end of the runway. A careless or lazy work crew that had been trenching had left the job site at quitting time without marking or removing the hazard. To compound the problem, the control tower had not advised the aircraft crew of the condition.

As I recall, the aircraft only lurched slightly. A fairly heavy landing ensued, but then the "juddering" of the wheels and the level of excitement in the cockpit bespoke more serious difficulties. We came to a halt part way down the runway and the order was given to leave the aircraft as soon as possible as the brakes were burning. Then came the announcement or the awareness — I can't now remember which — that the wings were spilling fuel. That really speeded up the exodus. Trouble was that the only conventional way out was down a ladder, some 10 feet or so from the main cargo door to the ground. Standing as I was near the end of the line, that group of passengers and fellow crew members seemed to move like snails! By this time we had removed the emergency escape hatches above the wings so that if something did ignite we could attempt to flee along the top of the wings, vaulting off the wingtips if necessary. However, opening these doors added considerably to the psychological pressure, as we could now hear very clearly the splashing of the fuel as it cascaded onto the runway. Furthermore a glimpse out any window provided the alarming sight of an ever-enlarging lake of gasoline flowing out in all directions, with us right in the centre.

17508 bores in to land at Trenton while training with No. 4 (T) OTU. Cpl. Barry Herron took this lovely photo on March 22, 1956. (DND PL88329)

About this time, my heart went up to my throat as I heard a loud "whoosh;" I thought she was starting to blow. As I learned afterwards, the sound had been a fire extinguisher fired by the flight engineer. He had lowered himself down a rope thrown out of the little-used crew door (next to the navigator's seat) and was dousing the still-smouldering brakes.

All this happened in minutes, but the episode still occupies as many of my memory cells as do whole months of my life.

Don Connolly

No. 4 (T) OTU

No. 4 (T) Operational Training Unit was activated March 1, 1952, "to provide conversion and continuation training in support of ATC's heavy and medium aircraft crews." Based at Dorval, the OTU operated North Stars, C-119s and Dakotas. The unit was responsible for, among other things, training all North Star aircrew. Besides training, it provided air transport as required. For instance, during 1953 a large backlog of air freight accumulated at Sept Iles which was urgently needed in the construction of the railroad to Schefferville, where iron ore deposits were being developed To assist, ATC loaned several North Stars and mounted Operation Overload. The OTU provided three North Stars, which during August moved over 600 tons of material up the rail line.

In January 1954 the OTU moved to new quarters at Trenton, where courses for pilots, navigators, radio operators, flight engineers and movement controllers were run for North Star personnel. The period December 31, 1954-May 31, 1955, included the following OTU activities:

Flying hours:	North Star	1260:50
	C-119	744:10
	Dakota	905:20
Total courses	32	
Graduates	160	

Training North Star flight crews went on continuously into the early 1960s, but the advent of turboprop transports signalled the end of North Star training. The OTU began its first Yukon course at Trenton in July 1961. Earlier that year, 412 Squadron had run its first Cosmopolitan course at Ottawa. The Yukon rapidly assumed long-range airlift duties with ATC, while the "Cosmo" took over VIP duties, though the C-5 was retained for some time after the remaining 412 North Star was phased out.

In the fall of 1962 the OTU acquired four North Stars from the disbanded 426 Squadron. These were used both for training and transport. The usual

North Stars share the scene at Menihek with a C-46 and a Lodestar. (John Timmins)

A North Star touches down on the gravel strip at Menihek during Operation Overload as technicians take time out to admire the pilot's efforts. (John Timmins)

On June 29, 1960, 17504 prepares to touch down on Runway 32 at Malton after a flight from Windsor, Ontario. (Larry Milberry)

variety of operations continued, including troop rotation to the Middle East. The OTU now carried the unofficial designation "213 Squadron" (half of 426). With the first Hercules OTU, run in August 1964, there is no more mention in the OTU diary of the venerable North Star.

Thunderbirds Bow Out

In January 1962 426 Squadron moved from Trenton to St. Hubert. This was necessary as the facilities at Trenton, now mainly supporting the new Yukon fleet, could not accommodate two major transport fleets. After brief operations from St. Hubert, 426 Squadron was disbanded on September 9, 1962. Activities during the St. Hubert interlude included shepherding duties on Operation Western Weal in April through July. This involved the overseas ferrying of 10 Beech Expeditors from Canada to France and two lots of 12 T-33s to Turkey. Some Congo flying was also done.

Flying rates for the first six months of operations from St. Hubert were: January 1228 hours, February 1220 hours, March 1387 hours, April 1081 hours, May 1085 hours and June 1537 hours.

The final Congo trip by North Star appears in the squadron diary for September 9 when an aircraft left Pisa for Leopoldville. Hereafter the Yukon handled that commitment through to the end of the operation on June 30, 1964. The big Yukon carried four times the load of the North Star. *The Roundel* later wrote of both the North Star and Yukon in the Congo:

In November '61 the RCAF pulled the North Stars off the Congo run and replaced them with Yukons. To F/L P. Hope, who flew both types to Leopoldville, the switch to Yukons was a decided improvement. "We could," he said, "take as much as 40,000 pounds, quite a contrast to the 10,000-pound payload of the North Stars." It is more than a matter of payload. The Yukons fly from Canada to the Congo in two giant steps, going from Trenton to Pisa, Pisa to Leopoldville in quiet, comfortable above-the-weather operations. The Yukons are back in Trenton in five days. North Stars lumbered their way to the Congo and back with more frequent refuelling stops.

But the Yukons got off to a bad

start on the Congo service. While the first Yukon arrived without incident, the second one was not so fortunate. The Congolese suddenly realized that there was a strange new aircraft landing on their airport. For some reason, they thought it was Russian. As the RCAF crew climbed down the ramp from their aircraft they were surrounded by gun-toting guards. The aircraft was impounded and searched. There was a tense moment as one of the Congolese soldiers started to unpack the safety equipment. If he had discovered the shotgun, which is a normal part of the Yukon's survival equipment, the Congolese would have been convinced that they had intercepted contraband. Fortunately, however, the curious soldier lost his zeal when he was told that if he unpacked the gear he would have to repack it himself. Only after the timely appearance of A/C C.G. Chapman and a senior Congolese official secretly summoned to the airport from Leopoldville, was the crisis ended and the Yukon and its shaken crew released.

This was but one of the many incidents which have plagued UN operations since that cosmopolitan force of civilian and military personnel first arrived in the Congo in the summer of 1960. But, in spite of all the trials and tribulations the ONUC, which is now entering its fourth year of operations, has done much to bring stability to that troubled land. And the RCAF officers, NCO's and airmen who have served, are serving or will be serving in the Congo can take justifiable pride in their role as peace-makers.

With 426 Squadron disbanded, its aircraft were dispersed. Four were transferred to the OTU, two to CEPE (Central Experimental and Proving Establishment) at Ottawa, six went to storage at Mountainview, and one was still in the hands of the CAIR (Calendar Aircraft Inspection and Repair) contractors. The last North Star left St. Hubert on October 19, 1962, and flew to Mountainview.

Rescue North Stars

In the early 1960s No. 107 Rescue Unit was operating from Torbay, Newfoundland, on search and rescue duties. Three Lancasters were in use, but in August 1963 two North Stars arrived as replacement aircraft. A third one arrived later, the aircraft being 17502, 17507 and 17510. The

In storage at Mountainview, 17510 awaits disposal. It bears the markings used while in service with 107 Rescue Unit. (K.M. Molson Collection)

unit diary notes, "All Lanc crews were checked out on the North Stars and all R/Os were checked out as drop masters for dropping MA1 Search and Rescue Drop Gear." In November 1963 the Lancasters were flown into storage at Dunnville. By year's end the North Stars had flown 398:10 hours from Torbay exclusive of training time. Unit activities in 1963-64 included:

Sept. 24, 1963. Lancasters FM104 and FM213 plus one North Star search for MATS C-133 missing between Boston and Lajes.

Oct. 27. North Star 17502 to UK on SAR training exercise.

Nov. 9. North Star 17507 escorts MATS transport with one engine out.

Dec. 10. 17502 from Greenwood to Kinloss, transport flight.

Dec. 16. 17510 escorts Seaboard CL-44 en route London-McGuire AFB with one engine out.

Dec. 18. 17510 on medevac from Gander to Toronto forced to make two-engine landing at Sherbrooke, Quebec.

Dec. 20. Search for French trawler *Douala.* Survivors later found by Argus.

Dec. 22. 17507, 17510 search for MV *Gladys Wiscombe.*

Jan. 22, 1964. 17510 search for MV *Colin II.* Ship found and ice breaker dispatched to rescue it from pack ice.

Feb. 14. 17507 from Greenwood to Bermuda, transport flight.

About this time the federal government revised its search and rescue policy and on February 29, 1964, 107 Rescue Unit was disbanded. Its North Stars were ferried to Mountainview to join others awaiting disposal there.

Prime Minister Louis St. Laurent, 412 Squadron pilot Stu Cowan and Canadair chief pilot Al Lilly with the C-5 at Calgary on its first VIP trip. (via A.J. Lilly)

Crème de la Crème—the C-5

Number 412 VIP Squadron received its first of two Merlin-powered North Stars in April 1949. In early 1950 it conducted its first round-the-world flight, carrying the Hon. L.B. Pearson to the Commonwealth foreign ministers' conference in Colombo. Between leaving on January 2 and returning February 8, the aircraft logged 125:20 hours.

The TCA crash of CF-TEL at Sydney in 1948 left the RCAF short one North Star, and, encouraged by an insurance payoff, the government authorized construction of one more aircraft on the North Star production line. The aircraft was to be for VIP use at 412 Squadron and it was decided to modify it suitably. Chief among the mods was the replacement of the noisy Merlins with Pratt and Whitney R-2800s. The aircraft was also pressurized. Its production may also have encouraged Canadair with hopes of further production, perhaps of its C-7.

The C-5 made its first flight on May 15, 1950, bearing the RCAF serial 17524. Al Lilly and Bill Longhurst took it up for a 35-minute shakedown

Crew of the C-5 during the visit of the Crown Prince of Japan in April 1953: LAW Grace Honkawa, LAC Joe Lanteigne, F/S Jerry Mignault, W/O E.J. "Benny" Benoit, F/O Al Kirk, Prince Akihito, F/L Hank Enns, F/L Jack Reid, F/L Dave Adamson, G/C J.G. Stevenson. (via D.R. Adamson)

flight. A second flight of 1:50 hours was made May 22, and the third flight of 1:45 occurred May 24. On July 7-11 it took Prime Minister St. Laurent to Calgary to open the Stampede. The C-5 was taken on strength by the RCAF on July 29.

For the next 15 years the C-5 was to be the premier VIP aircraft in the RCAF. Royalty and other civil and military VIPS of all sorts and from countries around the world were to enjoy its luxury. *The Roundel* paid tribute to the C-5 with this article entitled "Victor Charlie 10000" (referring to the C-5's later serial number):

If an aircraft is entitled to a measure of fame because of celebrated people it has carried, the number of countries it has visited, or the distances it has travelled, then No. 412 Squadron's C-5 is well established as a famous aeroplane.

Since this aircraft was assigned the task of transporting VIPs, its passenger list reads like an international "Who's Who." Its stewards have served meals and made up bunks for more eminent people than a waiter or chambermaid in Ottawa's Chateau Laurier or New York's Waldorf Astoria. Through the years, prime ministers and presidents, parliamentarians and princesses, have travelled on it, and, on four occasions, the C-5 has proudly flown the Royal Standard.

Not only is the C-5 the only one of its kind in the RCAF, but it is also a unique aircraft in itself, having an enlarged DC-4 fuselage, a DC-6 under-

The C-5 alongside its VIP partner 17518 at Downsview on April 15, 1957.

The C-5 at Dorval on October 31, 1951, upon arrival of Princess Elizabeth and Prince Philip. (*Montreal Gazette*/Public Archives of Canada PA129601)

Princess Elizabeth signs the C-5 log book while on the royal visit to Canada in October 1951. (DND PL52800)

carriage, and four Pratt & Whitney engines. This hybrid flying-machine, appreciably quieter than the time-tested North Star (and somewhat faster: 30 knots at 18,000 feet), came into service in July 1950 to fill the need for a pressurized VIP aircraft.

Although the largest aircraft in the Service, it carries only 27 passengers in a style suitable for official guests of Canada. Meals are prepared on board to suit the wishes and tastes of distinguished passengers. To guard against the possibility of spilling soup on some august lap, the steward receives reports of impending turbulence en route and plans his meal-hours accordingly. Since the C-5 generally flies at approximately 18,000 feet, however, turbulence is cut to a minimum and the likelihood of such unhappy accidents is negligible.

In its 5½ years of service, the C-5 has logged some 3,000 hours of globe-trotting flight. Its three major trips have been: an extensive tour of South America with Mr. C.D. Howe in 1953, a 24,000-mile trip around the world with Mr. St. Laurent in 1954, and Mr. Pearson's three-continent, twelve-country visit in 1955. On flights within North America, the C-5 normally carries a seven-man crew, but for extended trips, where maintenance may become a problem, a crew of as many as sixteen is carried. The latter precaution has already paid off. During Mr. Pearson's trip to India, when an engine failure occurred at Kuala Lumpur in Malaya, the aircraft maintenance team on board the C-5 promptly fitted a new engine, flown from the RCAF base at Langar, England, and the C-5 was able to rejoin Mr. Pearson in India in time for his scheduled departure. Another engine failure and subsequent engine change occurred in Korea during a visit by Mr. Brooke Claxton, the then Minister of National Defence, to the Canadian troops. The C-5 has never, however, been involved in an accident.

Only the most highly qualified personnel, both aircrew and groundcrew, are selected for service on the C-5. Pilots must have spent a period of at least two years in a transport squadron, operating on trans-continental and trans-oceanic routes, and they must have logged a minimum of 3000 hours (including 1000 hours on four-engined aircraft) in order to become eligible for duties as First Officer on the C-5. To qualify as a VIP captain,

Crew on the Prime Minister's world tour, February 4 - March 17, 1954: F/O Bill Middler, F/O Al Pulfer, F/L Russ Chalk, F/O Bob Doncaster, F/L Hank Enns and F/L Dave Adamson. Missing are W/C H.A. Morrison and W/O E.J. Benoit (DND PL59221)

ITINERARY

PRIME MINISTER'S TOUR

FROM	DEPARTURE DATE	DAY	LOCAL	TO	ARRIVAL DATE	DAY	LOCAL	N. MILES DISTANCE	FLYING TIME
ROCKCLIFFE	FEB 4	THU	1445	GOOSE BAY	FEB 4	THU	1900	761	3:15
GOOSE BAY	5	FRI	0700	LONDON	5	FRI	2000	2218	9:00
LONDON	7	SUN	1515	PARIS	7	SUN	1730	185	1:15
PARIS (ORLY)	10	WED	1030	BONN	10	WED	1145	260	1:15
BONN	12	FRI	1030	GROS TENQUIN	12	FRI	1115	105	:45
GROS TENQUIN	12	FRI	1400	ZWEIBRUCKEN	12	FRI	1430	50	:30
ZWEIBRUCKEN	12	FRI	1530	ROME	12	FRI	1930	690	4:00
ROME (CIMPIANO)	16	TUE	0915	BAHREIN	16	TUE	2000	2075	8:45
BAHREIN	17	WED	1015	KARACHI	17	WED	1530	862	3:45
KARACHI	19	FRI	1415	PESHAWAR	19	FRI	1715	605	3:00
PESHAWAR	20	SAT	1600	LAHORE	20	SAT	1700	206	1:00
LAHORE	21	SUN	1515	NEW DELHI	21	SUN	1730	300	1:15
NEW DELHI (PALAM)	25	THU	0900	AGRA	25	THU	0930	100	:30
AGRA	26	FRI	0900	BOMBAY	26	FRI	1200	550	3:00
BOMBAY	27	SAT	0930	MADRAS	27	SAT	1230	552	3:00
MADRAS	28	SUN	1500	COLOMBO	28	SUN	1700	380	2:00
COLOMBO	MAR 4	THU	0800	DJAKARTA	MAR 4	THU	1900	2000	9:00
DJAKARTA	6	SAT	0930	MANILA	6	SAT	1630	1400	6:30
MANILA	7	SUN	0900	SEOUL	7	SUN	1730	1500	7:30
SEOUL	10	WED	1000	TOKYO	10	WED	1300	500	3:00
TOKYO	13	SAT	0900	MIDWAY	12	FRI	2200	2250	10:00 Date Line
MIDWAY	13	SAT	1000	HONOLULU	13	SAT	1630	1150	5:30
HONOLULU	15	MON	2359	SAN FRANCISCO	16	TUE	1145	2190	9:45
SAN FRANCISCO	17	WED	0915	OTTAWA	17	WED	2130	2200	9:15
								23,089	106:45

DEPARTURE TIMES ARE THE TIMES AT WHICH THE AIRCRAFT SHOULD BE AIRBORNE AND IN CONSEQUENCE THE MEMBERS OF THE PARTY SHOULD PROVISION TO EMPLANE 15 MINUTES PRIOR TO THE INDICATED DEPARTURE TIMES.

THE ARRIVAL TIMES INDICATED ABOVE ARE RAMP TIMES, I.E., THE TIME AT WHICH THE PRIME MINISTER WILL DEPLANE.

ALL TIMES SHOWN ARE LOCAL TIMES.

W/C H.A. Morrison DSO, DFC, AFC, DC while commander of 412 Sqadron. At the time he was Captain of the C-5 during Prime Minister St. Laurent's world tour, February 4 to March 17, 1954. (via D. Adamson)

C-5 crew: officers in front are W/C W.K. Carr, F/L D.T. Thompson, F/O R.T. Brown, and F/L J.E. Perron. In the second row from the left are F/Sgt A.J. Cooper, F/Sgt E.C. Grose, Sgt R.E. Cameron, Cpl J.C. Lefebvre, Cpl P.R. Atkinson, Sgt J.O. Mignault and Cpl L. Toth. (PL107808)

January	5	Rockcliffe-San Juan	8:45 hours
	6	San Juan-Belem	8:50
	7	Belem-Rio de Janiero	7:10
	11	Rio de Janiero-Sao Paulo	1:25
	15	Sao Paulo-Buenos Aires	4:50
	18	Buenos Aires-Montevideo	1:00
	19	Montevideo-Rio de Janiero	4:45
		Rio de Janiero-Belem	6:30
	20	Belem-Port of Spain	5:00
	22	Port of Spain-Caracas	2:00
	27	Caracas-Maracaibo	1:30
	28	Maracaibo-Bogotá	2:30
February	1	Bogotá-Baranquilla	2:10
		Baranquilla-Trujillo	2:40
	2	Trujillo-Port au Prince	1:00
		Port au Prince-Havana	1:25
	5	Havana-Mexico City	5:20
	10	Mexico City-Rockcliffe	9:25

The log of the C-5 during a trade mission to Latin America in 1953 headed by C.D. Howe. Crew included F/L Adamson, F/L Reid, F/O Martin, F/L Webb, F/L Saunders, F/O Merrifield, F/O Penfold, F/S Benoit, Cpl Sjolin, Sgt Roy.

Sgt Jerry Mignault and Cpl Percy Atkinson, C-5 cabin crew, in the C-5's galley (DND PL113934)

Dec. 1950	C-5	Hon. Louis St. Laurent	Ottawa-London, England
March 1951	C-5	Hon. Louis St. Laurent	Ottawa-New York
March 1951	C-5	Hon. Louis St. Laurent	Ottawa-Bermuda
Oct. 12-31, 1951	C-5	Princess Elizabeth and Duke of Edinburgh	Toronto-Winnipeg-Rivers-Lakehead-North Bay-Montreal-Washington
Dec. 1951	C-5	Brooke Claxton	Ottawa-Tokyo return
April 1952	C-5	Lord Trenchard	London-Ottawa
June 1952	17520	Gov. Gen. Earl Alexander	Ottawa-Tokyo
May 1954	C-5	Gov. Gen. Vincent Massey	Ottawa-Washington
June 1954	C-5	Sir Winston Churchill	Washington-Ottawa-New York
Dec. 1954	C-5	Lester B. Pearson	Ottawa-Paris return

A sampling of VIP flights from L.W. Hussey's log.

The stately C-5 on the ramp at Uplands on April 5, 1961. (Larry Milberry)

the candidate must come up to the standards set by the squadron, and receive a recommendation from the Commanding Officer. The C-5's navigators and radio officers have correspondingly high standards, which include an annual pass-mark of 85% on theoretical and practical work. No crew is selected exclusively for the C-5, all "VIP personnel" being eligible; but, with such high qualifications required, the number of available crews is naturally limited, and, at the time of writing, only three captains and four first officers are permitted to fly the aircraft. Names of eligible crew members are kept on a roster and the crew is selected on a rotational basis for each flight.

The crew which flew Mr. St. Laurent on his globe-circling tour of eleven countries in 1954 totalled 25,000 hours of flying experience. At least two of the C-5's stewards, Sgt. J.O. Mignault with 2200 hours, and Sgt. J. Roy with 3000 hours, have more flying time than many aircrew members. Occasionally one or two of the crew belong to other Services than the RCAF. A USAF exchange pilot with No. 412 Squadron may sometimes go on a trip; and, on the journey to Moscow, a Russian navigator and radio officer joined the aircraft at Berlin to guide the aircraft through the Iron Curtain.

The crew of the C-5 strive to make each trip one of flawless performance, and a bulging file of complimentary letters from former passengers bears testimony to the success of their endeavours. Nonetheless, there *have* been times when this enviable record has been threatened for a few anxious moments. Such was the case during the Duke of Edinburgh's Canadian tour, when the hydraulic system failed on the approach to Vancouver airport. While the captain pressed on, the first officer and engineer took turns at madly wobbling the hand-pump to lower the undercarriage and flaps — and even after landing they continued their activities in order to keep the steerable nose-wheel operating. As the aircraft rolled to a stop they breathed a sigh of relief, confident that their royal passenger hadn't noticed anything out of the ordinary. They looked up to see the Duke smiling. "Nicely rowed, chaps!" he said. That is probably one of the most cherished compliments in the squadron's records.

The C-5 has a "ramp-time" to make at each destination, since there are often guards of honour and dignitaries on the ramp, waiting to greet the aircraft's passengers. This ramp-time is almost invariably made on schedule by the crew, with the able assistance of control-tower operators who, when they hear the call-sign "Victor Charlie 10,000" (in Canada) or "Canadian Air Force 10,000" (at foreign airports), help considerably by giving the C-5 landing priority. If the weather is "duff" and aircraft are holding, the C-5 is inserted in the stack so that it will be at the ramp on time. When the C-5 approaches a destination ahead of schedule, it circles the area, using up time until it can land exactly on ETA. There have been occasions, of course, when unforecast headwinds or storms have caused the C-5 to arrive late, but on the average the aircraft has a record for punctuality that would do credit to any airline.

The crews of the C-5 undergo experiences generally not encountered by other flying personnel in the RCAF. Wandering as they do to far-off lands where yellow fever and malaria are prevalent, they get more than their normal share of needles. When they start off on a trip in winter, they carry both summer and winter uniforms in preparation for the warm climate of the tropics. On one trip the crew changed uniforms three times as they came down from the frozen north into more temperate zones. Special passports, good for a particular trip, are issued to them by the Department of External Affairs.

Although the C-5 has no regular commitments, each of its trips being a special case, the most frequent run is from Ottawa to London, and Ottawa to Paris, for NATO and other conferences. In 1953, when the then foreign secretary, Sir Anthony Eden, went to Boston for an operation, he was flown by the C-5 at the request of Sir Winston Churchill. Commenting on the fact that the aircraft had been

A fine air-to-air view of the C-5. For looks, this first cousin of both North Star and DC-6 was in a class all its own. This photo was taken on September 2, 1955. (PL102424)

sent all the way from Canada to pick him up, Mr. St. Laurent expressed the high esteem in which the C-5 is held when he said: "Why not? There isn't another 'plane in the world like ours."

The Roundel

L.W. "Les" Hussey spent years with 412 Squadron flying on the C-5 and North Star and recalls of the North Star that it was a pilot's airplane in every way: "It responded well to pilot inputs, carried a reasonable load at a fair speed, and could be depended upon to perform as advertised in virtually all environments. In Arctic operations the Rolls-Royce engines would start in the extreme temperatures whereas the P & Ws would invariably present problems. But the North Star was noisy and not pressurized as was the C-5. The crossover exhaust system as used by TCA was not installed; some extra insulation was used instead. The propellers used by 412's North Stars were designed to alter the tip speed harmonics and thus reduce noise.

"Not all 412 Squadron North Star pilots flew the C-5 and there was always considerable rivalry between the two. The C-5 was generally used on VIP flights such as one on which I flew the Hon. Brooke Claxton, Minister of National Defence, to Tokyo in December 1951 and which was to take him to Korea for New Year's Day with the Canadian troops.

"Unfortunately, after about 38 hours with landings at Edmonton, Anchorage and Shemya we lost an engine on approach to Haneda Airport. Our arrival on the ramp turned out to be a horror show as the No. 1 engine was dripping large amounts of oil. A Cathay Pacific Airways aircraft then started up in front of us. Its prop wash blew oil and fluids all over the light-coloured suits and uniforms of the civil and military dignitaries meeting the minister!

"The engine had to be replaced but we didn't have a spare in Tokyo. We had to fly one out from Ottawa aboard, needless to say, a North Star. This was indeed the ultimate insult for the C-5 pilots — to be rescued by a North Star!"

Two Comets served 412 Squadron as VIP transports and on special duties, including many NORAD exercises. They shared space in 412's hangar at Uplands with two VIP North Stars and the C-5. (Wm. J. Balogh Sr. via D. Menard)

On Lester Pearson's round-the-world flight of 1954 the C-5 lost an engine at Kuala Lumpur in Malaya. Pearson went on to Calcutta on another aircraft, but the C-5 still had other passengers headed for the opening of a Canadian development project in India. The problem was solved by shipping an engine out from Dorval in a 426 North Star, arriving at 1:00 a.m. on a Sunday. WO2 Edgar "Benny" Benoit related, "We were given a deadline of noon Monday by which the aircraft had to be ready, so we got to work right away. There was no hangar large enough for the C-5 so we had to work in the open using a small mobile crane — all that the RAF had available. It was during the monsoon period so that it rained every day between 2 and 11 p.m. like you don't understand until you've seen it. At times we were standing in water half way up our legs. When it wasn't

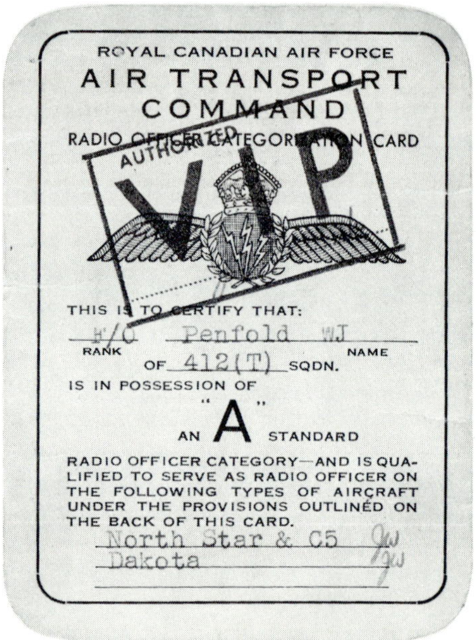
ROYAL CANADIAN AIR FORCE
AIR TRANSPORT COMMAND
RADIO OFFICER CATEGORIZATION CARD
AUTHORIZED
VIP
THIS IS TO CERTIFY THAT:
F/O Penfold WJ
RANK NAME
OF 412(T) SQDN.
IS IN POSSESSION OF
AN "A" STANDARD
RADIO OFFICER CATEGORY—AND IS QUALIFIED TO SERVE AS RADIO OFFICER ON THE FOLLOWING TYPES OF AIRCRAFT UNDER THE PROVISIONS OUTLINED ON THE BACK OF THIS CARD.
North Star & C5
Dakota

raining, the heat was scorching. We worked continuously for 26 hours without leaving the tarmac and beat the deadline by 5 hours."

In May 1953 412 Squadron began operating the first of its Comets, becoming the world's first operator to fly scheduled transatlantic jet transport service. The Comets vied for the

17506 at 6 Repair Depot, Trenton. The lower fuselage has been painted with anti-corrosion grey and carries the new Canadian flag, placing the time at 1965. In the foreground is another famous Canadair product, a CL-41 Tudor. (Harry Tate)

A sad sight. The once glorious C-5 being picked apart at Long Beach, California. (S. Kraus via John Wegg)

limelight but the C-5 remained the primary VIP aircraft at Uplands. The Comets were retired to Mountainview in 1965, where one was cannibalized and the other briefly civilianized, but the C-5 remained on active duty until retired from service on August 28, 1966. It too went to Mountainview from where it was sold to a US operator.

In typical Canadian fashion, the C-5 had just been refurbished prior to being sold for a give-away price. The Toronto *Globe and Mail* reported on this in a February 1968 item:

A C-5 North Star aircraft that had cost the Department of Defence more than $1.6 million, including $343,000 for refurbishing it to VIP standards in 1966, was sold in June, 1966, for $49,000, Auditor-General Maxwell Henderson has reported.

The North Star cost $1,226,000 in 1950. It was reconditioned in 1963 to keep it flying until 1965. Its use after that date was to have been reviewed annually, but the review was not made and its three spare engines were overhauled in 1965-66 at a cost of $40,000. Early in 1966, the Defence Department spent $343,000 to refurbish the plane to VIP standards and for inspections and repairs to gain an extra 1,500 hours of use.

The sale of the North Star for $49,000 included the three spare engines, which had not been used since their overhaul, Mr. Henderson said in his annual report tabled in the Commons on Monday.

The report soundly ticked off the Defence Department for its waste of public funds.

The C-5 was flown from Mountainview to Santa Monica by its American owner, but after apparently failing to obtain US certification the once-proud C-5 was ignominiously stripped of useful parts and cut up for scrap.

North Star Phase-Out

By late 1965 the North Star was redundant to RCAF requirements and was phased out in a ceremony at RCAF Station Trenton on December 8, 1965. The last service flight had arrived back at Trenton from a UNEF run to El Arish on November 14. For the phasing out ceremony, North Star 17506 flew from Downsview to Trenton for the ceremonial last flight, carrying guests and media people, Larry Milberry included. The half-hour flight was made in poor weather and, ironically, the aircraft arrived at Trenton with a hydraulic snag. This meant no "beating up" the field and upon landing the aircraft was handed over to the ground crew as unserviceable.

Hundreds were on hand for the last engine run-up later in the morning, with the mighty Yukon looming in the background. Afterwards a ceremony took place inside the 437 Squadron hangar where two shiny North Stars, a pipe band and a guard of honour were drawn up. In a nostalgic moment, the log books of 17506 were handed over to A/M W.A. Curtis by two 426 veterans. One was F/L John A. "Bud" McNair. By this time McNair had 5500 hours on the North Star and had flown it to the Congo, El Arish, the Arctic and among other things had recently captained the last trans-atlantic North Star run for ATC.

The other veteran of 426 Squadron was Sgt. Paul Pawliuk, with 4700 North Star hours. Pawliuk had been on the first flight out of Dorval on the Korean airlift and his experience included three royal tours, three world tours, and service in the Middle East, Congo and the Arctic.

Following the day's festivities, it was back to business. The Yukons continued their comings and goings to points around the globe, while the three North Stars were soon ferried the short hop over to Mountainview to await disposal. An era in Canadian military aviation was over, but thousands in the RCAF would not soon forget the grand old North Star. Their tributes could fill the pages of this book.

G/C Z.L. Leigh, Air Transport Command commander, knew the North Star from its earliest years. In his memoirs, *And I Shall Fly*, he records, "In March 1945 A/V/M Curtis and I flew down to Cartierville to take part in discussions regarding a new transport aircraft for both the civil and military operations which was to be built by Canadair Ltd. — the famous North Star. It was planned to use Rolls-Royce liquid-cooled engines in

Its Honour Guard presents arms as the North Star is retired from RCAF service at Trenton, December 8, 1965. (DND Z10578-1)

Sgt Paul Pawliuk and F/L Bud McNair present the ceremonial log book to A/M W.A. Curtis at the North Star stand-down ceremony on December 8, 1965. (DND Z-10578-4)

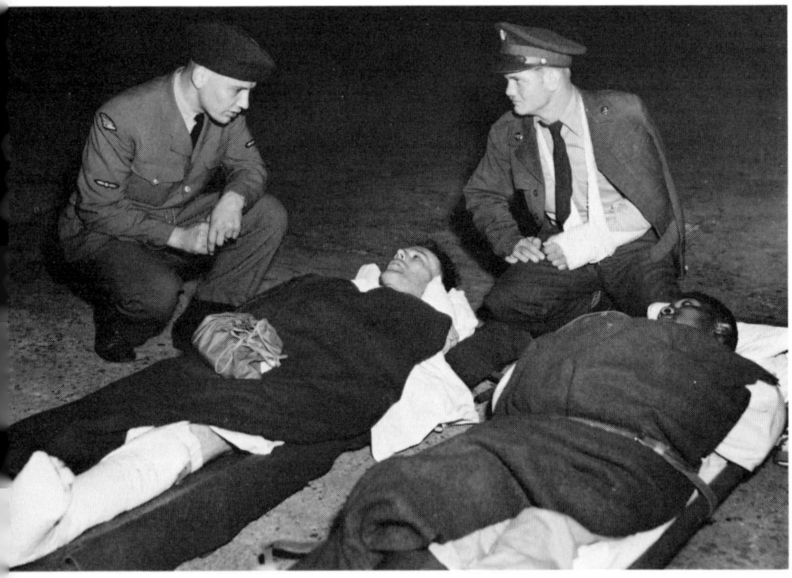

LAC Pawliuk chats with US Korean War evacuees at Itami, Japan, before their trip back "Stateside" in a 426 North Star. The North Star was preferred as a medevac aircraft as it was considerably faster than the C-54 on the long trans-Pacific trip. (DND PL52303)

F/O Bob Edwards in the captain's seat of a 426 North Star on the Korean airlift. Obviously not a "retread," Edwards was one of many 426 members who were fresh out of school yet functioning like old pros on the gruelling Pacific routes. (DND PL50553)

the aircraft. I objected, saying I would like to have Pratt and Whitney air-cooled engines installed. Both A/V/M Curtis and my old friend of TCA days, Mr. J.T. Bain, told me that I was wrong and that the Rolls-Royce engines would be better. Time proved that they were right and I was wrong. The Rolls engines did a terrific job, particularly under extremely cold conditions in our Arctic operations."

Bob Edwards, veteran of 426 and 412 Squadrons, also remembers: "With regard to my impressions of the North Star, I can only say that it was much more dependable than most and a delight to fly, providing one enjoyed better than average strength in his arms and legs. The Merlin engines made it a much greater success in the Arctic than the C-54 and, at a slight loss in range, gave it a good 30-knot speed advantage. It was also much more noisy than most aircraft, and those of us who have logged several thousand North Star hours invariably suffer some form of hearing loss.

"Of greater importance, though, throughout its history, no serious North Star accident or incident can be attributed to a fault in the aircraft itself. There are many recorded cases of error on the part of people flying and maintaining it, but there were no basic deficiencies like those that eventually showed up in aircraft like the Tudor, DC-6 and Comet.

"We flew the North Star in the days before weather radar and pressurization, and standard procedures for handling severe icing or thunderstorms were simply to 'press on' and depend on the aircraft to get you through — and it always did."

The Yukon (and civil CL-44) proved one of the greatest successes of Canada's aircraft industry. In RCAF service, the Yukon was a great advance over the North Star in all areas—speed, range, payload, economy. (Larry Milberry)

Turbine-powered Canadair CC-109 Cosmopolitans joined 412 Squadron in 1959 and eventually supplanted the squadron's remaining North Star and the C-5. This "Cosmo" visited Windsor on September 19, 1959. (Larry Milberry)

The RCAF began operating its C-130B Hercules in 1964. Here 10303 makes a pass during the July 1, 1961, airshow at Trenton. The Hercules, Yukon and Cosmopolitan brought Air Transport Command right up to date and signalled curtains for the veteran North Star. (Larry Milberry)

North Star 17515 as it has sat deteriorating at Rockcliffe for nearly 20 years. It is hoped this important piece of Canadian history and the rest of the National Aeronautical Collection will one day have a proper home. (via David Menard)

Old North Stars Never Die . . .

After over a decade of use, both the TCA and BOAC Canadair fleets were being replaced by the new turbines. North Stars and Argonauts were disposed of to a variety of independent operators. The aircraft enjoyed a new, if brief, career, mainly serving the tourist trade throughout Europe, though also in other regions.

Overseas Aviation

Formed in 1957 with Vickers Vikings, Overseas Aviation began flying Argonauts in 1958. These were needed as the airline expanded its IT (Inclusive Tour) services. G-ALHT, delivered in November that year, was the company's first Argonaut. The IT market was buoyant and Ron Myhill of Overseas was optimistic. At this time he also started Trans African Air Coach, introducing new low fares between Europe and Africa, using Argonauts 'HP and 'HT.

An interesting anecdote is told of 'HT. It was accepted from BOAC at Heathrow by Captain J. Burridge following a maintenance check. Because of the late time of day, a sign writer from Southend was brought along to paint the logo and cheat line on, as 'HT was scheduled to enter service the next day. Walking into BOAC's main hangar, the sign writer eyed his subject, borrowed a set of steps and went straight to work with his paints. Continuing steadfastly into the night, he finished the job and the Argonaut emerged the next day in the Overseas colours. Never was such a fast paint job seen in that hangar, and completed by one man at that.

The Argonaut fleet grew quickly and, besides flying ITs, was kept busy on general charter work as seventy-seaters to such far-flung places as Africa, Greenland, Japan and North America. On one trip, crews and equipment for a shipping company were carried to Hong Kong.

Charters for British tour operators were generally from Gatwick, but a series of holiday flights operated from Ringway to serve the Manchester area. To serve West German tour operators, Overseas set up a Hamburg-based office, primarily handling Mediterranean flights. A milestone was reached on June 14, 1960, when Overseas opened a big new hangar at Gatwick. With this, its Southend maintenance activities were closed.

Over the summer of 1960 Overseas operated IT flights from Belfast, Birmingham, Gatwick, Manchester, Glasgow and Southend, though some of these were just stops on flights originating elsewhere. For instance, flights from Belfast were joined up with flights from Glasgow to make up a full load. Thus, on June 12 G-ALHK operated Belfast-Glasgow-Palma, departing Renfrew airport, Glasgow, at 0656. One hesitates to think at what time the Ulster people had to leave their beds to meet their aircraft! However, the roar of four Merlins may have brought some joy to those at Rolls-Royce just south of Renfrew, where the Merlin was being overhauled by this time. The last of this series of summer vacation flights terminated at Belfast and Glasgow on October 2, with 'HD providing the service.

Through 1960 the Overseas Argonauts flew a varied route structure from Manchester, beginning June 4 when 'HD left Ringway for Perpignan. The same week 'HG left Ringway for Palma. Operating for Overseas from Ringway, OY-AAH of the Danish firm Flying Enterprise departed July 16 for Keflavik, thence to North America, returning next morning! The Manchester season ended September 26 as 'HD arrived from Pisa.

Meanwhile, Gatwick was humming with Merlins. Overseas dispatched flights to Basle, Copenhagen, Gothen-

North Stars for sale. This advertisement appeared in the British aviation press in 1961.

An ex-BOAC Argonaut at Newcastle in Overseas-Trans African Air Coach livery, August 6, 1959. (Ian MacFarlane)

burg, Lisbon, Luxembourg, Nice, Ostend, Palma, Perpignan, Rome, Tangier, Vienna and Zagreb. Ad hoc charters went to Amsterdam, Athens, Beirut, Brussels, Cardiff, Malta, Stockholm, Tel Aviv and Turkey. ITs from Germany were leaving from Berlin, Hamburg, Hanover and Munich. Also in 1960, Overseas Argonauts did trips for the Dutch airline Martin's Air Charter, and many TAAC flights were conducted to Lourenço Marques and Johannesburg.

Business was good the following year as well. Glasgow operations moved to Prestwick, likely to take advantage of better weather and longer runways. That summer Argonauts departed from Belfast/Prestwick to Barcelona, Perpignan, Rimini and Valencia. Flights from Birmingham ran to Naples, Nice, Palma, Rimini and Valencia. Four Argonauts worked out of Manchester to Malaga, Naples, Nice, Palma, Perpignan, Rimini, Tangier and Turin. From Gatwick, the Argonauts plied the airways from Copenhagen to Lyons to Zagreb.

As business seemed to be increasing, Ron Myhill and his directors felt that more aircraft were needed. BOAC had sold all its Argonauts to Overseas and other companies, but TCA was just selling off its North Stars. Overseas arranged to acquire 15 North Stars and spares for $660,000. The North Stars began arriving in the UK on June 2 and by August 12 the following had shown up at Gatwick: CF-TFE, 'J, 'K, 'L, 'M, 'N, 'O, 'Q, 'T, 'U and 'V. (Overseas never took delivery of the other four aircraft.) This was an impressive growth rate for just a few weeks!

The arrival of the North Stars posed a few compatibility problems. The North Star and Argonauts had dissimilar crossover exhausts, auto pilots and seating, and overhaul schedules for engines and propellers were different.

The British Airworthiness Registra-

Overseas Argonaut 'HT showing the starboard crew door open. Later OY-AAH of Flying Enterprise, it was eventually scrapped at Redhill (Air-Britain)

G-ALHT at London on September 9, 1957. It carries the Overseas/Trans African Air Coach titles, but still has its tail in the basic BOAC scheme. After serving with four different airlines, this Argonaut was scrapped at Redhill in 1966. (Peter Keating)

One of the longest line-ups of Canadairs, 11 aircraft all from the Overseas fleet. An Overseas Vickers Viking sits in the foreground. (via Stephen Piercey)

This Argonaut became famous for returning Queen Elizabeth to England following her father's death. It is seen here at Istanbul on March 11, 1960. Later OY-AFB of Flying Enterprise, it was scrapped at Copenhagen airport in 1965. (Peter Keating)

CF-TFO, used by Overseas on a Canadian C. of A., seen at Baginton, May 19, 1965, where it was scrapped by Keegan Aviation. (Ian MacFarlane)

tion Board had validated the North Star Certificates of Airworthiness for just a few months pending the establishment of guidelines which Overseas had hoped would allow them to be certified in England. Thus the first North Star to arrive was in service within seven hours of setting down at Gatwick. In fact the ex-TCA North Stars were never certified in the UK.

On July 31, 1961, Overseas inaugurated the first scheduled passenger service between Prestwick and Gatwick, stopping at Manchester. CF-TFM was used and carried 28 paying passengers plus guests. This was a "walk on" service operating three days a week. Unfortunately the idea did not catch on, and on a total of 12 flights only about 100 passengers were carried. The service was dropped August 11. But worse news was threatening, for at a hearing of the Air Transport Licencing Board in August, Overseas' financial status was revealed. Things were not good and the company's application to serve Gatwick-Porto Santo (Madeira) was rejected. A few days later British Petroleum cut off all fuel supplies to Overseas due to outstanding debts.

Over the weekend of August 12-13 Overseas continued operations, with CF-TFK and CF-TFT carrying passengers to and from Birmingham and Nice and Palma, but within two days most aircraft were too low on fuel to go very far. On August 14 CF-TFL operated Gatwick-Basle, followed by the return from Zagreb of G-ALHS. CF-TFV departed for Copenhagen, followed shortly by 'HS. A North Star touched down at Gatwick from Munich, followed by CF-TFK coming in empty from Birmingham. Later in the day CF-TFT and CF-TFO also showed up and by day's end Overseas had virtually ceased flying.

In October the North Star fleet was sold to J. Gaul for resale. They were ferried to Coventry for storage until resold or scrapped.

At its peak Overseas operated 16

Paul Tomlin took this view of eight Overseas Aviation Argonauts and North Stars at Gatwick on August 28, 1961, soon after the demise of the big UK tour operator. (via Stephen Piercey)

Argonauts and North Stars and at one time or another possessed 9 Argonauts and 11 North Stars, making it the fourth largest operator of the type just behind TCA, BOAC and the RCAF. Over the 1961 season Overseas had been Britain's second largest independent airline, next to British United Airways.

Derby/British Midland Airways

In the fall of 1961 Derby Airways purchased the five ex-Overseas Argonauts, G-ALHG, 'HN, 'HP, 'HS and 'HY. At the time Derby was a small airline flying DC-3s. The deal gave the Lombard Bank a sizable interest in the airline, which seemed wise at the time as it enabled the airline to buy the Argonauts very cheaply. Later the company would regret its enthusiasm as it learned how expensive the aircraft were to maintain.

So it happened that on October 5, 1961, the first really big aircraft to land at Burnaston airport, Derby, was G-ALHS, with Captain Van den Elst and F/O "Bluey" Wilson aboard. What the captain's thoughts were as he approached the pocket-size grass strip for the first time can only be guessed, but a beautiful landing followed and with reverse pitch the Argonaut came to a quick stop.

Early in November a start was made on the first proper training school set up within the company. The Link Trainer room of Air Schools Ltd. was revamped as the Argonaut ground school, with Merlin cutaway components and various other Argonaut instructional aids. Aircraft G-ALHG, 'HS and 'HY were soon being readied for service, while 'HN and 'HP were stored.

Formerly *Astra* of BOAC, 'HS was sold to Derby Airways in 1961, serving there and with BMA until withdrawn from use in October 1967. (A.J. Jackson)

Derby personnel photographed with 'HS: W/C Roxboro (chairman and joint managing director), Mary Agar (financial director), Denis Sullivan (commercial director) and R.R. Paine (technical director). Denis Aldridge (charter manager) is in the doorway. (via B.G. Cramp)

On February 27, 1962, Argonaut crew training began. G-ALHS under command of Captain Van den Elst flew Birmingham-Jersey return. Also aboard were Captains Pike (chief pilot) and B.G. Cramp (chief training captain), carrying out upper air work on their type conversion en route. Jersey was a convenient "aiming" position, but also had duty-free fuel available!

During training, certification checks of DC-4 flap jacks were undertaken. As Canadair jacks were scarce, the ARB allowed experimental fitting by Derby of the DC-4 parts. Tests required the flaps to be extended and to remain so well beyond normal limiting speeds. They worked perfectly, solving one spares shortage.

On March 17 Captains Pike and Cramp were on their way to Aden via Cairo in G-ALHS which had been dry leased to Aden Airways. Under command of experienced Argonaut captain Spenser of Aden Airways, line training was given to the Derby captains. Over the Alps, near Mont Blanc, the outside air temperature was so low that the control column could be moved a good 1½ feet either way from neutral before the aircraft would respond!

Pike and Cramp returned to England aboard VR-AAT, an Aden Airways Argonaut commanded by Captain Peter Williamson, that was on its way to Wymeswold in Leicestershire for maintenance at Field Aircraft Services. It happened that a BOAC Comet had ingested a stone at Khartoum and had been ferried to Aden where there was a spare engine. The replacement engine then suffered the same damage so the plane was sent to Nairobi where there was another spare. Amazingly, the same mishap occurred for the third time. To look after their Aden-Cairo and Aden-London passengers, BOAC booked these long-suffering souls on the returning Argonaut. This was likely the last time that BOAC passengers were carried on an Argonaut.

On the first sector to Cairo, Pike completed his Argonaut training. Cramp operated Cairo-Malta for a night stop, then on to Heathrow where the BOAC passengers deplaned. VR-AAT flew on to Wymeswold next morning, by which time Cramp was also fully qualified on the Argonaut.

Training now resumed using G-ALHY, but on one trip there was serious trouble. Captain Van den Elst was doing type conversion from Coventry (Baginton) with F/O Dave Cowell as trainee. Soon after one takeoff a starboard propeller went out of control and began overspeeding. It resisted feathering so the captain was faced with a discing airscrew which greatly increased drag. Straight and level flight was maintained and the crew noticed a runway ahead. Fearing that the propeller might break away and seeing that the runway and circuit were clear, they decided to land. The runway was at RAF Station Gaydon, a V-bomber base, and this Wednesday afternoon was sports time. One can imagine the surprise of the many people out "sporting" on the runway when, hearing the screaming engines, they looked up to see an aircraft, gear down, bearing down upon them. Fortunately everyone got clear and 'HY landed, at which point the engine seized and a broken connecting rod came through the side of the crank case.

On April 30, 1962, VR-AAS set off for Aden from Heathrow with a Derby crew which would ferry back G-ALHY, by then needed for Derby's summer charters. With full tanks, some Aden Airways staff and a load of spares aboard, the Argonaut climbed laboriously towards Dieppe. It was approaching 15,000 feet when the cabin heater fire warning light began to flicker. Engineer Norman Scott went aft to check. Then the warning light lit up brightly and the appropriate fire drill was carried out. Scott reappeared in the cockpit saying, "Punch the fire button. You can fry an egg on the cabin floor." With a fire aboard, fuel could not be dumped, but Captain Cramp made an emergency descent and warned Heathrow of his predicament. An over-weight landing was successfully carried out. Oddly, no sign of fire could be found when the aircraft was inspected.

Next day the flight reached Cairo uneventfully. Uneventfully, that is, until approach. A sand storm was blowing and just before touchdown, oil drums were seen to be littering the runway threshold. An overshoot was made, then a landing on another runway. Never a dull moment! On May 3 VR-AAS arrived in Aden and Captain Van den Elst with F/O Cowell ferried 'HY back to England.

As the runway was so short at Burnaston, Derby could not operate commercially from there, so facilities were arranged at Birmingham. Initial charters in 1962 were troop flights for the Ministry of Defence to Germany, as well as to the rocket range in the Outer Hebrides. On one troop flight G-ALHS landed at Glasgow. There an oil leak in No. 1 engine meant that a

"Hotel Yankee" at rest at Newcastle. (Ian MacFarlane)

three-engine ferry flight had to be made to Prestwick for repairs, the troops following by road, their commanding officer none too pleased. A minor part had to be flown in from Derby, and by this time a night stop for everyone was inevitable. Unfamiliar with the Argonaut, service people at Prestwick caused further damage. Now a three-engine ferry flight back to Burnaston was necessary and the C/O was more annoyed than ever. Repairs were completed, the aircraft returned to Prestwick, and the troops arrived at Dusseldorf the same day. Later in the year G-ALHY participated in ferrying troops from Bodø in northern Norway. The descent to land there was an anxious one due to severe icing, but at 3000 feet the ice began to fly off the aircraft in big chunks. A York passing through the same ice conditions lost all its aerials.

The Derby crew found accommodations scarce in Bodø and were rather dismayed to find the best available meal to be tough reindeer meat. But the locals were hospitable and a good night stop was enjoyed. A load of Royal Marines was flown to Exeter the next day.

Derby Airways had discovered early on that its Argonauts had a trim problem: they were tail heavy. To rectify this, the company's third aircraft, G-ALHG, was modified. The horseshoe-shaped lounge at the rear pressure bulkhead was removed and replaced with toilets. Two more windows were cut in the pressure hull to allow for an extra row of seats at the front of the cabin. The trim problem was relieved and seating rose from 75 to 78 passengers. At this time 'HG was painted in Derby's new livery, an attractive scheme that became the basis of today's British Midland Airways colours.

G-ALHG's first work was a number of flights operated for BEA along the Berlin Corridor over the 1962 Christmas season. On these flights, passengers complained that the cabin was cold. The cause turned out to be a flaw in the air conditioning. When the pilot selected a higher temperature, a lower one resulted.

Independent airlines such as Derby Airways never enjoyed the lavish facilities and services of the big operators. This became quite clear to Pike and Cramp at Le Bourget one snowy February day in 1963. When they arrived at their Argonaut they found it covered

G-ALHS with No. 1 feathered over Windermere en route Prestwick-Derby piloted by Captain Bert Cramp. (B.G. Cramp)

VR-AAT, formerly G-ALHV, scrapped at Burnaston by BMA. (via Peter Keating)

"Hotel Sugar" wonderfully pictured in its new-style Derby livery while flying near Birmingham, August 28, 1964. Note the extra cabin window near the inboard propeller. This was cut in when Derby was expanding seating forward. (via B.G. Cramp)

with snow. Airport staff were unwilling to assist. Unperturbed, the pilots borrowed brooms, clambered through the over-wing emergency windows, and swept their wings clear. Rarely had so much gold braid been so hard at work! The job done, ground crews de-iced the wings and 'HY got away, albeit two hours late.

In April 1963 it was announced that Derby would be carrying out a major series of charters from Cardiff to Barcelona and Majorca. Nine flights a week by Argonaut were expected. This bonanza unfortunately was not to be. On one early flight to Barcelona two engines failed. A landing was made at Hurn and two Dakotas completed the charter. In an incident later that July, 'HY had an exhaust system failure, forcing Cramp to make a tricky three-engine landing at Burnaston, complicated by the failure of reverse pitch. The aircraft stopped at the very end of the tiny field. These incidents were harbingers of greater difficulties to come.

Because of such flight delays, the Cardiff-based travel agency transferred its tours from Derby to Cambrian Airways. Derby now took a close look at its Argonauts. That their Merlin engines were deliverying progressively less power was evident from the engine instruments. One could discern that something was amiss by listening to the engines at shut-down. There was a chuff-chuff-chuff sound and, in fact, crews would comment that they had another "chuffer" on their hands. The noise was caused by individual cylinders on the compression stroke, when the exhaust valves in that cylinder had had part of the valve head burnt away by the hot gases escaping during the exhaust stroke. During the compression stroke the mixture, instead of being compressed, was simply pushed out of the cylinder through the sometimes quite large hole in the exhaust valve. Not only was there a severe power loss but the unburnt mixture would ignite in the exhaust manifold when it mixed with the exhaust gases from the other cylinders, causing disintegration of the exhaust system as had happened to 'HY on July 26.

The company studied its procedures to be sure that they were not at fault, and Rolls-Royce analyzed its manufacturing and overhaul techniques. Hundreds of instrument reading sheets were studied to see if the Derby crews were operating the Merlins within the Rolls-Royce parameters. Captain Cramp made many trips to Rolls-Royce at Hillington to meet with the manufacturers.

By this time it was apparent that the power control charts acquired by Derby were not adequate. Cramp asked Rolls-Royce for better data, but in the end had to conduct his own research program. Within three months he had produced the necessary chart, which was approved by Rolls-Royce and was used by Derby as long as its Argonauts were in service. Meanwhile Rolls-Royce was conducting an exhaustive quality-control program on Merlins yet finding that some engines were lasting a mere 35 hours before serious problems arose. Then, in September 1963, a Rolls-Royce boffin came up with a possible answer. He noticed that on the base of each exhaust valve stem, where it made contact with the rocker arm, the shape left by the rocker was oval instead of round. He suggested that an oil additive was forming a shellac on the valve stem which, when the engine was hot, became sticky and prevented the valves, especially the exhaust valves, from rotating due to "valve bounce" when the rocker had completely lifted off the valve at the end of the exhaust and beginning of the induction stroke. Further investigating would take months and Derby could not wait. Chief Engineer Frank Marshall ordered fleet-wide draining of the suspect detergent oil and its replacement with a non-detergent type previously used. The results were dramatic as the trouble ceased almost overnight in the Argonauts and the company's Pratt and Whitney-powered Dakotas. The next Merlin snag did not occur until three weeks later, and this was a minor problem. In all, the use of detergent oil was blamed for 42 premature Merlin changes at Derby.

As a result of the research done by Derby and Rolls-Royce, a joint legal action on their parts against the oil manufacturer was satisfactorily settled out of court, but not before Derby's reputation for reliability had been damaged severely, especially in south Wales and the West Country, where lost business was never recovered. Where Derby Airways had pioneered, other airlines were to reap the benefits.

On August 17, 1964, G-ALHS was used in the filming of *The High Bright Sun* at RAF Northolt. The aircraft was sprayed with a thin rubber solution over which the old BOAC livery was painted. The filming was done with Captain Cramp, F/O Neil Fagg and E/O Jed Bowker crewing. At day's end the BOAC markings were to be removed as the aircraft was required for Derby service the next day. The plan to strip the temporary markings off at Northolt didn't work out and when 'HS was due back at home base for a minor check, men were still trying to get the BOAC livery off. Abandoning this operation, the captain took off for Derby, masking tape and rubber strips flapping all over the aircraft and creating the din of a hailstorm. Once there the aircraft was finally cleaned up, though not without the help of several employees' wives. The Argonaut was back on the line next morning.

On October 1, 1964, Derby Airways, the operating division of Derby Aviation and part of the Air Schools group, changed its name to British Midland Airways Ltd. Soon the company's Argonauts began appearing in a new livery.

For six and a half years the Argonauts of Derby/British Midland provided good service, plying routes throughout Europe and the Middle East. By 1967 they were nearing the end of their useful lives and BMA was re-equipping with Viscounts. Then the company's fine record for safety was interrupted. On June 4, 1967, G-ALHG, coming in to Manchester from Palma, crashed disastrously at Stockport. In all, 69 passengers and 3 crew died; 12 others survived.

The Argonaut was about to land on Runway 24 at Ringway Airport when No. 4 engine failed, then No. 3. The aircraft began a right-hand 360-degree turn but by the time the turn was nearly complete, so much height had been lost that Captain Marlow had to make a crash landing. He selected the only clear space available, a bit of waste land in a built-up area, thus avoiding some large apartment blocks. Even so, disaster ensued.

The main cause of the accident was found to be a problem in the fuel tank selection system which allowed the rotary fuel cocks to be misaligned by as much as 15 degrees so that fuel could be transferred from the main fuel tanks to the auxiliaries without the crew realizing it. Exhaustive tests were conducted by the Board of Trade and the Royal Aircraft Establishment

G-ALHS at Manchester in British Midland Airways livery. (via Peter Keating)

with the full cooperation of BMA, using Argonaut 'HY as needed. Although the airline was cleared of responsibility in the accident, the remaining Argonauts were withdrawn from use, 'HS on October 16 and 'HY on November 6. Both were sold to Chartwell Aviation on October 19, 1968. Neither aircraft flew again, though during the Biafran war, 'HY was purchased for use, it was later realized, as a gun runner. The aircraft was actually about to take off from East Midlands airport for Zurich when it was successfully called back to the ramp and held on the ground.

Before leaving the Derby/BMA Argonaut story, one last anecdote bears telling. In an effort to find an inexpensive replacement for its Argonauts, BMA did extensive research into the possibilities of re-engining its Argonauts with the Rolls-Royce Dart turboprop. The airline was amazed to find that the Merlin and Dart pick-up points were identical. However, this idea was dropped, even though the airframes, especially the wing spars, had plenty of time remaining.

Flying the Argonaut

Captain B.G. Cramp writes this first-hand account of a typical Derby Airways Argonaut flight:

The Derby-based crew have been night-stopping in the Excelsior Hotel at Manchester Airport, only a short ride from the crew report station. The aircraft has not yet arrived and is running 20 minutes late, so there's no pressure on the crew, as yet.

The flight deck crew consists of captain, first officer and engineer officer. The E/O, or "flying spanner," is not a requirement on the Argonaut but many operators include him especially for operations into airfields without a base engineer or where limited engineering facilities exist. For today's flight no navigation officer is needed.

Once at the airport, the captain receives passenger and freight information from the handling agent and quickly calculates his dry operating weight. He orders the E/O to load full fuel (1634 Imp. gal.) once the aircraft comes on chocks. The pilots examine and discuss the meteorological situation and decide on operating altitude. As the later stages of the flight are over the Pyrenees, the Argonaut will have to fly above 12,000 feet, so that a change of supercharger gear will be required at some stage. The flight could stay low for the first two-thirds or so of the flight or go high straight away. For today's flight the latter course is chosen. The F/O goes off to work out his flight plan for 15,000 feet.

Once the aircraft arrives the passengers disembark, baggage is unloaded, and Customs is cleared. The E/O checks the technical log entries if required and discusses any snags with the inbound crew. Servicing and fuelling proceed, then the E/O begins his pre-flight. Meanwhile the pilots arrive at the aircraft and begin the comprehensive pre-flight check. About this time the passengers begin boarding and, when ready, the senior hostess reports that all cabin checks are completed and the passengers strapped in. All doors are closed and locked. The E/O passes the technical log to the captain for signature and hands a copy to the traffic officer who has boarded via the crew entrance with the Customs papers and Load/Trim Sheet. The captain passes him a signed copy of the required papers and he leaves the flight deck. The E/O then hauls in the small red bag containing the undercarriage ground locks, counts the locks, stows them and reports "All locks on board and

Derby/British Midland Airways Argonaut Operations: Summary

Inclusive tours
Cardiff and Bristol to Palma, Barcelona, Valencia, Rimini, Alicante, Tarbes, Ostend, Perpignan.
Manchester to Genoa, Ostend, Klagenfurt, Perpignan.
Birmingham to Barcelona, Ostend, Palma, Perpignan, Tarbes, Venice.

Scheduled services
East Midlands Airport to Dublin, Belfast, Leeds/Bradford, Glasgow, Jersey, Ostend.
Manchester to Ostend.
Birmingham to Jersey.
Luton to Jersey.

Captain B.G. Cramp at the controls during his pre-Argonaut days. (via B.G. Cramp)

Date	*Argonaut*	*Route*	*Time en route*
Sept. 3	'HY	Manchester-Dublin	1:15 hours
		Dublin-Tarbes	4:00
		Tarbes-Dublin	3:35
		Dublin-Derby	1:10
7		Cardiff-Bristol	:20
		Bristol-Barcelona	3:30
8		Barcelona-Bristol	4:05
		Bristol-Cardiff	:20
9		Cardiff-Ostend	1:25
		Ostend-Cardiff	1:35
10		Cardiff-Ostend	1:20
		Ostend-Bristol	1:35
		Bristol-Cardiff	:20
		Cardiff-Derby	:40
22	'HS	Manchester-Perpignan	3:45
23		Perpignan-Manchester	4:15
Oct. 19	'HY	Derby-Manchester	:35
20		Manchester-Bodø	6:10
21		Bodø-Exeter	7:20
22		Exeter-Derby	:55
23		Derby-London	1:00
		London-Elmdon	:40
27	'HS	Elmdon-Tarbes	3:00
28		Tarbes-Elmdon	3:45
Nov. 2	'HY	Birmingham-Hanover	2:25
		Hanover-Birmingham	2:40
Dec. 7	'HS	Birmingham-Hanover	2:30
		Hanover-Birmingham	3:10
14	'HY	Birmingham-Derby	:30
21	'HG	Hanover-Berlin	1:10
		Berlin-Munich	2:15
		Munich-Berlin	2:30
22		Berlin-Hanover	1:05
		Hanover-Berlin	1:00
23		Berlin-Munich	2:10
		Munich-Birmingham	3:15

A sample of Derby Airways Argonaut operations in 1962 from the log of Captain B.G. Cramp. During this period Cramp alternated between the Dakota and Argonaut.

stowed." The flight deck door is closed and locked.

The captain now calls, "Starting check," and having received ATC clearance starts No. 3 engine followed by Nos. 4, 2 and 1. Taxi check complete and clearance received, the Argonaut moves to the runway holding point and calls for the run-up and takeoff checks. The aircraft is now lined up on the runway, and with clearance received, the brakes are released. The captain operates the throttles and nose-wheel steering; the F/O pushes the control column forward to ensure good nose wheel traction and so the takeoff begins.

As the aircraft gathers speed the F/O calls out the speeds and at V_1 takes over the throttles while relinquishing the control column to the captain. At V_R the control column is pulled back, raising the nose wheel, and the Argonaut is in the air. As height is gained, the captain orders, "Gear up. After takeoff check," and, at 400 feet, "Flaps up." The F/O then reports, "After takeoff check complete," and climb power is called for and set.

As the aircraft climbs, the throttles are gradually opened to maintain the required manifold pressure until at around 9000 feet the change of supercharger gear from moderate supercharge to full supercharge is required, so long as the aircraft will be exceeding 12,000 feet. To go to FS, power must be reduced to lessen the load on the clutch mechanism springs. The captain informs the passengers then throttles back to 25"hg/2000 rpm. A deathly hush surrounds the aircraft at this point. When all four engines have changed gear, the master propeller control lever is set to 2650 rpm and the throttles opened to 48" and climb is recommenced. During the gear changing the aircraft loses speed and perhaps as much as 900 feet of altitude. The climb to altitude will be long and laborious. Once at height the aircraft levels off. The throttles are set full open and rpm reduced using the master lever to give 950 bhp.

The aircraft is responsive to elevator and rudder, but powerful biceps are developed when manually banking as the ailerons are quite heavy. Normally, though, the auto pilot is used.

The aircraft is not equipped with weather radar and a rough ride across the Pyrenees is always a possibility. Hail may be met with in this area, sometimes bending radiators and intercooler matrices so as to cause overheating.

As the aircraft nears Barcelona, ATC is called up for descent clearance.

A BMA Argonaut at Newcastle as as Ambassador departs. (Ian MacFarlane)

Once it is received, the captain calls for descent checks, the engines are throttled, the supercharger gear changed again, and descent begins. Approach and landing checks come in turn. The landing gear is lowered, the rpm (and noise) increased, and the aircraft prepared for landing. The captain has briefed the F/O about setting up the radio navigation aids and about reverse power needed on landing. On final, the captain has both hands on the control column and calls out the desired engine settings, sometimes increasing these if speed falls off too much.

Depending on conditions, the captain, upon reaching the runway, will call "Cut" or "Slow cut" and the F/O will pull off all power appropriately.

With all wheels down, the electrical circuit is made by pressure on the nose wheel to permit the reverse flap to be lifted, whereupon the F/O will lift the flap and await the order to go into reverse. At this point the Merlins are giving their characteristic pop and crackle in the exhaust system; and when the "Reverse power" order is given a distinct "plop" is heard as the propeller blades go into negative pitch, followed by a roar as the throttles are pulled through the reverse gate and into reverse power.

In reverse there is a large reduction of cooling air through the radiator matrix so rapid overheating is possible. Reverse is therefore quickly cancelled and the captain calls, "After landing check." Once on the ramp, the parking check is read off and the engines stopped. All necessary log entries are completed and, as the aircraft is scheduled to go on to Palma, the captain tells the E/O what fuel he wants and instructs the F/O to carry out the external and internal transit checks. Next, the whole process is repeated, sometimes for as many as four sectors in a duty day.

Captain B.G. Cramp

Air Links/Transglobe

Air Links was formed in 1958 and began operations the following year with Dakotas from Gatwick. Soon the Dakotas were sold and, hard pressed to supplement its lone Hermes, Air Links acquired the Argonaut OY-AAH from Flying Enterprise in January 1964. Brought up to standards at Gatwick, it was reissued its original registration G-ALHT. On March 24 it was flown to Bournemouth (Hurn) for a day of crew training and more training followed the next day at Stansted.

Air Links' first Argonaut revenue flight occurred March 28 when 'HT flew Gatwick-Ostend return. On its next charter, April 18, it flew Gatwick-Cherbourg return. Later in April it also flew to Milan and Rotterdam. The firm's second Argonaut, G-ALHI, was acquired from the Royal Rhodesian Air Force and arrived in June. Because of maintenance and modification work needed, it did not

G-ALHI in Air Links markings at Gatwick after its return from service in Rhodesia. (John Wegg Collection)

enter service until August 15 when it flew a Gatwick-Algiers charter. Through the rest of the summer these two aircraft were busy flying to Barcelona, Basle, Brussels, Le Bourget, Marseille, Nice, Ostend, Palma, Perpignan, Rimini, Tarbes, Valencia and Zurich from Gatwick. Some flights originated from Glasgow and there were others for the Ministry of Defence from Gatwick to Gibraltar and Hanover. There were also flights to the Middle and Far East, usually using Brindisi on the heel of Italy as a technical stop.

Also in 1964 the following Argonauts were acquired: G-ALHW (ex-RRAF, delivered in July), G-ALHM (ex-EAAC, delivered in August), G-ALHC (ex-RRAF, delivered in September) and G-ALHH (ex-RRAF, delivered in August). Of these, 'HC and 'HH were not to see further use. They were ferried from Gatwick to Redhill where they were scrapped. 'HM and 'HW were converted as 78-seaters, and 'HW was also modified as a freighter.

In 1965 Air Links was an all-Argonaut operation, as the Hermes had been withdrawn from use. British aviation history researcher Tony Merton Jones has chronicled the Air Links story and part of his work, reprinted here, outlines Air Links' Argonaut operations in 1965:

European sea ports had become familiar stopping off points for the Argonauts during the winter, and destinations like Genoa, Gothenburg, Athens, Venice and Antwerp all appeared in the aircraft log books for this period. With the approach of summer, familiar destinations began to reappear, with the first departure of the season to Tarbes taking place on April 15, and over the Easter weekend G-ALHI flew several Gatwick-Beauvais passenger services alongside a Shreiner Airways Friendship. Also on April 15, G-ALHW began a weekly freight service from Gatwick to Metz, France, and this service continued in operation until June 11.

On Saturday, April 24, 1965, G-ALHM flew the first of several day trip charters from Gatwick to Rotterdam. These were flown in conjunction with Dan-Air, who supplied two Dakotas and two Skymasters. First away on April 24 was Dan-Air's Dakota G-AMSS with the Argonaut taking off in second place. The 78 passengers aboard the Argonaut, though, were generally the first to arrive thanks to a good turn of speed provided by the four Merlins. These flights continued until May 9, with Britair's Constellation 5Y-ABF joining the operation for the last two days. This aircraft also operated one of Air Links' freight services to Metz during May. The spring also saw G-ALHT operating a series of Southend-Ostend schedules for British United Air Ferries in place of Bristol Freighters and Carvairs.

The summer inclusive tour charter program was impressive, and the three passenger aircraft opened the season in high spirits. In addition to the contract tour work, a great deal of ad hoc work was also flown, and Ministry of Defence troop charters provided welcome additional revenue. The Defence work mainly consisted of Gatwick to Düsseldorf, Frankfurt and Hanover charters, together with a number of Gutersloh-Manchester flights, and the now traditional Düsseldorf-Benbecula runs. Sometimes these flights were made direct, although on most occasions stops were made both northbound and southbound at Prestwick in order to clear customs.

Inclusive tour work began to build up in May, and on May 23 G-ALHT inaugurated a fortnightly Exeter to Barcelona service which was flown until October 10. Regular holiday flights linked Gatwick with Barcelona (commencing May 29), Nice (May 26), Guernsey and Jersey (July 6), Palma (May 29), Perpignan (June 12), Tunis (May 30) and Venice (May 20). Other destinations irregularly served during the summer include Basle, Beauvais, Luxembourg, Lyons and Rimini. Pilgrim flights to Tarbes operated at intervals throughout the summer, and departures to this airport were now made from Gatwick, Glasgow, Leeds, Manchester, Prestwick and Teesside. Ad hoc charter work continued to be varied in nature with G-ALHW flying a freight consignment from Tirstrup to Manchester on June 22. G-ALHT flew several services to the Italian Air Force base of Villafranca in late May and June.

The freighter Argonaut G-ALHW did not fly as often as any of the other three, and its main source of revenue was from the operation of flights between Gatwick, Dublin and Zurich and the occasional ad hoc charter. Two flights from Gatwick to Benina were made by G-ALHT early in June, with the aircraft staging outbound via Nice, and returning direct

'HM with Air Links seen at Newcastle, August 27, 1965. It was later purchased by BMA and used for spares. (Ian MacFarlane)

Following service in Rhodesia, G-ALHW operated with Air Links and is seen here at Gatwick on August 24, 1965. (Peter Keating)

to Gatwick. A Shannon-Milan flight was undertaken by 'HW on July 23, and amongst the other charters was a Gatwick to Sofia (Bulgaria) charter flown on July 10 outbound, returning three days later.

On August 1, 1965, Air Links was renamed Transglobe and the company began operating its first Britannia. This meant the end of the Argonauts. G-ALHT was sold in July, leaving three aircraft in service during August. Then another Argonaut was withdrawn and only 'HI and 'HM were in use. In October 'HW went to Redhill, thence to East Midlands where BMA used it for spares. On October 3 'HI flew its last Transglobe charter, a flight to Tunis. Ten days later it too made the 6-mile hop from Gatwick over to Redhill, leaving only 'HM in use. On October 23 'HM made the last Transglobe Argonaut revenue trip, flying to the Far East via Brindisi. The aircraft was back at Gatwick November 2. At 1430 on November 4, 'HM departed for Redhill and the wreckers' hammers.

Tony Merton Jones

Flying Enterprise

In January 1960 a new charter airline was formed in Denmark. Based at Kastrup, Copenhagen, Flying Enterprise was helped in its formation by Overseas Aviation and Ron Myhill, but had a Danish managing director, Hans Linde.

Flying Enterprise began by leasing Argonauts from Overseas, and the first to arrive, on January 20, 1960, was G-ALHT, now registered as OY-AAH and named *Robert Svendsen.* Three months later, on April 30, 1961, G-ALHU arrived at Kastrup as OY-AAI, *Alfred Nervo.* Charters were soon under way to the usual Mediterranean inclusive tour stopping-off points and to other destinations, including some for Overseas. On one of these latter trips, OY-AAH left Manchester, July 16, 1960, for North America via Keflavik, returning the next day. During this period, Flying Enterprise became one of the few West European airlines allowed to operate in East German air space.

On January 10, 1961, G-ALHN, now OY-AFA, *John Foltman*, was delivered to Kastrup. At month's end G-ALHK, now OY-AFB, *Ludvig Bjarkow,* arrived to complete the Danish fleet. On April 20 OY-AFA returned to Overseas to be replaced by OY-AFC, bearing the same name as 'AFA. When Overseas ceased operations in 1961, a large supply of Argonaut spares was shipped from Gatwick to Copenhagen and Overseas' chief pilot, Jack Burridge, joined the Danish airline.

One unusual trip flown at this time by Flying Enterprise was to a tiny strip on the east coast of Greenland. The arrival had to be timed precisely because of the brief daylight period and there were no navigation aids at the strip. Nonetheless the Argonaut got in and out safely.

In September 1962 the four ex-Overseas Argonauts were sold to Flying Enterprise by the liquidators. By

G-ALHI as it ended up at Stansted in 1976. (via John Wegg)

OY-AAI of Flying Enterprise in an alternate colour scheme, seen at Nairobi on January 21, 1963. (Peter Keating)

January 1964 Flying Enterprise had been reduced to three aircraft, OY-AAH having been sold to Air Links and reverted to G-ALHT. By this time the company was feeling the pinch from flying and maintaining the Argonauts. So it happened that in March 1965 Flying Enterprise wound up its operations and in July that year its fleet was scrapped at Kastrup.

OY-AAI and OY-AFC after being reduced to scrap at Kastrup, Denmark. (A.J. Jackson)

From BOAC to Overseas to Derby, G-ALHP provided reliable service along the airways from September 1949 until withdrawn from use over 12 years later. Here it bears the combined Overseas and Trans African Air Coach titles. (Peter Keating)

Glory to Dishonour

Former TCA and Overseas North Star CF-TFM became the subject of various enquiries while in the UK and it eventually ended up most unfortunately. On January 4, 1963, Keegan Aviation at Panshanger Aerodrome, Hertford, wrote to the Canadian DOT in the hope of speeding proceedings to get the aircraft flying again. In part the letter reads, "We would be sincerely obliged to you if you could peruse this document and advise us whether in your opinion if the work specified were carried out competently . . . if upon completion of said work you would then consider the aircraft to be in a fit state of serviceability for the issuance of a Canadian Certificate of Airworthiness." It goes on to explain that CF-TFM and other North Stars were about to be sold in Spain and that "unless we can receive such a statement from you these aeroplanes . . . will have to be scrapped completely which, apart from being an extremely sorry ending to the North Stars, would soon become common knowledge in Europe and possibly adversely effect the resale in the future of any Canadian registered aircraft."

The DOT replied on January 16 that certification would be no problem had the aircraft "been maintained in accordance with our approved TCA continuous maintenance system and had the aircraft been certified by TCA Engineers . . . however, not knowing the present physical condition of the individual aircraft, it is impossible to say whether this work would be adequate or not." The following week the DOT received a telex requesting permission to ferry CF-TFM to Spain for Trabajos Aereos del Sahara with a reply to be sent to a Captain V.G. Parkinson. Whatever the reply, the North Star did not go to Spain.

A further letter of October 19, 1964, from the Ministry of Aviation to the DOT requested registration information on CF-TFM for a solicitor and added that "this solicitor is acting on behalf of a Mr. Levy who is stated to have recently purchased CF-TFM at an auction held by the Sheriff of Northumberland at Newcastle Airport." The reply informed the Ministry that CF-TFM had been struck from the Canadian register on April 18, 1962.

The next entry of note from the file of CF-TFM is a DOT one: "On February 15, 1965, a Canadian North Star had been seized by the Dutch authorities at Amsterdam airport. Appears Panamanian registration painted on the side washed off in a rain storm revealing the old Canadian markings, CF-TFM." On June 21 the Chief Constable at Heathrow wrote the DOT asking for registration data for CF-TFM as he had been required "to undertake an investigation on behalf of the Director of Public Prosecutions in respect of a flight made by a DC-4M2 aircraft on 9th February 1965 from Newcastle Airport, United Kingdom, to Limburg, Holland, bearing the registration CF-TFM." The aircraft was impounded in Holland but along the way some maintenance was done on it there. It next appeared as BR-HBP and a news item reported it as sold to the King of Burundi. On December 19, 1965, it left Holland on a supposed test flight to Frankfurt, but ended up at Albenga, Italy, and reportedly conducted some clandestine flights over the next few months listed as I-ACOA. Its last flight was on October 11, 1966. While it was apparently carrying a load of machine guns reportedly from Prague via Palma to an African destination, it crashed in Cameroun.

These excerpts from "Gun-running via Birmingham" published in *Insight* give details of the clandestine operations of CF-TFM. The excerpts begin with reference to arms dealer Paul Favier.

. . . Paul Favier, now in his late forties, got an excellent training in the trade as a senior employee in the French Sureté Nationale. And ever since he decided to leave the public service for the private sector, he has, to judge from his mode of living, grown fat on its proceeds . . .

Favier's Nigerian shipment came from a mixed lot of British and American weapons given to the Dutch in 1945 which he bought from an Israeli entrepreneur named Arazi. Contacted in Tel Aviv last week Arazi—who had already profited from a complicated

CF-TFM seen at Malton in its heyday. This proud airliner was to end its career as a pile of rubble in West Africa. (Larry Milberry)

About to taxi for takeoff at Gatwick, July 1, 1961, CF-TFM served briefly with Overseas Aviation in 1961. (Peter Keating)

CF-TFM bearing its bogus Panamanian registration at Newcastle on February 5, 1965. (Ian MacFarlane)

barter deal between Israel and Holland—could remember only that the deal with Favier took place "some time ago." . . .

Soon after Favier completed the deal with Arazi, he started to ship 500 American Browning machine guns to one of Britain's largest "independents," Major Robert Turp, a 48-year-old ex-regular army officer, who achieved passing notoriety last spring when he rejected an offer to sell arms to Rhodesia. Turp reconditioned the guns at his Bexley Heath factory, sold them with British Government approval and split the profits with the suspicious Favier.

(When Favier arrived in London four months ago to collect his share, he curtly rejected Turp's cheque. So they went together to the National Provincial Bank in Oxford Street and drew £2,000 in cash. But Favier's caution rebounded, it seems; he went straight to the London branch of the Lebanese Intra Bank ["it's the only one I trust"] and deposited the cash: 10 days ago Intra closed its doors.)

But Favier was impressed enough by the Browning deal to suggest that they do a similar deal with the 3,600 sub-machine guns which were slowly rotting away in Dutch naval warehouses. Turp agreed, though the deal, he said, was naturally conditional on a Board of Trade import licence. (All weapons imports are administered by the Board of Trade, and controlled by the Ministry of Defence.)

Favier startled Turp by admitting that he already had lapsed licences for the guns. All they had to do was to renew them, he said. Favier's confidence was justified. Without any trouble, Turp had obtained within a month import licence number EX 193057/1, and the essential import certificate for presentation to the suspicious Dutch authorities, numbered NE 2253.

Not until the beginning of September did Turp discover that Favier had a client. Favier returned to London stating that he wanted to buy German M G 42 machine guns "for Nigerian clients in Geneva." Turp had the guns, but the clients wanted to see them first. The guns could not, however, leave the country without a licence, and the Nigerians could not come to Britain. "Why not show the military attache at the Nigerian High Commission?" Turp suggested. Favier looked horrified, phoned Geneva, and reported that the Nigerians would take the guns sight unseen. Advice from the Nigerian Central Government was the last thing the Geneva group wanted.

Favier's next request was for a British registered aircraft which could fly to Port Harcourt, in the Ibo dominated Eastern Region of Nigeria. Turp said it was impossible: the British Government, which has a way of finding out about unauthorized charters, would not be willing to look the other way, and there was no chance of an independent getting a straight-forward export licence to supply the dissident Nigerians. So Favier—with clients and weapons, but no transport—returned

CF-TFM with its Burundi registration. (via Stephen Piercey)

CF-TFM was operating as I-ACOA when it made its last flight on October 11, 1966. (John Wegg Collection)

to Paris. Turp was not to hear from him again until last week.

In the evening of October 8, a DC 4, with the Italian registration I-ACOA landed at Zestienhoven airport in Rotterdam, piloted by a 50-year-old German-American named Captain Henry Warton. The Dutch airport security police were less than enthusiastic about its arrival; they knew this DC-4 all too well.

For 10 months in 1965 it had stood on the tarmac at Schiphol airport in Amsterdam, impounded by the airport police who had watched fascinated as falling rain washed away the Panamanian registration it was flying under to reveal a previous Canadian registration. It was (as *Insight* reported on August 1, 1965) about to take part in another illicit gun running operation to Algeria organized by an intrepid American named Lucien Pickett.

In October 1965 a Swiss named Heuer said he had brought the ailing DC-4 to found an airline in the minute African State of Burundi. But he faded from the scene when he was charged in Germany with recruiting for a foreign army, fraud, embezzlement, unfair business dealings and arson.

Then Warton appeared, and after some persuasion from the Burundian Embassy in Paris, the Schiphol authorities were persuaded to release the aircraft to him. Nothing was seen of Warton or the aircraft for almost a year. Suddenly, he returned to Zestienhoven two weeks ago, penniless, stating blandly that he had come for 1,007 sub-machine guns. The security police naturally feared the worst.

But when a Piper Aztec, containing four people, one of them almost certainly Favier, flew in from Geneva just after midnight on the morning of October 9 and Warton was given the money to meet his bills, the police could not interfere. When Warton presented a valid British import certificate for the cargo of arms, however, "we were sceptical," says A.W.F. van der Hoek, the airport's director. But Warton could not be stopped.

Warton claimed the guns were going to Birmingham. "Well, you never take guns to England, especially Birmingham. But the papers were in order and so there was nothing anyone could do; and Mr. Warton had a flight plan for Birmingham," the director says with a good deal of embarassment.

The Dutch authorities had no way of knowing that the DC-4 had been impounded by police at a small Italian airport between Genoa and San Remo called Albenga. There is, the Italian police alleged, considerable doubt about the aircraft's ownership; the registration number, they added for good measure, is a fake.

Nor were the Dutch authorities to know that Warton would not land at Birmingham on October 9, even though the export licence was only for Britain. "The certificate was grossly misused," Turp said yesterday. The Dutch feel the same way.

By October 10, the day after he left Rotterdam, Warton's journey had taken the DC-4 beyond Palma to Hassi Messaoud on the edge of the Sahara in Algeria. When the aircraft left at 7.44 a.m. the following day, Warton told air traffic control that the destination was Fort Lamy in Chad, which lies close to both Cameroun and Nigeria. The next thing that is known about the ill-fated aircraft is that it crashed near Garoua that same evening. It certainly did not appear at Fort Lamy.

So the bizzare plot ended unhappily for everyone. Favier's reputation for delivering the goods is badly dented; the Nigerian rebels still want guns; and Captain Warton—with his three colleagues (all of whom survived the crash)—has been comprehensively charged by the Cameroun Government with transporting arms without authority, illegally flying over Cameroun air space, and flying a plane lacking the necessary flight documents.

The only satisfied body involved would seem to be the Board of Trade, whose import certificate was not, after all, successfully used to transport illegal arms to a member of the Commonwealth. But they might not be so lucky next time.

Insight

Aden Airways

When BOAC decided to replace its Argonaut fleet, it was predictable that its associate companies would inherit some of the older aircraft. So it was that three Argonauts, 'HR, 'HV and

An Argonaut at London prior to becoming VR-AAR. Salt corrosion caused by Persian Gulf environmental conditions later forced its retirement from service in 1963. (Peter Keating)

G-ALHG's history in photos. First it is seen in this A.J. Jackson photo in its striking BOAC livery. Peter Keating caught it at London in March 1958 bearing West African titles. In November 1963 he again shot it, this time with Overseas Aviation at Dakar. Finally, Peter's view of it in Derby Airways colours at Manchester on July 29, 1964. It later wore the BMA livery, a slight digression from the Derby scheme. 'HG crashed in June 1967.

'HX went to Aden Airways as VR-AAR, VR-AAT and VR-AAS.

The Argonauts were no stranger in Aden, where BOAC had operated them for a decade, and they were welcomed by Aden Airways as fast and air-conditioned replacements for its Dakotas. However, the high temperatures in desert Aden were to play havoc with takeoff performance, especially in keeping Merlin coolant temperatures within limits. Midday takeoffs were definitely avoided here!

Another problem facing Aden Airways was the salt-laden atmosphere at their main base. Corrosion became a serious problem to the Aden Airways engineers. So badly did salt attack the Argonauts that VR-AAR was scrapped at Nairobi after expiry of its Certificate of Airworthiness in February 1963. The same fate awaited VR-AAS in Aden when its certificate expired in April 1964. VR-AAT was sold to Derby Airways in December 1963 and flown by Captain Pike to Burnaston on December 13, where it was broken up for spares. During their service with Aden Airways, the Argonauts operated to Bahrein, Cairo, Khartoum, Mogadishou and Nairobi on scheduled services, and elsewhere on charters.

East African Airways

East African Airways Corporation was a joint government airline sponsored by Kenya, Tanganyika and Uganda with BOAC participation. To expand its international routes, it acquired four ex-BOAC Argonauts the first of which arrived at Nairobi on March 29, 1957. This was VP-KNY (ex G-ALHO) and it was joined by VP-KOJ (ex G-ALHM), VP-KOT (ex G-ALHJ) and VP-KOY (ex G-ALHD). Numerous BOAC staff were seconded to EAAC, including Captain Vic Parker who notes that the Argonauts were used on services to Karachi and Bombay via Aden, and to Durban via Salisbury. They also operated Nairobi-Benghazi-London.

EAAC Argonauts were used to aug-

Aethra just after its withdrawal from the BOAC fleet in mid 1959. (Peter Keating)

ment RAF Transport Command aircraft in Aden and Bahrein during the 1961 Kuwait crisis.

VP-KNY crashed on takeoff at Embakasi (Nairobi) while crew training. The three crew aboard escaped but the Argonaut was a total loss. VP-KOJ was substantially damaged at Benina when the nose gear collapsed one day. By 1960 the EAAC Argonauts had been largely supplanted by Comet 4s, and the three remaining aircraft returned to England.

Two other Argonaut operators during this period were West African Airways Corporation and Kuwait Airways. WAAC was a BOAC associate serving Lagos-Accra-London and was superceded after the independence of Ghana in 1958 by Ghana Airways and Nigeria Airways, both using chartered BOAC Britannias on their services to London. Kuwait Airways, also an associate of BOAC, used Argonauts into London and to Pakistan and India.

Rhodesian Argonauts

In 1959 the Royal Rhodesian Air Force was looking for aircraft to supplement its Dakota fleet. The Argonaut was chosen and in December 1959 four were purchased from BOAC along with spares, ground equipment and tools. These were delivered in early 1960 via Nairobi, where East African Airways trained the RRAF crews.

One of the aircraft, G-ALHW/RRAF 179, was retained as a VIP aircraft while the others were delivered as 65-seaters. The Argonauts joined the Dakotas and Pembrokes of No. 7 Squadron and were to serve the RRAF for over three years on a variety of operations.

In 1961 three were stationed at Khormaksor in Aden, moving troops of the 24th Brigade, British Army, from Kenya to Kuwait during an emergency when Iraq was threatening Kuwait. During that airlift, 757 troops were carried and 230:20 hours flown. During the Congo crisis, the RRAF Argonauts were kept busy rescuing refugees from that war-torn nation. On other operations, they sup-

△
VR-AAR at Bahrein on February 28, 1960. Salt corrosion forced its retirement from service three years later. (Peter Keating)

Another Aden Airways Argonaut, also later grounded from the effects of salt corrosion. (via Stephen Piercey)
▽

Aden Airways continued using the familiar BOAC livery. This Argonaut was shot at London Airport. (via Peter Keating)

Beautifully maintained East African Argonaut. An RCAF pilot recalls seeing EAAC Argonauts at Nairobi and being amazed at how finely kept they were. EAAC patched up his North Star after a truck had damaged it, and the repair job was a masterpiece. (Peter Keating)

ported RRAF Canberras on CENTO exercises in Pakistan and RRAF Vampires training in Germany. On local trooping, they carried two pilots, a flight engineer and wireless operator, while a navigator was added for longer flights.

In January 1961 Central African Airways of Salisbury was forced to ground its five Viscount 748s due to spar corrosion. Faced with a serious crisis in its operations, the airline borrowed the RRAF Argonauts with crews and between January and March these flew 19 services for 129 flying hours. During this period the Argonauts bore dual military and civil registrations.

While visiting Rhodesia for the

G-ALHC in basic BOAC colours but with East African Airways details. (Peter Keating)

VP-KOY at London in what is basically the BOAC livery. (Peter Keating) ▽

A BOAC Argonaut in Kuwait Airways titles but retaining its British registration as photographed at London in June 1958. (Peter Keating)

opening of the Kariba Dam, the Queen Mother's party was carried by Argonaut. At the time of the enquiry into Dag Hammarskjöld's death in a plane crash in Northern Rhodesia, an Argonaut was used in an effort to establish the cause of the crash and to reconstruct the approach pattern. The VIP Argonaut also carried Lord Dalhousie to Nairobi upon Rhodesia's self-proclaimed independence. Shortly after this, the Argonauts were re-sold to Air Links in England.

International Air Freighters

International Air Freighters was started by former TCA North Star captain Colin Campbell, who organized some financing from an Edmonton construction company and in May 1961 purchased three TCA North Stars for $50,000. CF-TFC was to be flown while the other two were for spares. Campbell's plan was to fly the North Stars between Canada and Cuba hauling freight. Three experienced TCA pilots were hired: Stan Hegstrom, Al Burns and Willy Milne. Another pilot, Clif Wenzel, was hired later on. When the Edmonton backer was threatened with being cut off by some of its US suppliers for trading with Cuba, it withdrew its support and new financing was arranged with some Toronto businessmen.

Flights began soon after the North Stars arrived in Toronto. Cargo hauled to Cuba ranged from chicks to piglets to powdered milk and pharmaceuticals. Some return flights were empty but others carried such loads as seafood and even watermelon. An agent in Toronto would be on hand to receive these shipments and re-route them to the United States. So much for the Cuban trade embargo! Shipments from Toronto to Cuba were generally paid for in advance by Cuba, and as business built up, International Air Freighters was dispatching two or three flights a week. An ex-Cuban DC-4 was then acquired and registered CF-NWM. On occasion a Cubana Super Constellation freighter was also parked with the IAF aircraft at Malton.

Most IAF flights between Toronto and Havana were routine, but on one with pilots Hegstrom and Wenzel aboard, the North Star was intercepted by a USAF F-102 of the Alabama Air Guard. The North Stars occasionally carried passengers as well, including Cuba's UN ambassador. He would bring his family from New York to Toronto and fly home by North Star

G-ALHC, later RRAF 180, before delivery to the Rhodesian Air Force. The photo was taken at Heathrow on January 21, 1960. (Peter Keating)

RRAF 182, also at Heathrow awaiting delivery to Rhodesia. (via Peter Keating)

CF-TFC appears to have been delivered to International Air Freighters on May 20, 1961, and flew its first trip to Havana the same day, an 8:45-hour flight with pilots A.M. "Al" Burns and W.D. "Willy" Milne aboard. As a rule, flights were under seven hours. To January 10, 1962, Al Burns' log shows 46 return flights to Havana. This photo was taken at Malton on March 29, 1962. (Jack McNulty)

Captain Willy Milne welcomes passengers aboard an International Air Freighters flight to Havana. (Stan Hegstrom)

rather than go to Mexico for the weekly Mexico-Havana flight.

Throughout this period International Air Freighters was operating on a Cuban licence, not having one of its own. A government agreement stated that while the Cuban licence was in use, a Canadian operator be allowed to fly trips to Cuba on a one for one basis. Thus Don McVicar's Dorval-based World Wide Airways was able to conduct many freight flights to Cuba during this same time using its DC-4, CF-NNN.

After about a year of operating, IAF was in financial difficulty. There did not seem to be enough money available for basic aircraft maintenance and to cover other operating expenses and in July 1962 the company closed its doors. The DC-4 reverted to Cuba and the North Stars were left to languish at Malton for several years until they were scrapped.

California Conversion?

In November 1961 TCA sold North Star CF-TFF to World Wide Airways of Montreal, who then concluded a deal with Russell K. Lamm of Oakland, California. Lamm, a long-time aviator and operator of an aircraft maintenance and overhaul base, had conceived the idea of converting a North Star from Merlins to Wright R-2600 radial engines, which were available cheaply from US Navy surplus.

Lamm purchased CF-TFF from WWA, having already worked out the engineering details. The engine change was technically no big problem, though a C. of G. problem could be anticipated due to the Merlin being a longer and heavier power plant.

CF-TFF was ferried to Oakland from Dorval via Chicago with Captain Ray Bicknell of TCA in charge. For US operations, it was re-registered N7866C. Meanwhile Lamm had been doing some marketing of his idea but found no operators ready to lay down cash for a North Star or DC-4 converted to Wrights. In less than a year he abandoned the idea, sold all the engineering to another company, and scrapped the North Star, ending what could have been yet another page in the North Star saga.

Another of the International Air Freighters North Stars shown derelict at Malton, July 31, 1964. (Peter Keating)

Basking in the Florida sun, CF-TFQ bears its short-lived SAM titles. Harold G. Martin took this photo at Miami. (via Stephen Piercey)

North Stars for Colombia?

As researched by Stephen Piercey, the Colombian airline Sociedad Aeronautica de Medellin (SAM) had plans in 1962 to acquire several of the former Overseas Aviation North Stars. These were ferried to Fort Lauderdale and a demonstration tour was organized to show the North Star on the SAM routes. Piercey reports in *Propliner*, "On May 18, 1962, with small SAM titles applied where the old TCA logo had once been carried, a hopeful CF-TFQ, still bearing TCA colours and its original fleet number, 217, departed to Medellin from Fort Lauderdale on the start of its week-long demonstration trip in Colombia. During May 18 and May 19 the old North Star undertook flights from Medellin with SAM director Oscar Mejia and other company personnel aboard. CF-TFQ was later ferried to Colombia's capital, Bogotá, where on May 24 she underwent air and ground inspection by the Aerocivil, Colombia's answer to the British Civil Aviation Authority." While in Colombia the North Star's pressurization system failed, as did an engine on takeoff from Bogotá. These events did not impress the airline and, with its own financial difficulties and uncertainty about being able to maintain Merlin engines in Colombia, SAM decided to steer clear of the North Star. CF-TFQ returned to Florida and its fate at the hands of the wreckers.

Missile-tracking North Star

In 1962 the US government required an additional tracking aircraft on its South Atlantic and South Pacific missile ranges. AVCO, prime builder of missile nose cones in the US, had been handling most data collecting contracts using such aircraft as the C-54, DC-6 and C-121. In mid-1962 several North Stars, from the collapsed Overseas Aviation in the UK, made their way to Florida where, it was hoped, they could be sold to Latin American operators. Air engineer Ted Schneider was at Fort Lauderdale one day when one of these aircraft, CF-TFL, came in. Schneider asked around about the airplane, checked it out and, knowing about the AVCO contract, called his friend T.W. Smith in Cincinnati to tell him there was an airplane that suited AVCO's needs for sale at a good price. Smith soon arranged a deal and in August 1962 he and Schneider flew the North Star to Blue Ash airport near Cincinnati.

Re-registered as N5822R, the North Star was leased to AVCO and work began to refit it as an airborne data collecting station. Gene Holden of AVCO supervised electronic matters during conversion work, while Virgil Jones became general overseer of the project and, later on, was responsible for logistics relating to "Twenty-two Romeo."

An AVCO paper from this time describes the equipment carried on the North Star, and its specific role: "Avco Field Engineering acquired, modified, and operated a pressurized DC-4M2 aircraft for special missile range data collection. This modification included installation of optical glass windows, three meteor radar systems, optical instrumentation, and a complete darkroom facility. This platform measured electromagnetic characteristics associated with ICBM type re-entry vehicles by use of the specially designed meteor radar. Photographic and magnetic tape equipment was used to record the radar video data. Optical instrumentation and a camera timing and control system were also provided to track the vehicle and correlate to the re-entry data. Additional communications and special navigation equipment was added to meet the missile range and project operational requirements."

N8722R taxis at Cincinatti's Blue Ash airport just before its first test flight after refit as a nosecone re-entry data collecting aircraft. (AVCO International)

E.G. "Gene" Harris was hired as first pilot on the AVCO North Star and the aircraft was put to work soon after its two-month refit. Harris, a long-time C-54 pilot in the US Navy, had most recently been flying modified C-54s for Bendix on satellite tracking station calibration. He, Carson Malcolm, Walter Wright, Hank Lillybeck and Hal Foot were to share flying duties on "Twenty-two Romeo" in the months that followed.

Operations on the North Star data collecting contract took place from Ascension Island, in the South Atlantic, and Kwajalein, midway between Hawaii and Australia. The aircraft would fly to the South Atlantic from Miami via San Juan or Piarco, Surinam, and Recife; and to the South Pacific from Miami via San Francisco and Honolulu. It was on the latter 12-hour leg that the twin 500-gallon R5D long-range fuel tanks were needed the most. Pilots appreciated the North Star's speed advantage over the DC-4 but it was quite a bit slower than the DC-6 or DC-7. One pilot recalls leaving San Francisco westbound and hearing a DC-7 on the radio just leaving Los Angeles for Hawaii. The North Star was still droning on towards Hawaii when the same DC-7 passed it heading back to LA!

Nonetheless, the North Star performed extremely well for AVCO. It suited their requirements in speed, range, payload and reliability. It was pressurized, so could accommodate the onboard radar which had to be pressurized to sea level. Amazingly, too, it was found to be comparatively vibration-free, which was essential for the smooth running of the sensitive scientific equipment. Crews insist that it was better in this respect than the DC-6, DC-7 or Super Connie. One experienced pilot described "Twenty-two Romeo" as "a great airplane, really nice to fly, and the quietest big piston transport I've ever flown." Who would believe it!

The North Star was generally on detachment for two and three month periods and in less than two years flew over 200 data gathering missions. As a rule about a dozen technical crew made up the complement, besides the pilots. When not busy flying, the crew went fishing and sailing. Carson Malcolm, an expert sailor, taught many of the AVCO people the basics of sailing.

Of course, the North Star was a great attraction wherever it went. Gene Harris took it into Clarke Field in the Philippines one day and raised a lot of eyebrows. Nobody there had ever heard of a North Star! At Kwajalein, Carson Malcolm says that beating up the field was mandatory every time he was aloft. Everyone there loved the sound of the North Star's Merlins. As he puts it, where else could you hear the sound of a squadron of Mustangs?

Aircraft serviceability remained high during the AVCO period for "Twenty-two Romeo." In the first year three engines had to be changed, but these were in rough shape when originally acquired. Virgil Jones had to scrounge all over for parts but with help from International Air Freighters in Toronto, Spartan in Ottawa and Bristol in Montreal he was able to keep the North Star and its Merlins on the go. One annual check was done by Aloha Airlines during a six-week stay in Honolulu.

All tracking flights were made at

AVCO technicians busily at work converting the cabin of their North Star. In all, three North Stars worked as airborne scientific stations: AVCO's, the NRC's Ice Wagon, and the NAE's CF-SVP-X. (AVCO International)

△ Engine maintenance for "Twenty-two Romeo," this time at Miami. Once the AVCO North Star had completed its shakedown period, it functioned almost perfectly. (AVCO International)

night and on these the North Star performed like a real star — it was always on station exactly as specified. In the Pacific this was a spot near Eniwetok Atoll. On one or two occasions the aircraft had to hold so long for the reentry that it had to abort its mission and return for fuel, but reliability was extremely high. An AVCO document notes, in describing the company's world-wide support operation, "This back-up makes possible the high level of performance evidenced during a two-year period when the AVCO DC-4M instrumented aircraft logged 268 scheduled departures with only seven minor delays."

In August 1964 the AVCO North Star contract ran out and the USAF took over its duties with other aircraft. "Twenty-two Romeo" was flown to Middleton, Ohio, and stored by T.W. Smith. Attempts were made to sell it but in the end there were no takers. It was donated to a local church, which sold it for scrap in a fund raising campaign. So ended the unusual career of one North Star.

This camera angle shows the main external mods done to N8722R: three antennae, the main one being 16 feet long, and a pair of large windows for optical tracking. (AVCO International) ▽

Pilot Gene Harris looking satisfied with everything. To Harris and the other pilots who flew "Twenty-two Romeo" for two years, the North Star was as fine a big propliner as any pilot could ask for. High praise indeed from captains, some of whom are still flying DC-6s and the like today. (AVCO International)

The Caicos Stars

In this article from *Propliner* Stephen Piercey tells the story of the North Stars that served the Turks and Caicos Islands in the British West Indies.

For ten years the city of Sarasota in west Florida was associated with some old Canadair C-54GM North Star aircraft. Seven former Royal Canadian Air Force (RCAF) machines were purchased from the Crown Assets Disposal Corporation in Canada by Cavalier Aircraft of Canada, a subsidiary of the Cavalier Aircraft Corporation, based at Bradenton Airport in Sarasota. Cavalier's goal was to specialize in military and civil conversions of the North American P-51 Mustang, and its president, David Lindsay, happened to be the president of Lindsay Newspapers and editor of the *Sarasota Herald Tribune* newspaper. Lindsay is reputed to have wanted the North Stars because he was a firm believer in anything with a Rolls-Royce powerplant. It is said that he once paid to have some junked former TCA North Stars that were standing derelict at Miami cut up because he couldn't face the sight of the abandoned Rolls-Royce-powered transports.

In the early 1960s Lindsay had formed Sarasota-based Trans Florida Aviation, a fixed-base operator. At least one of the North Stars was initially registered in this name. Lindsay had become involved in refurbishing Mustangs as early as 1962, and Trans Florida had undertaken minor work in this connection. The company closed in 1965, and it was not until the formation of Cavalier in 1966 that the Mustang program really got under way.

With the purchase of the North Stars, Cavalier Aircraft acquired a vast quantity of Merlin engine spares to support its Mustang rebuilding program. It is probable that David Lindsay purchased the RCAF's total North Star spares inventory, as his seven aircraft were the only RCAF North Stars to have been purchased. But the North Stars were not all bought for their parts — it was once Lindsay's intention to have the Canadian-registered Canadairs placed on the US register, to use some for a proposed commercial operation (possibly as Trans Florida Aviation), perhaps sell others and use the rest for their valuable parts. Of the seven Canadairs, just three flew again after arriving in Florida. It was anticipated that further aircraft would be used for a proposed commercial operation, although nothing was to come of this and the aircraft remained grounded at Sarasota, the property of Cavalier.

The seven North Stars were bought during 1966 for an estimated $10-20,000 each, and were registered CF-UOY, 'UXA, 'UXB, 'VLW, 'VLX, 'VLY and 'VLZ. From storage at Mountainview and Trenton, Ontario, they were ferried over an eight-week period to St. Jean, near Montreal, for refurbishing before delivery to Florida. CF-UXB and CF-UOY were the first two aircraft to be civilianized, and were the first to arrive at Sarasota, during August 1966. A third aircraft, CF-UXA, was airtested at St. Jean on August 27, finally arriving in Sarasota via Trenton and Tampa five weeks later, on October 8, 1966. Cavalier accepted delivery of its last North Star, CF-VLZ, in February 1967 . . .

During 1966 the Government of El Salvador was negotiating with Cavalier Aircraft for the purchase of six Cavalier Mustang 2s for use by the Fuerza Aerea Salvadorena. It also required a heavy transport to carry military cargo and personnel, and as part of the deal North Star CF-UOY was "thrown in" for good measure with a quantity of spares. Other reports indicate that up to $150,000 was paid for this aircraft — certainly Cavalier had spent quite a large sum of money preparing it for delivery. Under the captaincy of John "Bud" McNair, CF-UOY had operated a number of crew training details from Sarasota with Salvadorean trainees aboard

Air Caicos' two ex-RCAF workhorses, CF-UXA seen at Sarasota, and CF-UXB at Miami. (John P. Stewart and Harold G. Martin via Peter Keating)

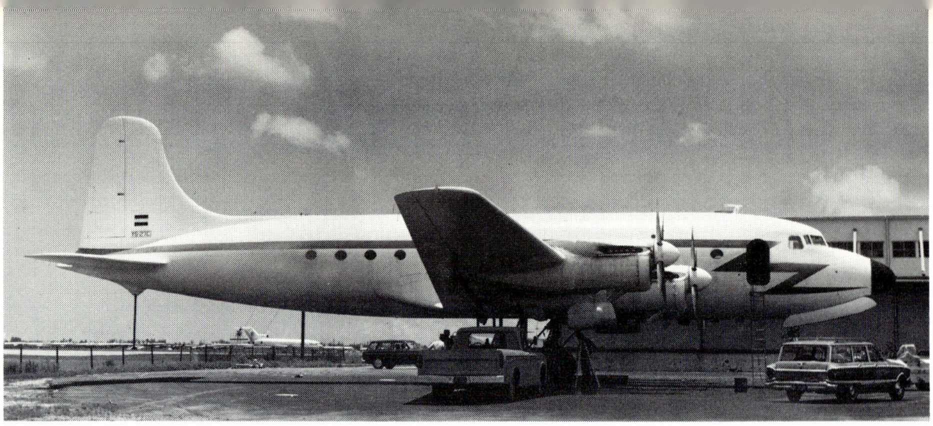

Central American North Star. YS-27C of the Fuerza Aerea Salvadorena photographed at Miami in 1963. (Hal G. Martin via Peter Keating)

during August 1966, and on August 31 the aeroplane departed Florida for St. Jean, where further work was undertaken. Bud McNair, a former North Star pilot with 426 Squadron, became Lindsay's personal pilot, and also chief pilot of the Cavalier Aircraft Corporation, flying both the C-54GMs and P-51s. He was destined to remain a part of one North Star operation or another until the type's last commercial days in the mid-seventies.

None of the North Stars were placed on the US register, and on paper they were leased to Cavalier in Florida by Cavalier Aircraft of Toronto. North Stars CF-VLW through to 'VLZ were never used, and languished at Sarasota until broken up in the early seventies.

David Lindsay was also a director of the Caicos Company, which owned acres of land down in the Turks and Caicos Islands in the British West Indies. The company owned the Admiral Arms hotel on South Caicos, which, at the time, was the island's only hotel (to this day it remains the most prominent on the island). "The islands," as they shall be referred to, are situated about 565 n.m. east of Miami. Trace a line from Miami down to the bottom of the Bahamian island chain and you can't miss them. North Star CF-UXB began operating for Cavalier and the Caicos Company on a private category C of A in August 1966. Loaded with supplies for the islands and the hotel, the first trip was "inaugurated" on August 11, when Bud McNair flew the Canadair from Miami to Grand Turk and South Caicos with David Lindsay acting as copilot. A return was made two days later. This was the beginning of a ten-year period of operation down to the islands by North Star aircraft, which was to eventually become the world's last commercial operation of the type. The islanders would never forget it.

While flying for Cavalier on business relating to the P-51 program, or while operating on behalf of the Caicos Company, CF-UXB averaged one visit to the islands per month, but she was often found away from base, calling at various airports and air force bases in North America with Cavalier representatives aboard.

During March 1967 a newspaper conference was held in Montego Bay, Jamaica, and, as the publisher of Sarasota's *Herald Tribune*, David Lindsay made arrangements to attend this important function. While the world's publishing magnates, editors and newspaper representatives were preparing to fly to Jamaica in their private Lear Jets or Gulfstreams, Lindsay was clambering aboard his old North Star at Sarasota, with pilot McNair. CF-UXB was about to operate her usual monthly run to the islands, and with her cargo and "special" passenger, flying as copilot, she left for Miami on March 27 to collect fuel. About three hours after departing from Miami, the North Star approached her first port of call, Grand Turk, the capital of the Turks and Caicos Islands. Here the aeroplane nightstopped, and the following morning, her cargo off-loaded, she set course for Montego Bay, some 550 miles south west, where she arrived amid a host of gleaming VIP executive jets in time for the convention. I suppose that a North Star would be the last type of aircraft from which one would expect to find a newspaper publisher disembarking — but what style. Once the conference was over, CF-UXB fired up and departed back to South Caicos, where she collected a cargo of fish and returned to Miami on April 2.

During early 1967 Air Caicos was formed by David Lindsay and partners (one of which was Douglas Dodds-Parker, a British M.P.) as a commercial foreign operator. As a British corporation, with three American board members and two Bermudan board members, the airline held a licence to fly cargo between Florida and the Turks and Caicos Islands — the only islands in the group that were capable of accommodating the North Star being Grand Turk, Providenciales Island (known as Provo) and South Caicos. Initially only CF-UXB was to be operated, on lease from Cavalier Aircraft of Canada, and sporting a new blue cheatline, airline titles and the Turks and Caicos crest on her nose, the company made final preparations for the new service. Air Caicos embarked on its first commercial service during August, becoming the first foreign air cargo carrier to operate from the islands.

Bud McNair became chief pilot of the newly formed Air Caicos company. With most of its personnel residing in the Sarasota area, the North Star continued to be based at this quiet residential airport, and was ferried on the hour trip down to Miami twice a week, where fuel and cargo was loaded for the run to the islands. The infrequent charters undertaken made welcome breaks for the North Star crew from the understandably monotonous routine of flying to and from the islands. Occasionally the North Star flew cigarettes to Hamilton Kindley Airfield in Bermuda from

The El Salvador North Star derelict at Ilopango, El Salvador, in 1977. (Stephen Piercey)

Miami on behalf of the Pitt Company, a large tobacco producer on the island. The owner, Jack Pitt, was one of the directors of Air Caicos. On one occasion, CF-UXB uplifted as many as six cars in one go from Hamilton Kindley Field and flew them down to South Caicos, where they were sold to the islanders.

North Star "X-ray Bravo" continued to clock up between 10 and 20 hours each week flying down to the islands, the routine being interrupted once a year when it was time for her to be ferried up to Toronto-Malton Airport, for an annual overhaul in the hands of Field Aircraft Services.

During February 1969 Air Caicos welcomed a former United Airlines mechanic, George Davy, to its operation. As copilot on the North Star, George was teamed with Bud McNair, and was later to become a co-captain. These two aviators have remained together and had the distinction of both flying in the left hand seat while operating the world's last commercial North Star service some seven years later.

As the only licenced cargo operator of the Turks and Caicos Islands, and the only company offering a reliable and regular service from Miami, the islanders became accustomed to the twice-weekly arrival of "their" North Star. She carried anything that would fit through her cargo door, and it was the North Star that was directly responsible for the development of these recently still under-developed islands. Indeed, the islanders owe just about everything they now have to the Canadian workhorse.

Air Caicos offered very low cargo rates to the islands. Consequently, everyone turned to them for business, and by early 1969 it was decided to place a second aircraft into service. The third North Star to be delivered to Cavalier during 1966 was chosen (CF-UXA), and following three years in open storage at Sarasota, she was prepared for the trip to Fields at Toronto for overhaul and refurbishing. On September 18, 1969, loaded with 16,500 lb of furniture and produce, North Star 'UXA thundered across Miami International Airport on her very first commercial service, down to Grand Turk. With the two Canadairs on line, the airline rotated them on the island runs from time to time.

Unfortunately, as the two aircraft were never in the air at the same time (one usually remained at its base in Sarasota), Miami Airport was never graced with the sight of Caicos' two North Stars parked together on the ramp. However, during 1969, another former RCAF machine was placed into full commercial service with Aerolineas El Salvador (AESA) and made occasional visits to Miami. This aircraft was Cavalier's former CF-UOY, which, upon her sale to El Salvador in 1967, had been given the air force serial FAS300. During 1969, while flying on behalf of AESA, she carried the civil registration YS-27C, uplifting general cargo and mail to and from the USA. Her career continued until 1970 when replaced by a Douglas DC-6B.

During 1970 Nassau-based Bahamas Airways was in a poor financial state, and was fighting a losing battle to survive. Its efforts to continue in operation failed, and after nearly 34 years of continuous inter-island operation the airline was closed. Air Caicos was immediately approached by the Governor of the Bahamas and asked if it would pick up the scheduled passenger route from Nassau to the Turks and Caicos Islands via intermediate points in the Bahamas. In addition to this passenger operation, Air Caicos was also committed to fly a once-weekly diplomatic flight on behalf of the British embassy from Nassau to Havana, Cuba, the initial commitment being to carry 20 passengers and up to 12,000 lb of cargo on each flight. Two DC-3s were purchased for the passenger schedules, while the trusty old North Stars were chosen as suitable aircraft for the diplomatic charter.

The first diplomatic flight by Air Caicos took place on November 6, 1970. As more than a dozen diplomats boarded the aging Canadair in the Bahamas, the newly employed stewardess dutifully handed out a set of ear defenders to each passenger in an attempt to make their two-hour flight more comfortable. The aeroplane was fitted with 20 seats forward, and the flights operated direct to Cuba — with a one hour turnaround time. Official complaints were often received through embassy channels, with the diplomats demanding a more up-to-date (or less noisy) form of transport. Fortunately nobody took any notice of these complaints (after all, there was no competition on the route), and old CF-UXB continued to ply between the Bahamas and Cuba, and her stewardess continued to hand out the ear muffs.

Bud McNair waves to photographer "Piston" Piercey as he heads for South Caicos from Miami. McNair assigned former RCAF North Star numbers as Air Caicos fleet numbers. UXA/Ft. No. 502 was formerly RCAF 17502; while UXB/Ft. No. 510 was RCAF 17510. (Stephen Piercey)

Handsome though it first appeared, the Bahamas passenger operation was not profitable. The problems encountered with a leased Convair aggravated the situation somewhat, and to make matters worse the geographical area was experiencing a recession during the early 1970s, forcing the load factors down still further. The North Star cargo operation was the only profitable division of the company, and it soon began to subsidize the passenger operation. It was clear that Air Caicos was in for a difficult time.

Friday, September 17, 1971, was a day that Air Caicos would not easily forget. Early in the morning Bud McNair, George Davy and Lee Edwards, an Air Caicos DC-3 pilot based in Nassau, boarded North Star CF-UXB for the short flight to Miami, where refuelling and loading was expected to commence mid-morning for an early afternoon departure to the islands. The North Star left her base at 0950. On approach to Miami International only two undercarriage lights illuminated in the cockpit, indicating that the port main gear had not lower-

Interior of Bud McNair's North Star, CF-UXA, looking aft. (Stephen Piercey)

Air Caicos chief pilot and North Star diehard Bud McNair prepares to replenish the oil in UXA. Note the spotless Merlin, obviously the pet of a very admiring captain. This photo dates to March 1975. (Stephen Piercey)

ed. Uncertain whether the alarm was genuine, the crew gently lowered the Canadair on to Miami's runway for a touch and go landing to check the situation. As soon as she made contact, they felt the left wing drop and realized that the warning was not false. The four Merlins surged with power as McNair aborted his landing. The unsafe indication persisted and the crew decided to return to base at Sarasota. All attempts to lower the port gear failed, and even a harsh bump on the runway at Bradenton Airport failed to release it. Air Traffic Control kept quiet about the troubled aircraft in an effort to avoid any publicity, although by this time an eager enthusiast had picked up the distress signal on his radio. The word got around fast, and soon what appeared to be the entire population of Sarasota had gathered around the airport perimeter to witness the spectacle.

Bud McNair decided that a wheels-up landing on wet grass would be better than attempting a two-wheeled landing on concrete, and advised the tower of his decision. Dozens of emergency ground crew waited patiently as the North Star circled the airport. At 1400 hours, more than four hours after taking off from Sarasota, one final attempt was made to jar the reluctant landing gear free. It was unsuccessful, and CF-UXB climbed away and circled to the south of the airport, where all but 100 gallons of fuel was jettisoned. At 1425 the tower advised McNair, "The grass is nice and wet. Good luck."

"You bet, Buddy," came the controlled, but not reassured reply.

Bud brought his North Star low over the field for a close look at his emergency strip, before approaching for the last time at 110 mph with the gear retracted. With the two inboard engines cut and propellers feathered, 'UXB bellied on to the grass in a spray of mud and water — coming to a level, straight-ahead stop a quarter of a mile further on. McNair, Davy and Edwards emerged from the crew door unharmed, the North Star seemingly undamaged.

Although 'UXB received only slight damage to her underside, she did not fly again, and with 21,674 flying hours to her credit, she was put out to grass, where her vital parts were removed to help keep CF-UXA airborne. The ensuing investigation revealed that an undercarriage shear pin (which should break with a force of 1-1/2g or more) had been replaced at some time by a stainless steel unit. On this occasion, when normal lowering refused to function, the pin which should have broken under load did not, effectively locking the left gear in its retracted position.

At the time of the accident, CF-UXA was in Toronto undergoing routine maintenance, and on her return to Florida she was pressed into service on both the regular freight runs to the islands and the diplomatic charter. Air Caicos was fast running out of money, and although there would have been little difficulty in finding extra funds from US sources, as a British concern more than 50 per cent of its finance had to originate from British backing. Its time was up. On January 27, 1972, CF-UXA turned finals into Miami International, on her return from the islands on the company's final flight. Cavalier Aircraft Corporation also closed its Mustang program during 1972, and the future began to look bleak for CF-UXA. At this time, the remaining four North Stars (CF-VLW, 'X, 'Y and 'Z) had deteriorated. . . and littered Sarasota Airport in their faded RCAF liveries.

The debts accrued by Air Caicos were too large to absorb and a reformation of the airline was out of the question. Bud McNair began to look for new ways in which to operate the redundant North Star, which remained grounded at Sarasota from January 27. He visited the islands to try and form Aero Caicos, which would essentially take over from where Air Caicos left off, although nothing ever materialized. The next few months were spent in finding new finance for the revival of CF-UXA. A consortium of ten people provided financial backing for the proposed operation, and during November 1972 the name of Turks Air Ltd. was incorporated in the Turks and Caicos Islands.

With the demise of Cavalier Aircraft of Canada, CF-UXA became the property of Canadian Aircraft Holdings. The Canadair was leased from this company and was prepared for re-entry into service by Bud McNair and George Davy at Sarasota, where all maintenance and overhaul was to be conducted. CF-UXA would be flown on the same routes as those served by Air Caicos, with the exception of the Bahamas contract. She took to the air on an airtest in November 1972, and on December 5, 1972, almost 11 months after the collapse of Air Caicos, she departed to Grand Turk with McNair and Davy at the controls on the first service under the new consortium.

As president of Turks Air, Bud McNair applied to the Turks and Caicos Islands' government on July 25, 1973, for a licence to carry cargo to the islands. Pending US approval and the go-ahead from the government of the islands, the North Star continued to be financed by the consortium. Turks Air finally received its operating permit on October 11, 1973, and was soon flying as many as ten round trips a month from Miami.

A typical week in the life of Turks Air saw the aeroplane being ferried to Miami from Sarasota on the Tuesday. She would load early on Wednesday morning for a mid-day departure to Provo Island, Grand Turk and South Caicos, where she would remain overnight. Returning to Miami on the Thursday, CF-UXA would be offloaded, and reloaded for another trip to the islands (just Grand Turk and South Caicos) later in the day. A return to Miami would follow on the Friday, and in the event of a cargo backlog, which was frequently the case, she would return once again, for the third trip of the week. Outbound cargo consisted of anything imaginable that could be loaded through her rear cargo doors, and on each return trip from the islands the North Star carried lobsters and conch from fisheries on Provo Island and South Caicos.

Turks Air undertook its own maintenance on the Canadair at Sarasota, and while airframe spares presented no particular problem, spares for the 1,760 hp Merlin 622 (which offered a 1,200-hour time between overhauls) were more difficult to come by. The main problem was that of engine time, and by 1974 Turks Air knew that with the limited amount of time remaining on the engines the North Star would not remain in service much longer.

Bud McNair had all known available spares for the Merlin 622, and he had conducted a wide search in places that North Stars had operated over the last 20 years for any possible lead to a stash of hidden parts. During 1973 and 1974 North Star engine exhaust stacks became extinct, and Turks Air therefore acquired Lancaster exhaust stacks (which could still be bought, if you knew where to go) for modification to fit the Series 622 engine. In 1974 the National Aeronautical Establishment of Ottawa offered its North Star, CF-SVP-X, to Turks Air, although, as Bud explained, the company was not interested in buying what was essentially an airframe with near time-expired engines and no spares. Ironically, the NAE's North Star was destined to outlive Turks Air's aeroplane, flying with the Establishment until 1976. During the late 1970s this last airworthy North Star was impounded on Great Inagua Island for an illegal flight — and remains at this location today.

Fortunately, CF-UXA rarely developed any serious engine malfunction in her civil career, thus prolonging Turks Air's spares inventory. Quite remarkably, in ten years of civil operation, Bud McNair could recall only four engine shut-downs with the North Stars — most of which were precautionary. That says something for Canadair, Rolls-Royce and Turks Air workmanship.

During June 1974 Turks Air was awarded a foreign carrier permit by the Civil Aeronautics Board, by which time the Canadair was clocking up a healthy 70 flying hours each month. Early in March 1975 I was extremely fortunate (and relieved) to have been able to travel to the British West Indies to sample Turks Air's operation with its unique North Star (simply referred to as "Star" by its crew and islanders).

Bearing in mind that Bud had been flying North Stars continuously for more than 20 years (including CF-UXA while she served the RCAF as 17502), it was hardly surprising that he had established a relationship with the aeroplane. Likewise, George Davy had flown the type since early 1969, and even though it is now long gone, he still recalls its almost legendary reliability. They both explained to me that the islands had no running water or electricity before the advent of the North Star (1966), and that thanks to Air Caicos' first aircraft, CF-UXB, and 'UXA in its role with Turks Air, the islanders now enjoy these amenities as naturally as we do. At Miami International early in the morning I witnessed Bud and George supervising the loading of their aircraft, doubling as loadmasters. CF-UXA's cargo included vegetables and fruit, and the loading operation took 4 hours. Shortly afterwards she fired up and departed for Grand Turk, just three hours away.

Like CF-UXB, Turks Air's machine had had her soundproofing removed. Cockpit noise level was too high to communicate by voice to the aircrew, but aft of the bulkhead the sheer roar of the Merlins and rattle from every nut and bolt in the cabin was intolerable — a couple of minutes was just about all I could take, and that's saying something. On arrival at each island, dozens of islanders lined the small airports, awaiting the North Star. To give an impression of how attached Bud and George had become to the islanders, Bud used to undertake personal shopping for the native islanders while at home in Sarasota. During the few days I travelled with CF-UXA, he made personal deliveries of swimming trunks, trousers, carpets and a bicycle to various people on South Caicos. How many airline presidents do you know who would do that for anyone?

South Caicos is a 15-minute hop from Grand Turk, and here the crew night-stopped at the Admiral Arms hotel, situated on the edge of the Caribbean Sea. Loading and offloading equipment (such as a fork lift truck) was unheard of on the islands (and may remain so today). Part of our cargo was a water filtration plant, destined to supply running water to the hotel and its environs. The plant was packed in eight crates, the heaviest of which weighed more than 1,000 lb. Try to imagine off-loading such a cargo without anything as simple as a fork lift. Other large cargoes carried by 'UXA during March included a new Chevrolet hatchback and an electricity generator, which, when installed, would supply all the electricity for South Caicos.

Back at Miami there was perishable cargo for the islands stacked to ceiling height in the Turks Air warehouse in the northwest corner of the airport, and the crew prepared for a late evening return flight to South Caicos. I stood out by the freight sheds near the taxiway, waiting to catch a last glimpse of the departing North Star, which by now had slipped away to the south side of the airport and disappeared among a handful of jets, holding off the runway in a long queue awaiting take-off clearance.

Half a dozen widebodies, Boeing

UXA at home base, Bradenton Airport, Sarasota. Nearby is the derelict CF-VLZ, formerly RCAF 17516. (Stephen Piercey)

The view from CF-UXA on approach to Grand Turk Island. (Stephen Piercey)

727s and DC-8s blasted off into the sunset, before the familiar shape of Turks Air's North Star appeared from behind an Eastern TriStar, poised just off the runway. From my standpoint, a little less than a mile away, any sound was drowned by the noise of a fuel-guzzling Colombian Boeing 720 starting up for its night run to Barranquilla and Bogotá. The Canadair rolled to the active runway. She came to a halt, as if talking herself into the decision of flight, and through the haze of jet exhaust was seen to start her take-off run. And then the sound hit me . . . the distinctive and magical roar of her Merlins at full boost, sounding like a squadron of powerful Spitfires, flooded the airport, drowning every surrounding noise. The growling departure must have rekindled forgotten memories for many casual standers-by, and even the Cuban aircraft loaders and the driver of the mobile dinner truck stopped in their tracks to listen to McNair's North Star getting airborne . . .

The North Star soldiered on throughout April and May, although with her remaining engine time dwindling to less than 100 hours, the situation had become quite critical. On the morning of June 5 the North Star uplifted over 14,000 lb. of furniture and produce for the islands, but on the return leg the following day, No. 4 engine began to play up, and lost its oil pressure. With no replacement Merlin, or sufficient spares remaining, Bud McNair and George Davy realized the end of the North Star was imminent. She arrived back at Miami early in the afternoon and

UXA being off-loaded at Grand Turk and in the Turks Air livery. (Stephen Piercey)

disgorged her very last consignment of cargo to the fishery trucks. Later she was taxied across to the centre of the airfield for temporary parking. In the meantime, Turks Air leased capacity on a Fort Lauderdale-based DC-4, N67147, in order to continue the vital flow of cargo to the islands.

For the last time the "Star" slipped away from Miami as quietly as her Merlins would allow on the afternoon of June 19, en route to her home of ten years. Indeed it was an extremely sad day for both Bud and George. CF-UXA made gentle contact with Sarasota's runway, although it was not to be her final landing after all. George Davy commented how sorry he was to have been in the right hand seat for the last landing. With that, Bud McNair opened the Merlins to half boost and the unladen North Star climbed in a sprightly manner from the airport for one last circuit. As Bud changed seats with his co-captain, the governor on No. 4 motor failed and the engine began to over-

Off to the blast furnaces. CF-UXA, carved up and ready to be recycled into who knows what, was unceremoniously hauled away from Sarasota in the spring of 1976. (Stephen Piercey)

CF-SVP-X in formation with other aircraft of the National Aeronautical Establishment: a Beech 18, T-33 and Harvard. (NAE)

speed. The Merlin was shut down shortly before the final touchdown. CF-UXAs log book closed on 22,500 hours.

The Merlins from CF-UXA were sold to a Californian boat company for use on power boats — so the spirit of the North Star is kept alive. Other parts of this remarkable aeroplane continue to fly on Turks Air's DC-4. Sadly the airframe of CF-UXA was given to a scrapman, and over a four-week period the remnants of this once-proud transport were trucked away from Sarasota.

Even today the old North Star is sorely missed, as testified by the crew's epitaph that few things in aviation have worked as well.

Stephen Piercey

Airborne Laboratory

In 1962 RCAF North Star 17514 was made available to the National Aeronautical Establishment for use as an airborne laboratory. The operation of this North Star has been detailed by Stephen Piercey of *Propliner*. Piercey, one of aviation history's most intrepid devotees, made a point of being on the last survey mission operated by the NAE North Star. His account of the 1976 mission entitled "North Star's Arctic Farewell" is reproduced here:

The aircraft concerned is ex-Royal Canadian Air Force 17514, which has been with the NAE at Ottawa since 1962 in spite of remaining officially on charge with the RCAF until 1965. Registered CF-SVP-X. . . the aircraft has clocked up some 20,066 flying hours since its first flight in 1947 . . .

CF-SVP-X was by no means operated conventionally during its working life with NAE, having been used mainly for research and development work on anti-submarine systems and for aeromagnetic and geophysical surveys. A number of unfamiliar modifications have been acquired over the past decade, the most obvious of which is a "sting" boom extending from the tail of the aircraft. This houses a magnetic anomaly detector (MAD) at its tip.

Used for a number of different projects, the prototype MAD, an extremely sensitive magnetometer system, was developed by the NAE in the early 1960s and was originally carried in a "bird" attached to a bomb rack under the starboard wing. During flight the bird was trailed on a 300 ft. line. But for a number of reasons (of which in-flight instability was the most important) this method was supplanted by the present installation. The boom keeps the MAD fully 25 ft. away from the airframe, thereby significantly reducing aircraft-induced magnetic interference. Any remaining errors are corrected by an on-board compensation system calibrated in flight over areas with known magnetic characteristics.

The NAE's North Star has taken part in a wide variety of projects, including Department of National Defence-sponsored research into the aeromagnetic detection of submarines, and research on such VLF navigation systems as the Global GNS 200 and 500. But much of the North Star's later life has been spent on experimental aeromagnetic surveys, covering vast areas stretching from the North Pole down to South America and across to Greenland and Ireland.

The boom-mounted magnetometer measures the Earth's total magnetic field and the data received are processed in flight by an on-board digital computer. VLF navigational data from a GNS 200 receiver also pass through the computer, and the resulting magnetic and navigational information is recorded simultaneously on magnetic tape for future analysis. The Geological Survey of Canada, which jointly sponsors the work with the NAE, hopes in time to produce complete magnetic profiles of the areas covered.

This work has had two main objectives. Top priority has gone to the identification of possible oil-bearing areas on the continental shelves of northern Canada. The aeromagnetic technique makes possible relatively cheap and easily applied geophysical surveys, and usually yields data accurate to 10 per cent on the thickness of potentially oil-rich sediments. The project has been remarkably successful and a number of discoveries have been made, mostly in the Labrador Sea and Baffin Bay. Sediment thicknesses in excess of 40,000 ft. were discovered on the outer part of the Labrador Shelf, leading to oil-company leases in the area during the mid-1960s. Sediment-filled troughs have also been found in Cumberland Sound, south-east Baffin Island, Melville Bay and north-west Greenland, and thicknesses of more than 16,000 ft. have been detected over most of Baffin Bay itself.

Objective No. 2 has been the mapping of the magnetic "stripes" characteristic of ocean rocks produced by sea-floor spreading (continental drift) in the North Atlantic and Labrador Sea. In this way the history of continental movements can be studied. The North Star mapped the stripes which occur along the axis of the Labrador Sea, conclusively demonstrating for the first time that continental drift had occurred between Greenland and the North American continent. Indeed, it was found that a triple-spreading junction (a point at which segments of the Earth's crust have parted) occurs south of Greenland. This work contributed greatly to the revolution in the Earth sciences which occurred during the 1960s and led to a complete change in our understanding of the history of the Earth's crust during the past 175 million years.

Although I had already sampled the pleasures of flying in the North Star, I was still very much looking forward to another flight. So when it was confirmed that I would be crossing the Arctic on the world's last North Star mission I was, needless to say, overjoyed. The object of the expedition was a low-altitude survey of the Gulf of Boothia and Admiralty Inlet, North West Territories, to obtain magnetic profiles of both areas. On the morning of April 21 [1976] the crew and I — each carrying a bulky set of Arctic clothing — boarded the North Star. There were 15 of us, led by Mr. A.D. Wood, head of the Flight Research Laboratory, who alternated with the other pilot, Mr. W.T. Chevrier, as aircraft commander. Capt. W.D. Sherk, a navigator seconded from the Canadian Armed Forces, and flight engineer O.H. Stevens completed the flight-deck crew. The geophysical team consisted of two members of the Geological Survey of Canada, Dr. P. Hood and Miss Margaret Bower (who has accumulated well over 2000 North Star hours). The remaining souls acted as nursemaid for the aeroplane and its equipment — research engineers, technicians and aircraft maintenance crew, with temporary experimental observer Piercey bringing up the rear.

The North Star, finished just as it was in its RCAF days (except that blue replaced red), looked immaculate. As it bathed in the sun on the apron, CF-SVP's undersurfaces still bore a mirror finish testifying to a wealth of care and attention over the past 14 years. The interior is still military all-green, although most of the seats have been removed to make room for the special equipment installed. Two racks house the electronic equipment: one at the rear accommodates the magnetometer gear, and the other, running back down the aisle from the front, supports most of the equipment needed on survey flights.

I sat in an aft-facing seat right between the engines — quite the noisiest place in the whole aircraft. At 10:30 a.m., about an hour behind schedule, we taxied out to the far side of the airport. There we sat for a minute before Research 8 (our callsign for the trip) was given the go-ahead and became airborne amidst the customary Merlin din.

We climbed to 7000 ft. over Ottawa and headed north towards Frobisher, our en-route staging point. Before take-off I had been told that I would be well advised to wear ear mufflers. I chose not to, and bitterly regretted my decision for the next six and a half hours! Flying at 220 mph, we made our way across the icy wastes of northern Canada, over the Hudson Strait, and onwards to Frobisher, 1200 miles north of Ottawa. After circling the town once we put down on an ice-layered runway. The door was opened and I was treated to my first impression of the Arctic — intense cold . . .

After an overnight stay we returned to the aircraft, only to find that one of the digital tape recorders aboard was not working. We resigned ourselves to a long delay, but by afternoon the aircraft was ready and the crew scheduled an early-morning departure the next day. I awoke to a clear, crisp morning, with the temperature hovering at about -5°C.

After take-off for Resolute Bay, our designated base for the next four days, we again flew north at 7000 ft. and as each minute passed the landscape below became whiter and whiter. As we were a day behind schedule the aircraft captain decided to go via Admiralty Inlet and survey the area on the outbound trip instead of on the way home. After about four flying hours we began the descent to the southern end of the inlet, which runs parallel to Brodeur Peninsula on Baffin Island. To obtain the data required for the magnetic profiles the North Star had to fly a saw-tooth pattern backwards and forwards across the area at 1000 ft. Unfortunately, visibility was not as good as I had hoped and I could see little but blinding whiteness stretching to the horizon. The survey was complete after 413 line miles and two flying hours. We then climbed away from the inlet after a satisfactory day's work and headed north-west for Resolute Bay, one hour away.

We had flown as many as 1500 miles that day, so the North Star had certainly earned a rest. After arriving at Resolute, a small "town" with a population of no more than 350, the North Star was bedded down for the night. Small heaters were placed in the engine intakes, which were then sealed with guards. Frozen, unstartable engines are a very real risk in the Canadian Arctic. Resolute Bay is only one-tenth the size of Frobisher, yet the airport is ten times busier, with up to ten DC-3s coming and going frequently throughout the day. We walked through deep snow to the small hotel at the edge of the airfield.

It was at this point that I really felt the cold for the first time — the temperature was -15°C — and I gave fervent thanks for my large sack of Arctic clothing. It was still light at 11 p.m. (the sun officially sets at 10:30 p.m. and rises at 2:09 a.m., but it never really got dark) and as I photographed the aircraft sitting amidst snow drifts at 1 a.m. I was struck by how crisp, clean and clear the atmosphere was.

I awoke next day to the sound of throbbing Merlins. Every morning,

CF-SVP-X lifts off from Frobisher on a mapping mission, April 1976. (Stephen Piercey)

The NAE North Star seen refuelling under Dakota C-FIAZ's nose at Churchill. (Stephen Piercey)

before the day's flying began, the engines were run up for half an hour to clear the oil that had congealed overnight. Departure time was 9:30 a.m., and we were bound for the Gulf of Boothia. I turned down this flight so that I could take some action photographs as the North Star took off and landed. I was free to wander out to the edge of the runway and was afforded some splendid photographs as the aircraft departed. Such was the stillness, the aircraft was still audible 15 minutes later. Research 8 arrived back safely that afternoon, having covered the mid-Boothia region in another eight flying hours, and was again prepared for a 9:30 a.m. departure the next day to survey the northern section of the Gulf.

The scenery I was to experience on this sortie was nothing short of spectacular. The day was clear, and at 1000 ft. operational altitude gave me a really splendid view of the ice-bound Arctic. The North Star flew lines 20 nautical miles apart across the survey area, and as we approached and flew parallel with the cliffs bounding the Gulf it was possible to study the steep rock formations in great detail. Polar bear tracks were also clearly visible and in the afternoon we spotted three of the creatures that had made them. We also flew over a caribou herd, and at 4:15 that afternoon a school of about ten 15-ft. white Beluga whales was sighted north of Brodeur Peninsula.

Seven hundred and fifty line miles were flown that day and, as the few remaining engine hours were steadily clocked up, the end of the North Star's career crept nearer. With 1800 line miles already covered in the two areas surveyed since our departure from Frobisher, the last 440 survey miles were to be flown across the southern Gulf of Boothia on our way back to Ottawa the next day.

Take-off time was set for 8:30 a.m., and shortly after this time the engines fired up for what was to be the North Star's last operational flight. Still sprightly after 29 years of hard flying, she became airborne halfway down Resolute's runway to begin the 2½-hour flight to the tip of the Gulf of Boothia. The weather on arrival was not too clear, and although we were flying at the usual 1000 ft. there was little to be seen from the cabin.

Fitted at the rear of the aircraft is an NAE-developed safety device known as the "crash position indicator." In the event of a forced landing a crash-proof radio transmitter is automatically released by impact switches in the nose and wings of the aircraft, thereafter transmitting for seven days to alert rescuers of the aircraft's position. It is a real comfort to the crew of aircraft which, like the North Star, fly over some of the most desolate areas in the world. Recently, during low-altitude flying, a bird hit the impact switch in one of the wings. The radio beacon was ejected and within 15 minutes its signals had been picked up and "rescue" was on its way.

With the last survey complete, Mr. Wood took the aircraft back up to 7000 ft. and instead of returning to Ottawa via Frobisher, decided to drop in at Churchill on the western shores of the Hudson Bay. Little snow was evident on the ground, indicating that we were nearing warmer climes again. After a perfect landing the aircraft was parked among a handful of DC-3s and a Lambair Otter.

My last flight in the North Star was set to begin at 10:30 a.m. We took off on time and climbed smoothly to 9000 ft. where at 230 mph we headed for home. After a 6-hour flight Research 8 put down for the last time in anger. It had just 10½ engine hours left. The passengers watching from the terminal had little idea that this North Star arrival marked the end of 30

The NAE North Star at Uplands. Note the underwing "bird." When in use, it would be strung out behind the aircraft on a long cable. (NAE)

years of dependable worldwide operations.

The North Star turned off the runway and headed back to the NAE hangar, where the Merlins were silenced for virtually the last time. The aircraft was originally to be withdrawn from use in 1968; this date was put back until 1974 and delayed still further until 1976. Will there be another stay of execution? Unfortunately the answer is "No." This is the final "end" and the time for the North Star's last bow has come.

Stephen Piercey

Not long after, in September 1976, Stephen Piercey received a letter from Marg Bower, a computer operator with the NAE and part of the scientific crew aboard CF-SVP-X. She wrote that "on May 8, 1976, we made our final working flight in the North Star, a doppler calibration and photo mission. A Harvard flew with us and we took pictures of each other. Rumour has it that the Air Force will be taking the North Star back to keep on display and to fly occasionally at airshows. Hope so, as I'd really hate to see the old bird chopped up into little bits . . . Now every vestige of project equipment has been removed from the North Star and it looks absolutely naked. Except for the tail boom, you would never know we had been there. When they weighed the plane, it was 7000 lb lighter. I have the computer and some of the peripherals set up in a lab right now and am doing nav 'refly' on this spring's mission. Oddly, the system doesn't work as well as it did in the North Star. Someone suggested playing it a tape of Merlin engines!"

On April 7, 1977, CF-SVP-X was flown to CFB Trenton for storage and disposal. Crew included Col. J.C. Henry (base commander, Trenton), and NAE personnel A.D. Wood, Bill Chevrier and O. Stevens. It was hoped that the aircraft might become a permanent display at Trenton but it was put on the block by Crown Assets for sale to the highest bidder. Various parties were interested, including the Canadian Warplane Heritage. Toronto-based freight operator Carl Millard bid on the North Star as its cargo doors and other parts would be useful as spares for his fleet of DC-4s. But the successful bidder was Aircraft Salvage of Dallas at a reported $26,000. Around July 10, 1977, a crew arrived at Trenton to ferry the North Star, now registered N8022L, to Tucson.

N8022L was no sooner airborne than problems forced an emergency landing at Rochester, New York, just across Lake Ontario from Trenton. The aircraft was grounded there for over a week, then flew on its way. But trouble again cropped up and a few minutes after takeoff from Rochester an emergency descent had to be made into Buffalo.

Four days later the North Star made another bid for Tucson, but somehow it was bound not to get there. No. 2 engine blew its coolant and seized up. The crew brought their lame duck down at Peoria, Illinois, where it was to sit until early September while the engine was overhauled in Los Angeles. The North Star finally reached Tucson where it was readied for use as a freighter.

On November 3, 1979, N8022L was sold to a Miami concern. It left Tucson and next turned up unserviceable on Greater Inagua Island off the eastern tip of Cuba. This was in December 1979. A mechanic from Tucson flew out to inspect the aircraft and found a minor foul-up caused by dirty fuel. He went home, then was called back in early January. This time there was serious trouble as someone unfamiliar with the Merlin engine had run the aircraft up and in the process seized an engine. Several Merlins were rounded up in Mexico and the plan was to get these overhauled in San Diego and have lots of engines to keep the North Star on the go. On the go for what, though?

About this time several people associated with N8022L were arrested. The US Drug Enforcement Agency had been watching their movements and had determined that they were using the North Star on "clandestine" operations. Since then, several people have been sentenced to prison terms in the case. At the time of writing, the aircraft, once part of the RCAF's proud North Star fleet, veteran of the Korean airlift and many another historic operation, lies basking in the sun on Greater Inagua, but perhaps it is not down and out. A number of Miami-based pilots are known regularly to discuss resurrecting N8022L. Only, of course, after a good few beers, but one never knows . . .

The Last Word

As so many of those who knew it have agreed, the Canadair North Star is now "ancient history." With few exceptions, the airplane has faded from the scene, and even from memory. Most of those who fly today for Air Canada, CP Air, British Airways and the Canadian Forces (known in "the good old days" as TCA, CPA, BOAC and the RCAF) wouldn't know a North Star or an Argonaut if they tripped over one. All the more reason for a book such as this which can give credit to a noteworthy Canadian venture, and to thousands of people for whom it was so important.

You've seen in this book how the airplane was criticized, even maligned. Some of this was natural enough, for, like many other airplanes, the North Star had its problems. Ask any TCA mechanic from the late 1940s. He's likely to get hot under the collar and give forth with a litany of horror stories! But people learned to cope

North Star veterans J.L. "Lindy" Rood, A.J. "Al" Lilly and George Lothian at George Lothian's place in Largo, Florida, in February 1982. Pilots of great experience, each made important contributions to the development of the North Star, and to the success of TCA and Canadair. They represent a much larger group of dedicated people who led the way on Canada's postwar scene. (Larry Milberry)

Lunchtime photo of TCA North Star "old timers" taken October 9, 1981 at the Cara Inn, Toronto International Airport. In the front row from the left are captains Jerry Lloyd, Art Bush, Bob Penrose and Harry Clarke. Standing are captains Ron Dennis, John Higham, Don Lamont, Lyle Greenlaw, writer Larry Milberry, captains Don Lowry, Don McLean, Gus Bennett, and purser-steward Glen Cawker. (G.S. "Monty" Montgomery)

Harold Dondenaz and Beth Buchanan of Air Canada's Historical Services greet Larry Milberry at Dorval airport in July 1981. To Beth's left is Don McVicar, wartime pilot, head of World Wide Airways and author of *Ferry Command* and *North Atlantic Cat.* (Matthew Milberry)

with the airplane and before too long most of the serious bugs were cleared up. By 1955, for instance, the Argonaut was the strength of the BOAC fleet and the airplane was making money. Company chairman Sir Miles Thomas noted at the time, "I would like to pay particular tribute to the way in which the Argonaut fleet has held the fort during the difficult period that we have gone through in the last 12 months. During the financial year just ended, they have achieved an aircraft utilization of 3440 hours, equivalent to 9.4 hours a day, and the standard of performance of the aircraft . . . has become increasingly attractive to passengers." Douglas test pilot Bob Brush recalls that "the North Star with its Merlins was a beautiful machine to fly," and, if you please, "the North Star was the lowest in vibration and the most pleasant to fly of all the transports I have tested."

World's last North Star M2. This former TCA aircraft still sits at Mexico City airport where it has served many years as a café. It's unlikely that this important piece of Canadian history will ever get back to Canada. (Peter Keating)

This North Star M1 is currently on Greater Inagua Island between Cuba and the Caicos Islands and appears to be in good condition. But at the moment its future is uncertain. If any North Star is ever to fly again this would be the one. (Stephen Piercey)
▽

Captain Peter Cane of BOAC writes of the Argonaut: "It was an easy, simple aircraft to fly, with positive controls on all axes, if somewhat noisy with its stub exhaust Merlins. Personally I preferred its handling characteristics to those of the 749 Constellation, my previous aircraft. I distinctly remember the Argonaut as a most stable platform for astro, the main method of long range navigation on the Eastern routes. It was a delightful aircraft to land, although a full load takeoff in really hot conditions tended to be a bit sluggish. All the pilots who flew Argonauts had the greatest confidence in the aircraft. It became the backbone of BOAC operations and, beyond any doubt, proved itself a success."

Captain John R. Johnson of BOAC had this to say about the Argonaut: "At the controls you get the feeling that, whatever happens, the plane will see you through . . . The Argonaut may be a bit slow by modern standards, but it's tough and has no vices. It's a gentleman's airplane."

So after all is said and done, Douglas, Rolls-Royce and Canadair, it looks like you built a good one!

Maps

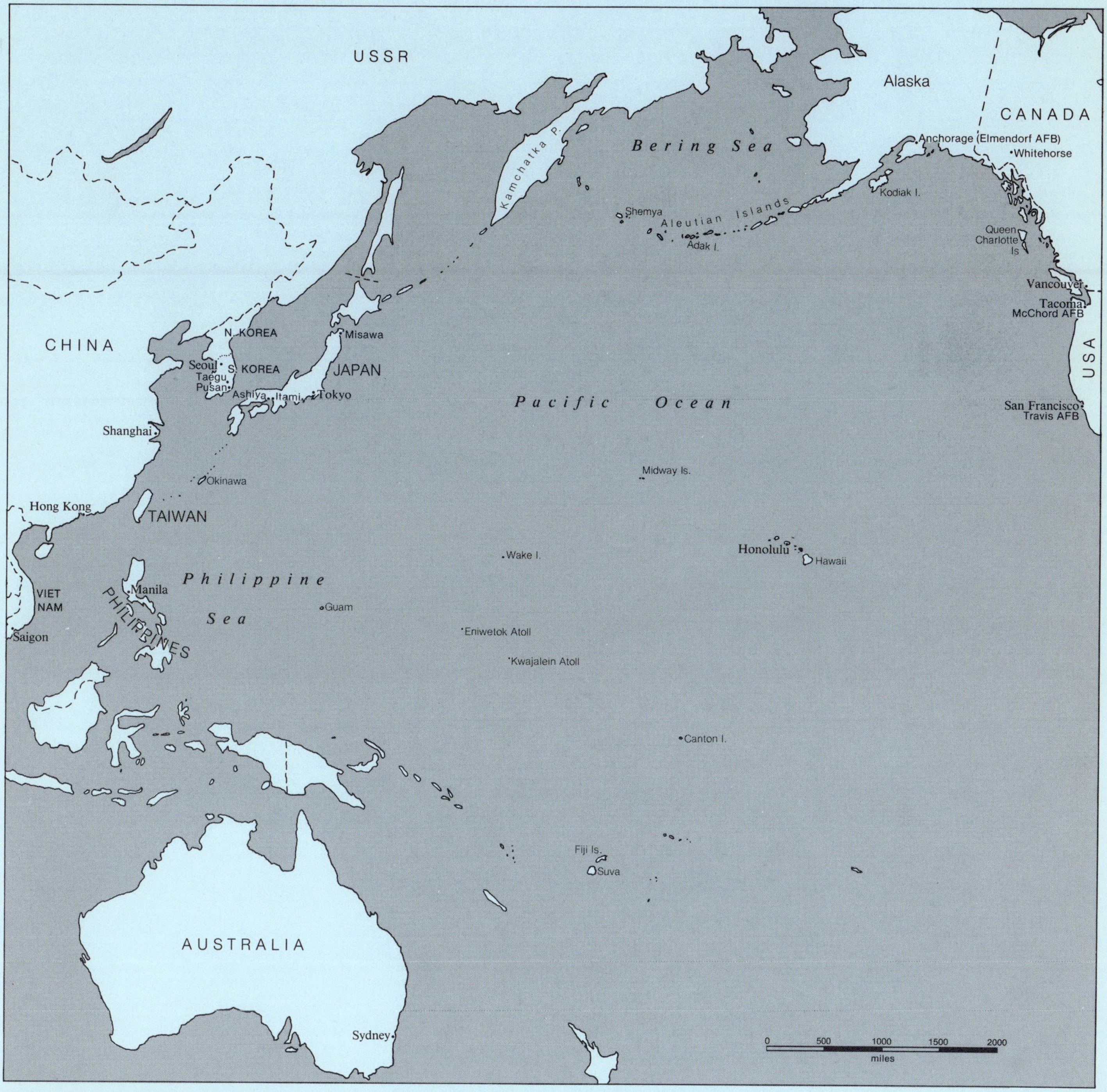

The Pacific Rim, showing ports of call for the North Stars operated by the RCAF, CPA and AVCO. (R. and M. Brass)

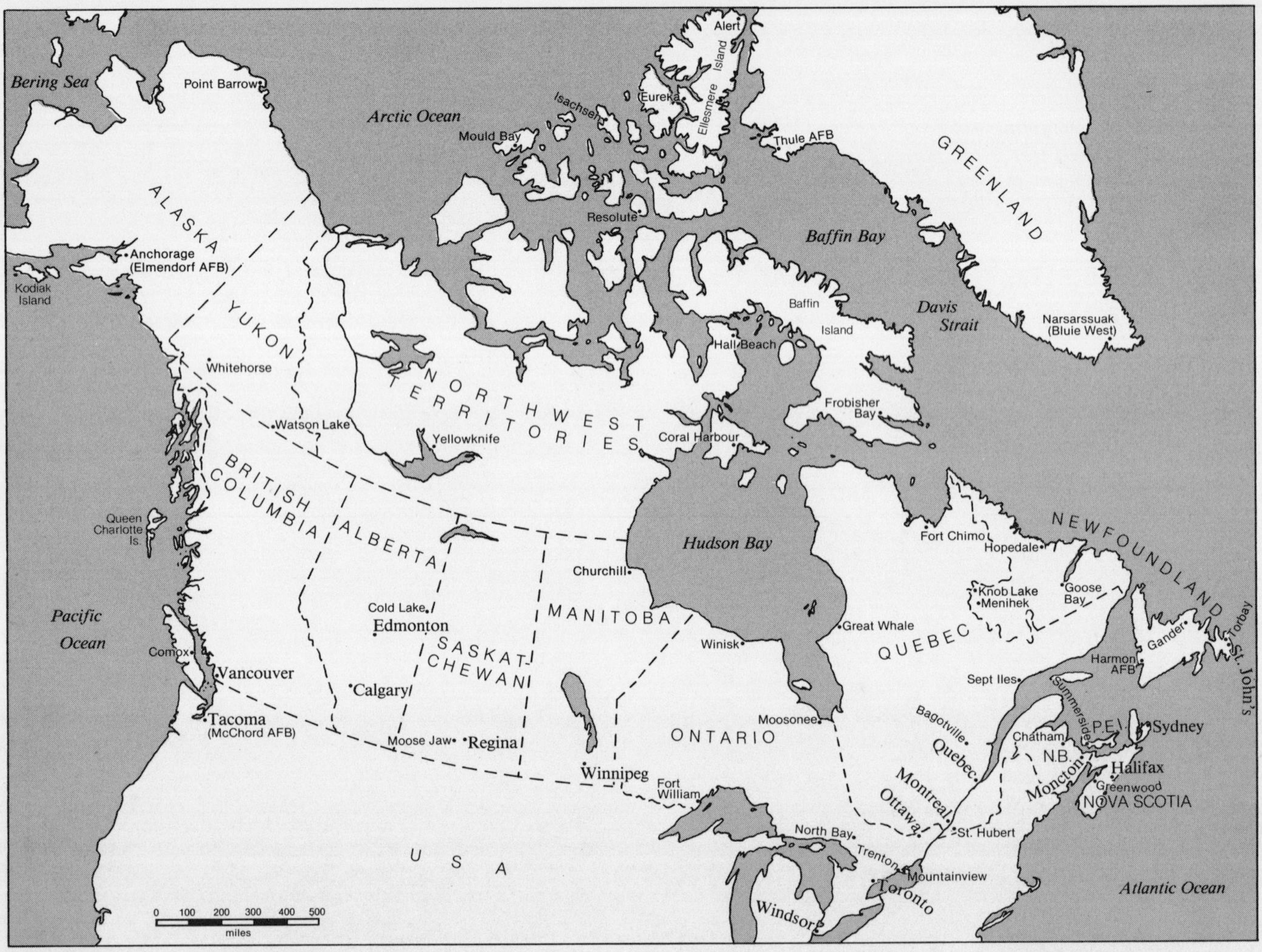

Canada, showing the principal centres for North Star operations as noted in this book. (R. and M. Brass)

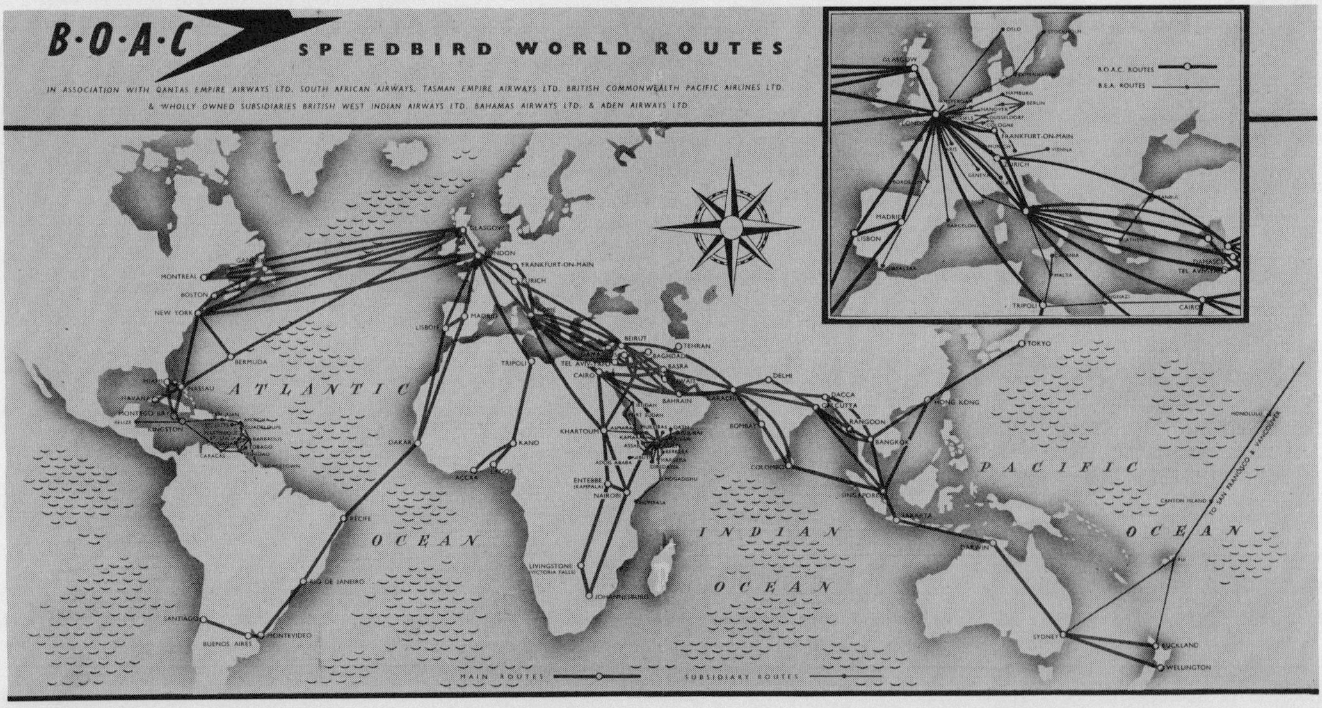

BOAC routes from the company's 1952 timetable.

Appendices

Reproduced here is a memorandum by J.T. Bain of March 14, 1944, to the Operations Manager of TCA. It explains much of the background to TCA selecting a new airliner for its postwar needs.

Winnipeg, March 14th, 1944.

Subject: Discussions Concerning York Aircraft & Douglas D.C.4. Agreement

Memo to Operations Manager

The following recount of my visit to Toronto, Ottawa and Montreal is prepared for your information and as a memo for file.

On Monday, February 28th, Mr. Dyment and I visited Victory Aircraft Ltd. to discuss the specifications of the Avro York. These discussions were prompted by advice from Mr. Symington that TCA would take over one York and, if investigations showed the York to be satisfactory airline airplane, that consideration would be given to the purchase of a total of three.

We were received at Victory Aircraft by Mr. D. Boyd and we placed before him the preliminary specifications prepared in Winnipeg subsequent to discussion with Mr. English and Mr. Symington.

After a brief review of the specifications we obtained Mr. Boyd's assurance that everything possible would be done to make the York a reasonable airline aircraft. It was pointed out to Mr. Boyd that there were many details of the York that made it unsuitable in its present form, such as lack of cargo space, lack of winterization, arrangement of aircraft systems, etc. Mr. Boyd fully appreciated the points we mentioned and assured us he would be very keen to make the York as good as possible for our purpose.

From Mr. Boyd's office we went into the details of the preliminary specification with Mr. Marchant and Mr. Shaw, Chief Engineers of Victory Aircraft. The remainder of Monday and most of Tuesday was spent in this detail discussion. Two main points arose from the examination of the preliminary specification. These were:

1. No design data is available in Canada for Avro York and there appears to be reason to doubt that the parent English Company can supply adequate data to secure a Canadian Certificate of Airworthiness at a gross load of 65,000 lbs.

   Without complete design data and calculations it will not be possible to incorporate TCA's requirements, nor have we any basis on which to deduce the economics of York operation.

2. The present Engineering staff at Victory Aircraft is inadequate to complete the necessary project engineering entailed in the TCA modifications. There is great enthusiasm to do a good and complete job but there also is only small realization of the amount of work involved and the difficulty of obtaining properly experienced engineering help.

Because of the above two points Mr. Dyment and I again visited Mr. Boyd's office and re-discussed the project.

The possibility of making a complete stress analysis in Canada was examined and Miss Elsie MacGill (Consulting Engineer, Toronto) was considered capable of handling the work. (Later enquiry showed that the time element would be too heavy, also the cost of such independent analysis was considered excessive. It was therefore considered that before putting forward any such recommendation that the possibility of obtaining any assistance from the parent company should be fully explored.)

Mr. Boyd believed that, with some assistance from TCA Engineering Dept., the necessary project engineering could be handled. (I do not share this view fully with Mr. Boyd because, whilst it is agreed that TCA Engineering could be of assistance, the penalty to present operations would be too heavy.)

Summing up the situation we have:

Insufficient data to complete the York as specified by TCA and,

Insufficient Engineering help or ability to produce the airline aircraft.

From Victory Aircraft Mr. Dyment and I proceeded to Ottawa where I was due to attend a meeting of the Post-War Aircraft Manufacturers' Committee in Mr. Bell's office; however, in view of the information and findings made at Victory Aircraft I was prompted to call Mr. English from Ottawa and outline the picture as we saw it. Remembering that Mr. Bell had been somewhat disturbed over our inability to produce the Avro York specifications over-night, I believed it would be to advantage to outline to Mr. Bell our discussions at Victory Aircraft. Mr. English concurred that a discussion of the situation with Mr. Bell might have very desirable results in unifying the points of view of everyone concerned, consequently, an appointment was requested with Mr. Bell, and we attended at his office on Wednesday, March 1st, at 9.45 A.M. During the interval between calling Mr. Bell and attending his office, Mr. Bell advised Mr. C.D. Howe that we were in town and the Minister expressed a desire to see us.

In Mr. Bell's office we outlined the York picture detailed above, and Mr. Bell appreciated the necessity of obtaining correct design data before proceeding further with the construction of the York.

During the discussion of the York Mr. Bell advised us that the agreement between the Canadian Government and the Douglas Aircraft Corporation had been signed, giving Canada permission to build, under licence, the Douglas DC4 aircraft. This subject was mentioned to establish whether or not it affected the suggested York program. Also in Mr. Bell's office there was some mention of the plant in which the DC4 should be built. Mr. Dyment and I expressed no opinion on this matter as we had no knowledge of any Canadian manufacturing plant with the exception of Boeing, Vancouver, and Victory Aircraft. Mr. Bell suggested that, if the opportunity presented itself, we should examine the Vickers plant at first-hand.

Mr. Bell then took us around to Mr. Howe's office where we again briefly outlined the York picture as we saw it. In this discussion, Mr. Howe mentioned that he could authorize the issue of a C. of A. for the York aircraft without the necessity of full collaboration with England. I suggested, however, that it would not be in the best interests of TCA to use on our system an airplane, the strength data of which was an unknown factor. Additionally, it would be undesirable to place in our service a Canadian built aircraft which did not give us first-class reliability.

Mr. Howe then mentioned the Douglas DC4 agreement and expressed the view that DC4 equipment would be available in a period of 15 months, consequently Mr. Howe was interested to know how this would affect the Avro York picture. I expressed the view that we would be fortunate to obtain Douglas DC4 equipment from any Canadian manufacturer in a period of time shorter than two years, also that, as the York could be available in a much shorter period of time, that it probably would still be correct to proceed with the construction of the Yorks. (A time element of 6 months to 1 year was given, depending on the amount of information available in England on the York.)

Mr. Howe asked some questions about the Lancaster operation, referring specifically to the Merlin engines. I outlined to Mr. Howe what had been done to correct the Merlin operation and he expressed a degree of satisfaction with the results obtained. Mr. Howe expressed the hope that the Merlin could be developed into a satisfactory commercial engine as it was his desire to "use" the Merlin engine.

In a further brief discussion with Mr. Bell, I intimated to him my desire to proceed to Montreal immediately rather than wait for the Post-War Aircraft Manufacturers' Committee meeting, the following day. He agreed that the committee meeting would not be of too important a character and that it would be better to acquaint Mr. Symington with the details of our conversations.

On Thursday morning, March 2nd, we were given an appointment by Mr. Symington and reviewed for him all of the discussions outlined above. The DC4 agreement was further discussed in its application to TCA and we expressed to Mr. Symington our lack of confidence in the ability of Boeing to build for us a satisfactory aircraft. Mr. Symington advised us to take the opportunity of examining Vickers' plant, should it present itself. In regard to the Yorks, it was decided that Mr. Dyment and I would proceed to England to investigate at first-hand the availability of design data and the establishing of the maximum permissible gross load.

On Friday, March 3rd, I received a telephone call from Mr. B. Franklin of Vickers Aircraft and he invited us to examine his plant. As a result we visited Vickers on Friday afternoon. From the outset it was quite clear that Mr. Franklin was seeking to impress us with the suitability of the Vickers plant to manufacture the Douglas DC4. He outlined for us the past history of Vickers in meeting commitments and schedules, and the reduction of man hours effected in the construction of the Consolidated Catalinas. We outlined for Mr. Franklin some of the problems entailed in the construction of an airline airplane. We endeavoured to show him that the basic DC4 required considerable alterations to make it an airline airplane suitable for operation by TCA. We gave Mr. Franklin a rough estimate that the time to produce a DC4 for TCA would probably take 40-50% longer than the basic airplane detailed by the Douglas drawings.

In this discussion with Mr. Franklin we were careful to avoid any commitment that TCA would support Vickers in their desire to secure the Douglas contract. In reply to direct question, however, I told Mr. Franklin that, from what I had seen, I believed the Vickers plant would be suitable for the fabrication of the Douglas. (Note: Amplifying somewhat on this statement, for your information, it is my opinion that of the three factories, Boeing, Victory and Vickers, Vickers is by far the most suitable for the fabrication of the Douglas.)

On Saturday morning, March 4th, I again called Mr. English long distance to explain the inference of our discussions with Mr. Franklin the previous day, also that, as the Post-War Aircraft Manufacturers' Committee meeting had been called for Monday morning (March 6th) in Ottawa, it was likely that some questions regarding the DC4 agreement would be asked by various parties who seem to be interested in the Douglas contract. Because I appreciated that I have no authority to make commitments in any such regard for TCA, I desired some guidance as to how such questions should be answered. Mr. English advised me to go and see Mr. Symington and tell him the same story. This was done.

At the outset of our discussion with Mr. Symington he very shortly told us that any questions regarding the Douglas contract should not be answered, excepting to advise the questioner that such matters would be decided by the Minister. We proceeded, however, to explain to Mr. Symington that our basic desire was to obtain the most suitable aircraft that was possible for TCA and, with this sole object in view, we believed it would be of interest to Mr. Howe to have a recount of our experiences with the Boeing Aircraft Company, during the conversion of our 14H2's to 14-08's. We also explained to Mr. Symington the type of revision we were requesting from the standard Douglas DC4 as we hoped thereby to produce an airplane slightly better in performance and considerably more economic to operate. At the close of the discussion, Mr. Symington asked that I go to Mr. Howe's office on Monday morning and recount the same story to Mr. Howe.

On Monday morning, March 6th, I attended at Mr. Howe's office as instructed by Mr. Symington and reviewed various factors concerned in the production of a TCA aircraft. I also recounted our history with Boeing Aircraft and, in reply to Mr. Howe's questions, advised him I considered that the Vickers plant would be satisfactory for the fabrication of the Douglas DC4. Mr. Howe agreed that, from his knowledge, the Vickers plant was probably the most suitable plant in Canada; however, he also pointed out that commitments had already been made to the British Government for the minimum of 40 Lancasters per month of Canadian manufacture. The Vickers plant, it was his belief, would be fully occupied to obtain such production. I assured Mr. Howe that my only desire was to give him some information to assist his decision regarding the awarding of the Douglas contract.

Later the same day Mr. Franklin of Vickers Aircraft called me in my room of the Chateau Laurier and advised me that he had had an interview with Mr. Howe regarding the Douglas DC4 contract. According to Mr. Franklin, Mr. Howe had made a commitment that, if he (Mr. Franklin) could show how 40 Lancaster aircraft per month could be produced in Canada, he would consider awarding the Douglas contract to Vickers. Mr. Franklin proceeded to outline that it was his belief that 40 Lancasters a month could be produced from Victory Aircraft if he was allowed to devote the energies of a portion of the Vickers plant to a sub-assembly subcontract. Knowing that I was in some degree familiar with Victory Aircraft, he asked me if I knew of any additional assistance he could give to Mr. Boyd of Victory Aircraft which would assist him to produce the necessary quota of Lancasters. In reply to Mr. Franklin, I advised him, that with the superficial knowledge I had of Victory, there was little I could tell him excepting that I knew Mr. Boyd had had considerable trouble with final assembly and flight testing. I suggested that the production of Lancasters was a subject for discussion between Mr. Boyd and Mr. Franklin and not one on which I could give any advice or assistance.

The following morning (Tuesday, March 7th) I again called Mr. English to acquaint him with the details of my discussion with Mr. Howe and the subsequent conversaton with Mr. Franklin.

(Sgd) "J.T. Bain"
Supt. Engineering & Mtce.

Excerpts from a letter of November 13, 1946, written by J.T. Bain to H.O. West in Seattle. The letter gives some insights into the inner workings of Canadair at the time.

Montreal, P.Q.,
Nov. 13, 1946.

Mr. H.O. West,
Seattle, Washington.

Dear Oliver:

I much regret not getting this letter away to you yesterday.

I am going to be quite open in describing the present situation in this plant in the hope that it may give you a fair idea of what is going on and act as a basis for your examination of any proposal Canadair may make to you.

Plant

The plant is a new one built during the war to handle an extensive contract for PBY's. Although mainly constructed of wood, it is quite modern in layout. For the present contract there is adequate plant capacity and complete coverage of equipment. The latter closely follows the pattern established in Chicago by the Douglas company for their extensive C54 contract.

Additional to the main plant a considerable portion of the factory area previously occupied by Noorduyn Aviation has been taken over by Canadair as a "conversion plant" in which an extensive programme of C47 to DC3 conversion is in process.

Work in Process

There are two main jobs in hand.

1. The lesser of the two jobs is the C47 conversion programme employing approximately 2,000 men. Converted C47's have been completed for many different countries. The quality of work compares more than favourably with the average of these jobs done elsewhere. There has been considerable disagreement between Canadair and the Government and within Canadair on the handling of the conversion programme. Ben Franklin, the president of Canadair, was smart enough to purchase all the work in process at Oklahoma City after VJ Day and as such has a corner on the only available civil source of supply of C47 materials in the world. By agreement with the Government, the conversion programme was started to keep up the direct labour in the plant and to retain the services of tradesmen for whom work did not exist in the initial stages of DC4 tool-

ing. Since that time, because of influences I will explain later, the tail has somewhat wagged the dog and the main project in my opinion has suffered very considerably from the pressure which has been applied to the commercial business of C47's.

2. The second main programme is, of course, the C54-Gm and DC-4M types being built for the R.C.A.F. and T.C.A. respectively. The maximum number of aircraft at present involved is 70, for which instructions to proceed have been received for 39. Subsequent orders to the total quantity will be given (unless policies are changed) during fiscal years '47-'48 and '48-'49. Some negotiations have been conducted between the Canadian Government and the Douglas company in an effort to expand the territory clause of the License Agreement. Attempts in this line to date have met with little success; however, we do have the rights to sell the aircraft in England and considerable interest has been shown by both the R.A.F. Transport Command and later by B.O.A.C. The present contract was given to this plant by the Government on a cost-plus basis.

Organization and Management

Before any discussion of the existing management and organization I would like to reiterate the salient points of the above.

1. The plant and all its equipment is owned by the Canadian Government.

2. The License Agreement giving Canada the right to build the DC4 was completed between the Canadian Government and the Douglas Company. The contract for DC4 construction was awarded by the Government to Canadian Vickers.

Since the contract was awarded by the Government to Canadian Vickers, Mr. B.W. Franklin, Manager Aviation Division, Canadian Vickers, completed an agreement whereby a new company (Canadair Limited) took over the contract and ran the plant on a management fee basis for the Canadian Government. At the time Canadair was formed a purchase option was granted by the Government to the Canadair company. Mr. Franklin, of course, remained as president of Canadair Limited holding all but two of the shares.

Ever since the formation of Canadair Limited, Mr. Franklin has been negotiating with the Electric Boat Company, New London and New York, to take up the option to purchase. At the present time Electric Boat have become involved only to the extent of investing $100,000. for a 10% share in Canadair Limited.

Ben Franklin is a remarkable man in many regards. He will invariably come out on the top side of any negotiation. He has remarkable ability to get the best out of any situation where money is involved. His ability as a negotiator can perhaps best be demonstrated by the fact that he is president of the only major aircraft plant in Canada, yet is a newcomer to the aviation business. He possesses singular mental agility but is certainly closer to being a promoter in the worst sense of the word than he ever has been a manager of an aircraft plant.

Within Canadair there are many very excellent men but the organization is such that they are not used to best advantage. The interrelationship of engineering and production is entirely incorrect and expensive, a fact that is indicated by the present excessive overhead. Much can be done to straighten out the present picture and must be done before any future programme is undertaken.

What follows may be a little incoherent as the situation is extremely complicated but it may serve as a basis for your own questions.

As I have mentioned before, Ben Franklin has been negotiating with the Electric Boat company to interest them financially in taking over the present contract, plant and equipment.

The management fee contract under which Ben Franklin is operating has been interpreted differently by the Government and Canadair Limited. The Government quite justly claims that all work completed in Canadair contributes to the DC4 programme. Ben Franklin chose to interpret his contract as permitting the Canadair company to complete and take a profit from commercial business not directly associated with the DC4. (It appears he has some very strong legal support.) A very handsome sum was accumulated by Canadair Limited from commercial business (mainly the sale of C47 parts from Oklahoma City) and this financial showing was used to good advantage with the Electric Boat company.

The difference of opinion between the Government and Ben Franklin only came to light recently during the first discussions of a purchase agreement. Mr. Howe roundly accused Ben Franklin of being a racketeer and was all for cancelling his contract and reverting completely to Government-controlled operation of the plant. Franklin's strong legal position, plus his side-stepping agility resulted in a more amicable arrangement wherein J.H. Berry (President of Canadian War Assets) was appointed as Managing Director of Canadair Limited, and Government auditors were put into Canadair to make detailed examination of the company finances. This arrangement was completed some weeks ago and Berry did in fact make an appearance in this plant; however, there is no visible evidence of anything he has accomplished.

Mr. Howe has softened his attitude towards the Electric Boat - Franklin combination to the extent that he is willing to negotiate directly with Electric Boat but has insisted that the management of the plant be put into competent hands. It was at this stage of discussion that your name was first mentioned by Mr. Howe.

In discussion with J.H. Hopkins of the Electric Boat company, I got the impression that they also would welcome competent management of the plant by someone of proven ability in the aircraft manufacturing industry, but they also were extremely positive that Franklin should be retained as president because of his particular ability in seeking out and promoting new business.

The first stages of discussion between the Government and Electric Boat have thus ended in agreement that in any event the plant should have an experienced manager, but there is no indication that agreement will ultimately be reached regarding the sale of the plant, or the taking over of the DC4 contract. Personally, I doubt very much whether the picture can be made attractive enough to interest the Electric Boat company. Franklin has declared his intention to immediately leave Canada if the Electric Boat company does not finalize a purchase agreement with the Canadian Government.

The costs of production in Canadair are running higher than necessary but, in all fairness to Canadair and Franklin, they are reasonably comparable to present sales prices for similar gross weight aircraft in the U.S.A.

The Government very strongly feels that two major aircraft plants should be retained in operation in Canada but, at the same time, are not prepared to make adequate financial contributions. From T.C.A.'s point of view, we want to be in a position to buy the best aircraft available anywhere in the world but are perfectly willing to purchase Canadian-built aircraft if they will measure up to our requirements and specification. It was only by Canadian manufacture that our present DC4 was possible.

The probability that a new type of aircraft for trans-ocean travel will be developed in Canada is very strong. The combination of the U.S. knowledge of transport aircraft and the British turbine engine developments offers possibilities for the manufacture in Canada of advanced types not available anywhere else in the world. Mr. Howe and Mr. Symington have indicated strong support of any such development project and a means does exist whereby financial assistance could be obtained.

As I advised you on the telephone, it was Mr. J.H. Hopkins of Electric Boat who asked me if I could get in touch with you. In the whole matter I am, of course, only an intermediary and would take no part whatever in any negotiations between yourself and Canadair. I have, however, indicated to Mr. Hopkins that men of your calibre and record do not come cheaply - a fact which did not seem to disturb him to any great extent. The future of Canadair and this plant is really impossible to predict at the present stage. Handled properly, it should be possible to develop and manufacture aircraft for export in addition to Canadian consumption. On the other hand, without proper financial backing and a clear cut programme, the handwriting may be on the wall within thirty days.

I hope very much to have the opportunity to discuss the matter more fully with you before you meet Hopkins or Mr. Howe. You know, of course, that you can count on me for any assistance.

Best regards.

Sincerely yours,
(Sgd) Jas. T. Bain.

Excerpts from the purchase agreement between Canadair Ltd. and Trans-Canada Air Lines, March 31, 1947:

Canadair shall manufacture, sell and deliver to the Buyer, and the Buyer shall purchase from Canadair 20 North Star Model DC-4M-2 aircraft . . . The aircraft shall be presented for acceptance as follows: the first in July 1947; two per month in August, September and October; and at least two per month thereafter . . . The price of each aircraft sold and delivered hereunder shall be $660,000 . . . Canadair agrees, represents and warrants that each aircraft and spare part delivered under this agreement shall, at the time of delivery thereof to the Buyer, conform with the detailed specification and be free from defects in material and workmanship . . . Before offering acceptance to the Buyer, Canadair shall conduct test flights on each aircraft sufficient to assure proper operation of the aircraft. Acceptance flights not exceeding 10 hours for the first aircraft and three hours for each subsequent aircraft shall then be conducted with the Buyer's representative aboard to demonstrate to the Buyer that the aircraft complies with the requirements of this agreement . . . Until each aircraft has been operated for 5000 hours or for a period of 24 months from the date of acceptance thereof (whichever first shall occur), Buyer shall not use any aircraft except on operations wholly within the Dominion of Canada or in international operations primarily originating in and based upon the Dominion of Canada . . . Canadair agrees, without cost to Buyer, to undertake immediately and to complete as promptly as possible, an investigation . . . to increase to 80,000 pounds the permissible gross weight for takeoff of the aircraft . . .

The new North Star airliner received a generally optimistic greeting from the press.

NEW GIANT TCA PLANE CHRISTENED

MONTREAL, July 22—"This is what we have been looking for. Every indication points to the fact that we have thoroughly achieved the aircraft specified by TCA engineers." This was the comment on The North Star, after its successful test flights, by J.T. Bain, superintendent of engineering and maintenance, Trans-Canada Air Lines, formerly of Winnipeg, who, at the request of B.W. Franklin, president, was loaned to Canadair Limited as executive assistant to the president, in charge of engineering and co-ordination between all departments in the manufacture of Canada's largest aircraft.

The North Star, four-engine 40-passenger transport, is the first of a fleet being built at Canadair by the Canadian government for TCA and the department of national defence. It was christened Saturday afternoon by Mrs. C.D. Howe, wife of the minister of reconstruction and supply.

"Jim" Bain, brilliant young aeronautical engineer, is modest about his part in the achievement, yet the basic idea was his and he followed it through with a faith and energy based on 26 years of hard aviation work. Bain pays wholehearted tribute to the engineering staffs of Trans-Canada Air Lines and Canadair and to the rank and file of the 8000 workers at the plant.

Mr. Bain joined Trans-Canada Air Lines in 1938 shortly after its foundation, becoming superintendent of maintenance and overhaul about a year later and superintendent of engineering and maintenance in 1942. He has been in Montreal since November, 1945, superintending the maintenance of North Atlantic aircraft and working on the development of the North Star.

—WINNIPEG FREE PRESS
July 22, 1946

BIG AIR TRANSPORT 'NORTH STAR' NAMED AT COLORFUL CEREMONY

40-Passenger Craft Marks New Era for TCA; Mrs. C.D. Howe Officiates

Aviation in Canada took a "notable" step forward on Saturday afternoon, when the largely Canadian-designed "North Star", four-engined, 40-passenger pride of Trans-Canada Airlines and the Dominion Government, was launched by Canadair Limited at St. Laurent.

The huge craft is among the largest transport planes built, and is claimed to be the most efficient in its class. The one launched Saturday will not see commercial use, since it is a testing prototype of similar ships yet to leave the assembly line.

The traditional bottle of champagne was broken over the ship's nose by Mrs. Howe, wife of the Rt. Hon. C.D. Howe, Minister of Reconstruction, who said, "I name thee North Star and wish thee Godspeed."

Following the ceremony, the plane was taken for a trial run. The thousands of spectators who had assembled under the warm sun to witness the event expressed admiraton as it banked and flew low over the runways.

"This plane will probably mean more to us than any other type of aircraft," said Air Vice-Marshal W.A. Curtis, representing the RCAF. "I am confident it is the exact type we need for transporting both men and material any place."

He praised the "foresightedness" of those who were responsible for production of the North Star. The plane represented a "great step forward," he stated, and should be especially appreciated by Canadians because it was designed at a time when most other nations had stopped pioneering in the field of aviation.

Mr. Howe lauded the utility of the ship, saying it would "bring glory to Canada and to TCA over the seven seas." The Air Force, he said, might use it as the first Canadian-built transport, and TCA "will like it for its comfort and practicality."

Other speakers included Hon. Ernest Bertrand, and His Worship Mayor Camillien Houde. Also present was Hon. J.L. Ralston, former Minister of National Defence. A reception was held in the Canadair cafeteria after the launching.

Officials pointed out that the plane is largely Canadian in design. While based on several Douglas ships, it has been "considerably modified" by TCA draftsmen. It is equipped with Rolls-Royce Merlin power-plants, and will be used both by the Government and TCA. The company expects to have several in use by the end of the year on national and international flights.

The craft has a maximum cruising-speed of 325 mph, a non-stop area of 3,500 miles, and a tricycle landing-gear, a wingspread of 117-1/2 feet, and is 97 feet in length.

B.W. Franklin, president of Canadair, officiated.

—THE MONTREAL DAILY STAR
July 22, 1946

DUBIOUS MAKESHIFT FOR AIR INDUSTRY

None of the stubborn dictations of Canada's air policy is more peculiar or puzzling than the course followed on the country's aircraft industry.

During the war years Canada developed a magnificent pool of aviation technique by building large numbers of various types of aircraft. At the insistence of a number of aviation enthusiasts, a representative committee of experts was set up in 1942 to investigate post-war possibilities and made recommendations to the government. That report, never published, urged government support for a small industry to produce Canadian-designed Canadian-built aircraft. It would be partly subsidized—the custom in all aircraft manufacturing countries—by government and RCAF orders.

Instead of encouraging Canadian initiative in aircraft design and manufacture, the government inappropriately settled upon a remarkable compromise. It set up a Crown company to build under licence an American aircraft in which was to be installed a British aircraft engine. The DC-4 is an admirable plane: the workhorse of the air, it has been dubbed. The Rolls-Royce Merlin engine is one of the world's best aircraft engines. But combining the two is another matter altogether, aeronautically; and no one can foresee the probable results until the new product has been fully tested under typical operational conditions.

But there are worse features to the scheme than the uncertainty of performance. Building under foreign licence reduces Canadian aircraft manufacturing from the status of an industry to that of mere unskilled assembly. In spite of the minor alterations that have been made here, the DC-4M is nothing more or less than a straight assembly-job. And it was put together at costs to the country that, while undisclosed for unstated reasons, are rumoured to be vastly out of proportion to the result achieved.

An additional cause of regret to those who are concerned with Canada's aviation future is that the Canadian Government should invest its faith and its people's hope in an aircraft that is at best obsolescent. The present is an interim period in aircraft design and manufacture. Planes far better than the DC-4 are being built at this time and will be at the disposal of Canada's air competitors at about the same time that the DC-4M is being put into operation by TCA.

Considering the heavy cost of Mr. Howe's experiment, and the several years it has taken to produce even a prototype by assembly, it must be hoped that the DC-4M will meet immediate requirements. It is evident, however, that Canada's air future is seriously narrowed by this unimaginative makeshift.

—THE GAZETTE,
Montreal, July 25, 1946

NEW PLANE IS WORLD'S BEST
C.D. Howe Tells of "North Star" and of TCA's Overseas Services

VANCOUVER, B.C., October 10, 1946. "This is the finest of all the post-war airplanes and a great triumph for our Canadian industry," said the Right Honourable C.D. Howe, Minister of Reconstruction, in describing the new "North Star" aircraft to a large audience brought together here today by the Board of Trade and Canadian Club.

"I came to Vancouver this time on a trial flight of the new 'North Star'," continued Mr. Howe, "and this aircraft will shortly go into service on the North Atlantic. Already this airplane holds the record for the fastest east-west crossing from Montreal to London, as well as the fastest west-east crossing from Ireland to Montreal. The pressurizing removes all the discomfort associated with changes in altitude. In short, it is exactly what the air traveller has been asking for.

"It seems to me that Canadians hardly appreciate the possibilities of our air service across the Atlantic. A few weeks ago, I found it necessary to be in London on a date that would not permit travel by ship. My wife did not favour travel by air, but agreed to go by air, provided I would agree to return by ship. We left Montreal at 5 o'clock in the afternoon, spent one hour at Gander airport, which is becoming a crossroads of the world at 6 o'clock Montreal time and 11 o'clock Prestwick time, stayed there for lunch, and arrived at our hotel in London in time for afternoon tea. In other words, only one working day had been lost. The trip across was so pleasant that my wife changed her ideas about plane travel, and asked me to arrange for return by plane, instead of by ship. On the day of departure from London, I worked at Canada House until 1 o'clock, then returned to the hotel for lunch and the trip to the airport. We left London airport at 2:30 p.m., had tea at Prestwick, arrived in Sydney, N.S. at 1:30 a.m., where we had a late supper, and reached our home in Ottawa at 5:30 a.m. I was at my desk at ten minutes of nine, ready for a good day's work. In other words, only one-half day of working time was lost on the return trip, and the comfort on the plane was such that not even loss of sleep was involved.

"The Canadian Trans-Atlantic service is now being operated with six unpressurized planes, which make 12 round trips each week. Before the end of the year, Trans-Canada will accept delivery of 20 pressurized planes. Most of these will be operated over the North Atlantic where travel seems to be limited only by the available space, particularly westbound, but it is also intended to introduce 2 round trips daily with pressurized four-engine equipment between Vancouver and Montreal, as well as a service from Montreal to Bermuda, Nassau, Jamaica, Trinidad and South American points. Trans-Canada Air Lines has an international reputation for reliable service, and the addition of this Canadian-built equipment enhances its position. Next summer, we intend to extend the service across the Pacific.

"In speaking of Trans-Canada Air Lines, I should mention the fact that its President, Herbert J. Symington, who is with us today, is making his last trip over the system in that office. Mr. Symington has had a great part in the growth of TCA and to him goes much of the credit for the world-wide reputation that it enjoys today. We all regret that pressure of his private business makes it necessary for him to retire as President at this time. Throughout the years of his association with Trans-Canada Air Lines, he has served the air line without compensation. I know of no one who has devoted more of his time and great ability to the public service, without monetary reward, over a greater number of years, than has Herbert J. Symington."

—TRANS-CANADA AIR LINES
Press Bureau, Vancouver, B.C.

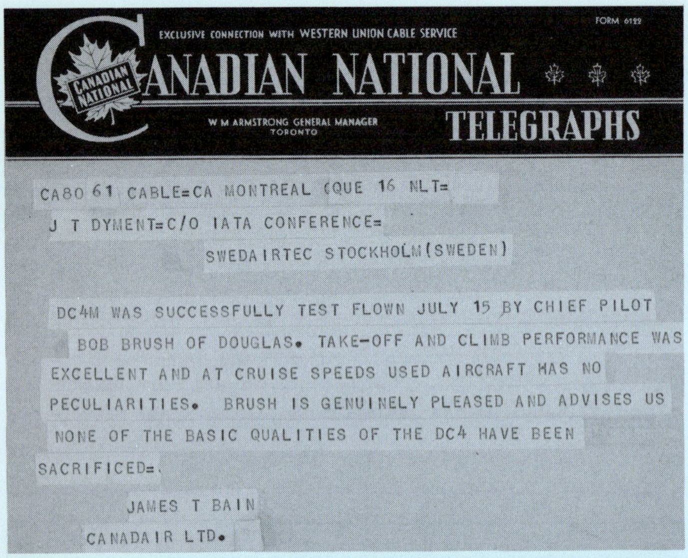

EXCLUSIVE CONNECTION WITH WESTERN UNION CABLE SERVICE

FORM 6122

CANADIAN NATIONAL TELEGRAPHS

W M ARMSTRONG GENERAL MANAGER TORONTO

CA80 61 CABLE=CA MONTREAL QUE 16 NLT=

J T DYMENT=C/O IATA CONFERENCE=

SWEDAIRTEC STOCKHOLM(SWEDEN)

DC4M WAS SUCCESSFULLY TEST FLOWN JULY 15 BY CHIEF PILOT BOB BRUSH OF DOUGLAS. TAKE-OFF AND CLIMB PERFORMANCE WAS EXCELLENT AND AT CRUISE SPEEDS USED AIRCRAFT HAS NO PECULIARITIES. BRUSH IS GENUINELY PLEASED AND ADVISES US NONE OF THE BASIC QUALITIES OF THE DC4 HAVE BEEN SACRIFICED=

JAMES T BAIN

CANADAIR LTD.

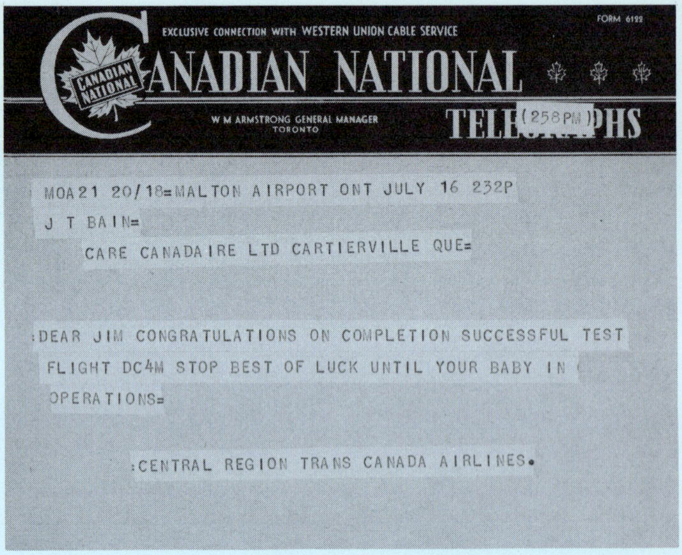

EXCLUSIVE CONNECTION WITH WESTERN UNION CABLE SERVICE

FORM 6122

CANADIAN NATIONAL TELEGRAPHS

W M ARMSTRONG GENERAL MANAGER TORONTO

(258 PM)

MOA21 20/18=MALTON AIRPORT ONT JULY 16 232P

J T BAIN=

CARE CANADAIRE LTD CARTIERVILLE QUE=

:DEAR JIM CONGRATULATIONS ON COMPLETION SUCCESSFUL TEST FLIGHT DC4M STOP BEST OF LUCK UNTIL YOUR BABY IN OPERATIONS=

:CENTRAL REGION TRANS CANADA AIRLINES.

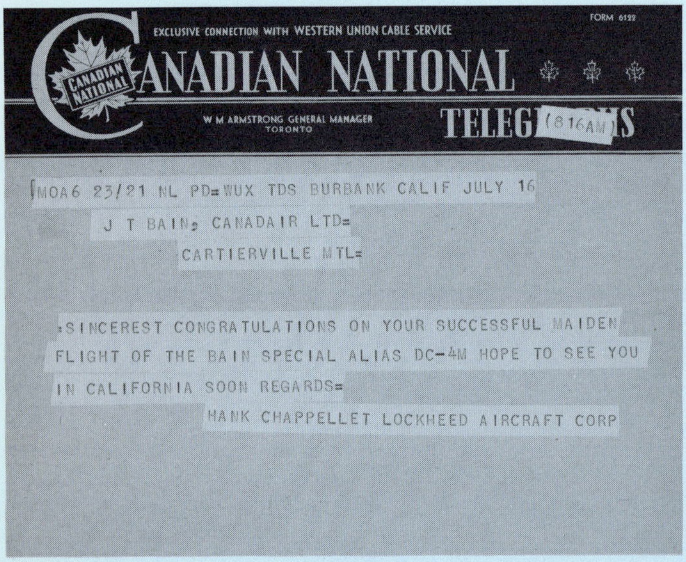

EXCLUSIVE CONNECTION WITH WESTERN UNION CABLE SERVICE

FORM 6122

CANADIAN NATIONAL TELEGRAPHS

W M ARMSTRONG GENERAL MANAGER TORONTO

(816 AM)

MOA6 23/21 NL PD=WUX TDS BURBANK CALIF JULY 16

J T BAIN: CANADAIR LTD=

CARTIERVILLE MTL=

:SINCEREST CONGRATULATIONS ON YOUR SUCCESSFUL MAIDEN FLIGHT OF THE BAIN SPECIAL ALIAS DC-4M HOPE TO SEE YOU IN CALIFORNIA SOON REGARDS=

HANK CHAPPELLET LOCKHEED AIRCRAFT CORP

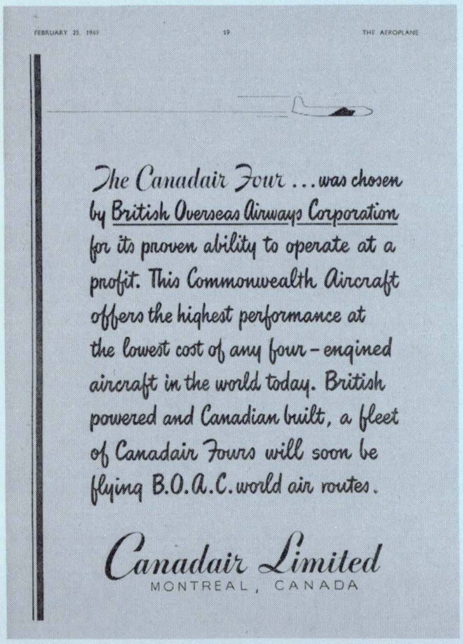

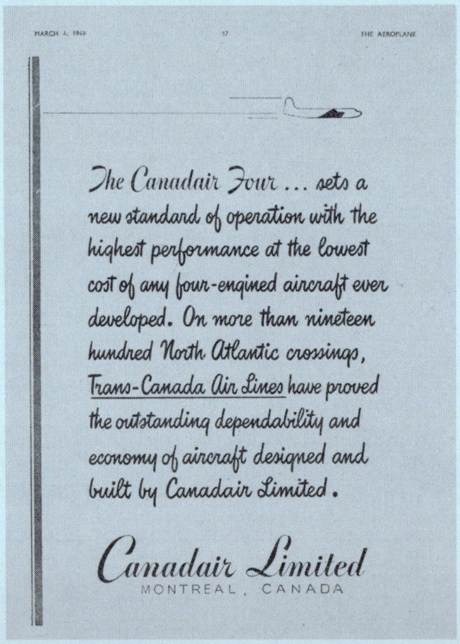

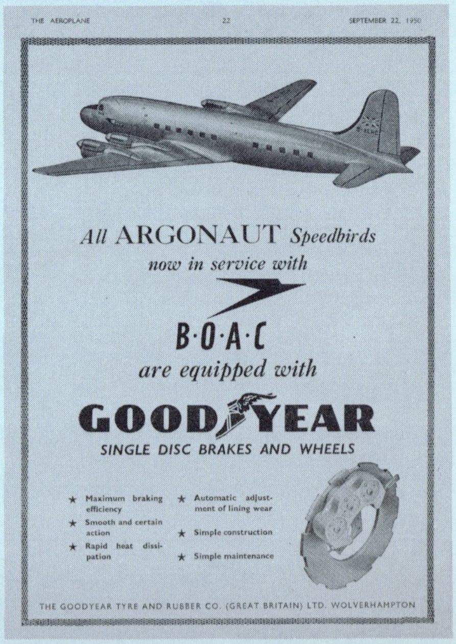

Production List

Canadair North Star, C-4 Argonaut, C-4-1 and C-5 production list. Canadair designations not given are: CL-2 (all North Stars), CL-4 (all other production aircraft except c/n 171), CL-5 (one aircraft only).

Type	*Serial*	*Operator*	*Registration*	*Subsequent Registration(s)*	*Subsequent Operator(s)*	*Notes*
C-54GM	101	Canadair	CF-TEN-X	17525	RCAF	Registered to Canadair 1.7.46.f/f (first flight) Cartierville 15.7.46. Used by Canadair as test, development and demonstration a/c. TOS (taken on strength) by RCAF 25.8.50, 426 Sqdn. W/O (written off) at Athens 21.6.59. SOS (struck off strength) 13.7.59.
DC-4M1	102	TCA	CF-TEK	17518	RCAF	RCAF but loaned to TCA 19.11.46. TCA f/n (fleet number) 84/184. TOS RCAF 16.3.48. Returned to RCAF 20.4.49. To 412 Sqdn. in 34-seat VIP configuration. Extra window in main aft cargo door. Same door permanently installed. APU (auxiliary power unit) installed in rear fuselage. SOS 12.9.62 at Aircraft Industries, St. Jean, Quebec.
DC-4M1	103	TCA	CF-TEL	17519		RCAF but loaned to TCA 17.6.47. TCA f/n 85/185. TOS RCAF 16.3.48 426 Sqdn. W/O in crash at Sydney, N.S. 12.8.48. Never carried RCAF markings.
DC-4M1	104	TCA	CF-TEM	17520		RCAF but loaned to TCA 21.1.47. TCA f/n 86/186. TOS RCAF 16.3.48. Returned to RCAF 4.3.49. To 412 Sqdn. as per 17518. W/O in crash at Hall Beach, NWT, 1.2.62. SOS 26.4.62. As CF-TEM made TCA transatlantic North Star inaugural flight 15.4.47.
DC-4M1	105	TCA	CF-TEO	17521	RCAF	RCAF but loaned to TCA 10.2.47. TCA f/n 87/187. TOS RCAF 6.3.48. Returned to RCAF 31.10.49, 426 Sqdn. SOS 18.3.65. Sold ex Mountainview to Samuel E. Sigal Enterprises Ltd. 30.10.63.
DC-4M1	106	TCA	CF-TEP	17522	RCAF	RCAF but loaned to TCA 20.3.47. TCA f/n 88/188. TOS RCAF 16.3.48. Used by CPA for Shanghai proving flight 1949. Returned to RCAF 15.8.49, 426 Sqdn. W/O with 426 Sqdn. at North Luffenham 8.5.54. SOS 29.11.54.
DC-4M1	107	TCA	CF-TEQ	17523	RCAF	RCAF but loaned to TCA 28.3.47. TCA f/n 89/189. TOS RCAF 16.3.48. Returned to RCAF 25.6.49, 426 Sqdn. W/O Resolute, NWT, 15.3.51. SOS 18.4.51.
C-54GM	108	RCAF	17501			f/f 14.7.47, pilots Lilly & Wickham. TOS RCAF 4.9.47, 426 Sqdn. W/O in ground fire Rockcliffe 22.7.53. SOS 4.8.53.
C-54GM	109	RCAF	17502	CF-UXA	Air Caicos Turks Air	f/f 10.7.47, pilots Lilly & Milne. TOS RCAF 2.9.47, 426 Sqdn. 107 Search and Rescue Unit with SAR mods. SOS 9.5.66. Disposed of ex Mountainview. Sold to Trans Florida Aviation 28.1.66 and registered to Cavalier Aircraft of Canada. Operated from Sarasota and Miami. Last flight 19.6.75. Scrapped at Sarasota May 1976.
C-54GM	110	RCAF	17503			f/f 8.8.47, pilots Lilly & Anglin. TOS RCAF 7.10.47, 426 Sqdn. W/O in crash at Vancouver 30.12.53. SOS 14.1.54.
C-54GM	111	RCAF	17504	CF-VLW		TOS RCAF 10.10.47, 426 Sqdn. SOS 3.10.66. Sold to D.D. Allyn of Sarasota 28.9.66. Ferried to Sarasota and stored there until scrapped.
C-54GM	112	RCAF	17505			TOS RCAF 7.10.47, 426 Sqdn. W/O in crash at Shemya 27.12.53. SOS 4.2.54.
C-54GM	113	RCAF	17506	CF-VLX		TOS RCAF 1.11.47, 426 Sqdn. SOS 3.10.66. Made last official North Star flight 8.12.65 flying from Downsview to Trenton. Disposed of ex Mountainview. Sold to Trans Florida Aviation Sept. 1966. Ferried to Sarasota and stored there until scrapped.
C-54GM	114	RCAF	17507	CF-UOY FAS-300 YS-27C	FAS	TOS RCAF 27.11.47, 426 Sqdn. SOS 8.10.65. 107 Search and Rescue Unit with SAR mods. Disposed of ex Mountainview. Sold to Trans Florida Aviation 10.66. Sold to Fuerza Aerea Salvadorena Dec. 1967 and delivered July 1968 as FAS-300 and later YS-27C. Withdrawn from use by 1970. Possibly still derelict at Ilopango airport, El Salvador.
C-54GM	115	RCAF	17508	CF-VLY		TOS RCAF 14.11.47, 426 Sqdn. SOS 3.10.66. Participated in North Star stand-down ceremony at Trenton 8.12.65. Disposed of ex Mountainview. Sold to D.D. Allyn 28.9.66 reportedly for $18,000. Ferried to Sarasota and stored there until scrapped.
C-54GM	116	RCAF	17509			TOS RCAF 10.12.47, 426 Sqdn. Mishap near Ashiya, Japan 19.4.51. W/O in ground fire at Trenton 24.3.60. SOS 11.5.60.
C-54GM	117	RCAF	17510	CF-UXB	Air Caicos Turks Air	TOS RCAF 10.2.48, 426 Sqdn. SOS 9.5.66. 107 Search and Rescue Unit with SAR mods. Disposed of ex Mountainview. Sold to Trans Florida Aviation 28.1.66 as CF-UXB. Landing mishap at Sarasota 17.9.71. Scrapped at Sarasota. Total time 21,674 hrs.
C-54GM	118	RCAF	17511			TOS RCAF 10.2.48, 426 Sqdn. SOS 18.3.64. Sold ex Mountainview to Samuel E. Sigal Enterprises, Ltd., New Toronto, 30.10.63. Lightning strike reported 22.3.55 and part of starboard aileron blown off.
C-54GM	119	RCAF	17512			TOS RCAF 27.2.48, 426 Sqdn. SOS 16.3.64. Sold ex Mountainview to Samuel E. Sigal Enterprises, Ltd. 30.10.63. Made first non-stop trans-Canada flight, Vancouver-Halifax 14/15.1.49 completing the 2785 miles in 8:32 hrs.

Type	*Serial*	*Operator*	*Registration*	*Subsequent Registration(s)*	*Subsequent Operator(s)*	*Notes*
C-54GM	120	RCAF	17513			TOS RCAF 27.2.48. SOS 27.4.56. Experimental and Proving Establishment research a/c "Rockcliffe Ice Wagon." W/O in ground fire at Dorval 19.3.56.
C-54GM	121	RCAF	17514	CF-SVP-X	NAE Aircraft Salvage	TOS RCAF 8.3.48, 426 Sqdn. SOS 27.9.65. To National Research Council and operated by NAE (National Aeronautical Establishment). Offered for sale from Trenton by Crown Assets Disposal Corp. Nov. 1978. purchased Feb. 1979 by Aircraft Salvage of Dallas. Sold to Miami interests and used on clandestine operations. Presently unserviceable on Greater Inagua Island.
C-54GM	122	RCAF	17515			TOS RCAF 8.3.48, 426 Sqdn. SOS 20.12.65. Participated in North Star stand-down ceremony at Trenton 8.12.65. Lost dinghy from port wing in turbulence 9.5.50. Lightning strike near North Battleford 8.5.53. To National Aeronautical Collection, Rockcliffe.
C-54GM	123	RCAF	17516	CF-VLZ		TOS RCAF 17.3.48, 426 Sqdn. SOS 3.10.66. Disposed of ex Mountainview. Sold to D.D. Allyn 28.9.66. Ferried to Sarasota and stored there. Scrapped.
C-54GM	124	RCAF	17517			TOS RCAF 6.4.48, 426 Sqdn. SOS 18.3.66. Sold ex Mountainview to Samuel E. Sigal Enterprises, Ltd., 30.10.63.
DC-4M	125	TCA	CF-TFA	XA-NUM	LAUSA	Del. to TCA 1.10.47, f/n 201. Christened *Cartier*. Threw propeller at Vancouver 27.9.54. Hours to 31.12.61: 39,326. Sold to Lineas Aereas Unidas SA 21.8.61. (Also noted as LAUMSA.) Believed scrapped in Mexico.
DC-4M2	126	TCA	CF-TFB	XA-NUU	LAUSA	Del. to TCA 10.10.47, f/n 202. Christened *Cornwallis*. Landing mishap Syhdney 10.4.57. Hours to 31.12.61: 35,781. Cargo conversion 2.11.57. Sold to LAUMSA 17.10.61. Believed scrapped in Mexico.
DC-4M2	127	TCA	CF-TFC		IAF/ Cubana	Del. to TCA 22.10.47, f/n 203. Christened *Champlain* Hours to 31.12.61: 38,992. Sold to International Air Freighters and operated on Cubana licence freighting Toronto-Havana. Last flight to Cuba 25.7.62. Scrapped at Malton.
DC-4M2	128	TCA	CF-TFD			Del. to TCA 28.1.48, f/n 204. Christened *Selkirk*. Crashed on Mt. Slesse, B.C., 9.12.56. Hours to crash: 27,196.
DC-4M2	129	TCA	CF-TFE	YV-C-LBU	Overseas	Del. to TCA 10.12.47, f/n 205. Hours to 31.12.61: 39,627. Sold to Overseas Avation 1.7.61. Del. through Gatwick 16.7.61. Sold to A.J. Gaul 13.10.61. Ferried to Florida by Keegan Aviation for Linea Expressa Bolivar C.A. 27.1.62 as YV-C-LBU. (LEBCA also noted as LAVSA.) Damaged by Hurricane Cleo at Miami 27.8.64. Scrapped at Miami 1967.
DC-4M2	130	TCA	CF-TFF	N7866C	WWA	Del. to TCA 19.12.47, f/n 206. Hours to 31.12.61: 38,916. Sold to World Wide Airways of Dorval 7.11.61. Ferried to Oakland, Cal., for intended conversion to Wright engines by R.K. Lamm. Scrapped.
DC-4M2	131	TCA	CF-TFG	XA-NUW	LAUSA	Del. to TCA 19.12.47, f/n 207. Landing mishap at Saskatoon 30.6.53. Hours to 31.12.61: 38,529. Sold to LAUMSA 9.11.61. Operated as the Wings Cafe at Mexico City airport since 1965.
DC-4M2	132	TCA	CF-TFH	XA-NUR	LAUSA	Del. to TCA 10.1.48, f/n 208. Explosive decompression on the ramp at Dorval 25.11.53. In cargo service 12.10.56 to 1.6.58. Hours to 31.12.61: 37,190. Sold to LAUMSA 3.10.61. Believed scrapped in Mexico.
DC-4M2	133	TCA	CF-TFI	XA-NUV	LAUSA	Del. to TCA 23.1.48, f/n 209. Hours to 31.12.61: 39,112. Sold to LAUMSA 23.10.61. Believed scrapped in Mexico.
DC-4M2	134	TCA	CF-TFJ		Overseas	Del. to TCA 6.2.48, f/n 210. Cargo conversion 19.9.55. Hours to 31.12.61: 37,541. Sold to Overseas Aviation 1.7.61. Del. through Gatwick 26.7.61. Sold to A.J. Gaul 13.10.61. Ferried to Florida by Keegan Aviation for LEBCA, Jan. 1962. Damaged by Hurricane Cleo at Miami 27.8.64. Scrapped.
DC-4M2	135	TCA	CF-TFK		Overseas	Del. to TCA 14.2.48, f/n 211. Nosegear inadvertently retracted, Sydney, N.S., 24.10.60. Hours to 31.12.61: 39,048. Sold to Overseas Aviation 1.7.61. Del. through Gatwick 23.7.61. Scrapped at Baginton.
DC-4M2	136	TCA	CF-TFL	N8722R	Overseas T.W. Smith	Del. to TCA 18.2.48, f/n 212. Hours to 31.12.61: 39,180. Sold to Overseas Aviation 1.7.61. Sold to A.J. Gaul 13.10.61. Ferried to Florida by Keegan Aviation. Sold to T.W. Smith of Cincinnati. Leased by T.W. Smith to Avco Electronics. Converted at Cincinnati as space program data collecting lab, Aug. 1962. Operated by AVCO till mid-1964. Scrapped at Middleton, Ohio, 1964.
DC-4M2	137	TCA	CF-TFM	HP-925 BR-HBP I-ACOA	Overseas	Del. to TCA 27.2.48, f/n 213. Total hours to 31.12.61: 36,020. Sold to Overseas Aviation 1.7.61. Del. through Gatwick 3.7.61. Sold to A.J. Gaul 13.10.61. Stored at Baginton until Nov. 1963. Various adventures hereafter until crashed in Cameroons 11.10.66.
DC-4M2	138	TCA	CF-TFN		Overseas	Del. to TCA 5.3.48, f/n 214. Hours to 31.12.61: 35,515. Sold to Overseas Aviation 1.7.61. Sold to A.J. Gaul 13.10.61 and ferried to Panshanger by Keegan Aviation. Ferried to Luton 27.7.64 and scrapped.
DC-4M2	139	TCA	CF-TFO		Overseas	Del. to TCA 18.3.48, f/n 215. Hours to 31.12.61: 37,338. Sold to Overseas Aviation 1.7.61. Sold to A.J. Gaul 13.10.61 and ferried to Baginton by Keegan Aviation. Scrapped May 1965.

Type	Serial	Operator	Registration	Subsequent Registration(s)	Operator(s)	Notes
DC-4M2	140	TCA	CF-TFP		WWA	Del. to TCA 8.4.48, f/n 216. Hours to 31.12.61: 35,827. Sold to World Wide Airways 7.11.61. Cannibalized at Dorval. Fuselage became club-house for Galaxy Flying Club north of Montreal, then scrapped.
DC-4M2	141	TCA	CF-TFQ		Overseas	Del. to TCA 29.4.48, f/n 217. Hours to 31.12.61: 35,777. Sold to Overseas Aviation 1.7.61. Sold to A.J. Gaul 13.10.61. Ferried to Baginton by Keegan Aviation thence to Florida. Demonstrated in Colombia for Société Aerien Medallin. Damaged by Hurricane Cleo at Miami 27.8.64. Scrapped at Miami 1967.
DC-4M2	142	TCA	CF-TFR		IAF	Del. to TCA 12.5.48, f/n 218. Hours to 31.12.61: 37,550. Sold to International Air Freighters 26.12.61 but not operated by them. Scrapped at Malton.
DC-4M2	143	TCA	CF-TFS		IAF	Del. to TCA 20.5.48, f/n 219. Hours to 31.12.61: 36,070. Sold to International Air Freighters 26.12.61 but not operated by them. Scrapped at Malton.
DC-4M2	144	TCA	CF-TFT		Overseas	Del. to TCA 4.6.48, f/n 220. Hours to 31.12.61: 36,069. Sold to Overseas Aviation 1.7.61. Del. through Gatwick 6.7.61. Sold to A.J. Gaul 13.10.61. Ferried to Baginton. Scrapped May 1965.
C-4	145	BOAC	G-ALHC	180 601-N VP-YZN	RRAF Air Links	f/f 6.3.49. Del. to BOAC 29.3.49 as *Ariadne*. Sold to Royal Rhodesian Air Force 14.1.60 as 180 and later 601-N and VP-YZN. Sold to Air Links 13.8.64. Ferried to Redhill Aug. 1964. Scrapped July 1965.
C-4	146	BOAC	G-ALHD	VP-KOY OY-AFC	EAA FE	Del. to BOAC 10.5.49 as *Ajax*. Used on BOAC's proving flight June-July 1949. Fitted for use on royal tours and VIP flights. Leased to East African Airways 25.2.59 as VP-KOY. Sold to Overseas Aviation 22.5.59 and later used by Flying Enterprise as OY-AFC *John Foltmann*. Sold to Flying Enterprise Sept. 1962. Scrapped at Kastrup (Copenhagen) in July 1965.
C-4-1	147	CPA	CF-CPI	CF-TFU YV-C-LBV	TCA Overseas LEBCA	Del. to CPA May 1949 as *Empress of Sydney,* f/n 401. Sold to TCA 27.11.51 asCF-TFU, f/n 221. Cargo conversion 15.4.55. Hours to 31.12.61: 31,232. Sold to Overseas Aviation 1.7.61. Sold to A.J. Gaul 13.10.61. Ferried to Baginton by Keegan Aviation thence to Florida for LEBCA as YV-C-LBV in Jan. 1962. Ditched off Caracas 6.7.62.
C-4-1	148	CPA	CF-CPR			Del. to CPA 1949 as *Empress of Vancouver,* f/n 402. W/O in crash at Tokyo 9.2.50 while making a night landing in bad weather.
C-4-1	149	CPA	CF-CPJ	CF-TFV YV-C-LBT	TCA LEBCA	Del. to CPA 11.7.49 as *Empress of Auckland*, f/n 403. Sold to TCA as CF-TFV and del. 5.2.52, f/n 222. Cargo conversion 28.3.55. Hours to 31.12.61: 30,059. Sold to Overseas Aviation 1.7.61. Sold to A.J. Gaul 13.10.61. Ferried to Florida Jan. 1962 for LEBCA as YV-C-LBT. Scrapped.
C-4-1	150	CPA	CF-CPP	CF-TFW	TCA	Del. to CPA 11.7.49 as *Empress of Hong Kong*, f/n 404. Sold to TCA 2.1.52 as CF-TFW, f/n 223. Destroyed in collision with RCAF Harvard 3309 over Moose Jaw, Sask. 8.4.54. Total hours to date: 7763.
C-4	151	BOAC	G-ALHE			Del. to BOAC 15.6.49 as *Amazon.* Later re-named *Argo.* Crashed in storm Kano 24.6.56.
C-4	152	BOAC	G-ALHF	VP-KOI	EAAC	Del. to BOAC 25.6.49 as *Atlas*. Sold to East African Airways 25.8.57. Acquired by Britair in 1964 and restored to British register. Thought broken up at Redhill. In West African Airways and Kuwait Airways service 1957-1958.
C-4	153	BOAC	G-ALHG		WAAC Kuwait Airways Overseas Derby/ BMA	Del. to BOAC 5.7.49 as *Aurora*. Sold to Overseas Aviation 14.4.60. Sold to Derby Airways Oct. 1961. Transferred to British Midland Airways 1.10.64. Crashed at Stockport in UK 4.6.67. This aircraft operated last BOAC Argonaut service 8.4.60.
C-4	154	BOAC	G-ALHH	181 602-0 VP-YZO	RRAF Air Links Transglobe	Del. to BOAC 15.7.49 as *Attica.* Severely damaged at Tokyo 29.1.52 and rebuilt at Hong Kong. Sold to Royal Rhodesian Air Force 23.2.60 as 181 and later 602-0 and VP-YZO. Sold to Air Links 13.8.64 and ferried to Redhill where scrapped July 1965.
C-4	155	BOAC	G-ALHI	182 603-P VP-YZP	RRAF Air Links/ Transglobe	Del. to BOAC 21.7.49 as *Antares.* Sold to Royal Rhodesian Air Force 28.3.60 as 182 and later 603-P and VP-YZP. Sold to Air Links July 1964 and entered service with them 15.8.64. Transferred to Transglobe 9.8.65. Flown to Redhill 13.10.65. Flown to Stansted 20.1.66 for use by Stansted Fire School. Burned 1971.
C-4	156	BOAC	G-ALHJ	VP-KOT	EAAC	Del. to BOAC 27.7.49 as *Arcturus.* Leased to EAAC in Feb. 1958 then restored to BOAC for use as apprentice training aircraft. Finally used for fire fighting training at Heathrow and derelict there to 1982.
C-4	157	BOAC	G-ALHK	OY-AFB	Overseas FE	Del. to BOAC 3.8.49 as *Atalanta.* Fitted for use on royal tours and VIP flights. Sold to Overseas Aviation 23.3.59. Transferred to Flying Enterprise 31.1.61 as OY-AFB *Ludvig Bjarkow.* Sold to Flying Enterprise 1962. Scrapped at Kastrup July 1965.

Type	Serial	Operator	Registration	Subsequent Registration(s)	Subsequent Operator(s)	Notes
C-4	158	BOAC	G-ALHL			Del. to BOAC 9.8.49 as *Altair*. Crashed at Tripoli 21.9.55 while landing at night in bad weather.
C-4	159	BOAC	G-ALHM	VP-KOJ	EAAC Air Links/ Transglobe BMA	Del. to BOAC 15.8.65 as *Antaeus*. Sold to East African Airways 30.8.57 as VP-KOJ. Sold to Britair thence to Air Links in July 1964 and entered service with them 1.10.64. Transferred to Transglobe 9.8.65. Flown to Redhill 4.11.65. Ferried to Castle Donington 9.9.65 for use as spares by British Midland Airways.
C-4	160	BOAC	G-ALHN	OY-AFA	Overseas FE Derby	Del. to BOAC 19.8.49 as *Argosy*. Sold to Overseas Aviation 1.2.60. Transferred to Flying Enterprise 10.6.61 as OY-AFA *John Foltmann*. Returned to Overseas 20.4.61. Sold to Derby Airways for spares and scrapped at Burnaston.
C-4	161	BOAC	G-ALHO	VP-KNY	EAAC	Del. to BOAC 29.8.49 as *Althea*. Sold to East African Airways 29.3.57 as VP-KNY. Crashed at Nairobi 11.4.62.
C-4	162	BOAC	G-ALHP		Overseas Derby/ BMA	Del. to BOAC 2.9.49 as *Aethra*. Sold to Overseas Aviation 12.6.59. Registered to Derby Airways Oct. 1961. Used for spares and scrapped at Burnaston.
C-4	163	BOAC	G-ALHR	VR-AAR	Aden Airways	Del. to BOAC 14.9.49 as *Antiope*. Sold to Aden Airways 17.2.60 as VR-AAR. Withdrawn from use at Nairobi 21.2.63 due to salt corrosion.
C-4	164	BOAC	G-ALHS		Overseas Aden Airways Derby/ BMA	Del. to BOAC 20.9.49 as *Astra*. Sold to Overseas Aviation 17.2.60. Sold to Derby Airways Oct. 1961. Leased to Aden Airways March-May 1962. Transferred to British Midland Airways 1.10.64. Withdrawn from use 16.10.67 and scrapped.
C-4	165	BOAC	G-ALHT	OY-AAH	Overseas FE Air Links/ Transglobe	Del. to BOAC 28.9.49 as *Athena*. Sold to Overseas Aviation 31.10.58. Transferred to Flying Enterprise 20.1.60 as OY-AAH *Robert Svendsen*. Sold to Flying Enterprise Sept. 1962. Sold to Air Links Jan. 1964 and entered service with them 8.4.64. Transferred to Transglobe 9.8.65. Ferried to Redhill. Scrapped Jan. 1966.
C-4	166	BOAC	G-ALHU	OY-AAI	Overseas FE	Del. to BOAC 24.10.49 as *Artemis*. Served with Kuwait Airways during 1958. Sold to Overseas Aviation 30.1.59. Transferred to Flying Enterprise 30.4.60 as OY-AAI *Alfred Nervo*. Sold to Flying Enterprise Sept. 1962. Scrapped at Kastrup July 1965.
C-4	167	BOAC	G-ALHV	VR-AAT	Aden Airways	Del. to BOAC 7.10.49 as *Adonis*. Sold to Aden Airways as VR-AAT. Sold to Derby Airways Dec. 1963. Transferred to BMA. Scrapped at Burnaston.
C-4	168	BOAC	G-ALHW	179 600-M VP-YZM	RRAF Air Links Transglobe	Del. to BOAC 18.10.49 as *Aeolus*. Used on East African royal tour April 1954. Sold to Royal Rhodesian Air Force 15.12.59 as 179 and later 600-M and VP-YZM. Sold to Air Links July 1964 and entered service with them 5.11.64. Transferred to Transglobe 9.8.65. Ferried to Redhill 6.10.65. Ferried to Castle Donington 11.3.66 for spares for BMA, then scrapped.
C-4	169	BOAC	G-ALHX	VR-AAS	Aden airways	Del. to BOAC 27.10.49 as *Astraea*. In service with Aden Airways from April 1960 on. Withdrawn from use April 1964 due to salt corrosion and scrapped.
C-4	170	BOAC	G-ALHY		Overseas Derby/ BMA	Del. to BOAC 11.11.49 as *Arion*. Last Argonaut in fleet and delivered 8 months ahead of schedule. Sold to Overseas Aviation 9.3.60. Sold to Derby Airways Oct. 1961. Transferred to British Midland Airways 1.10.64. Withdrawn from use 6.11.67 and scrapped.
C-5	171	RCAF	17524	10000 N17599		TOS RCAF 29.7.50. To 412(T) Sqdn. for VIP operations. Configured for 24 passengers. Renumbered as 10000 17.2.51. Withdrawn from use 28.8.66. Stored at Mountainview. Sold in US as N17599. Ferried to Long Beach, Cal., and scrapped.

North Star/C-4 Operators by Nationality

Country	National designator	Operator(s)
1. Aden	VR-A	Aden Airways
2. Canada	CF	Trans-Canada Air Lines, Canadian Pacific Airlines, Royal Canadian Air Force, Air Caicos, International Air Freighters, National Aeronautical Establishment, World Wide Airways
3. Denmark	OY	Flying Enterprise
4. Kenya	VP-K	East African Airways, West African Airways
5. Kuwait	n/a	Kuwait Airways
6. Mexico	XA	Lineas Aereas Unidas (Mexicanas) S.A.
7. Nicaragua	FS/YS	Fuerza Aerea de Nicaragua
8. Rhodesia	VP-Y	Royal Rhodesian Air Force
9. United Kingdom	G	British Overseas Airways Corporation, Air Links, Derby Airways/British Midland Airways, Overseas Aviation
10. United States	N	T.W. Smith Leasing, Aircraft Salvage, Trans Florida Aviation, R.K. Lamm
11. Venezuela	YV-C	Linea Expresa Bolivar C.A.

RCAF North Stars Involved in Category A Incidents or Ground Fires

Date	Aircraft	Unit	Location	Details
Mar. 16, 1953	17523	426	Resolute	While attempting takeoff from icy runway with strong cross wind and limited visibility, aircraft swung off runway and burned. As captain ordered "Pull it off" at the moment of takeoff, F/O pulled back on yoke while F/E pulled back on power! No injuries among 18 aboard.
July 7, 1953	17501	CEPE	Rockcliffe	Aircraft caught fire while being loaded for a northern flight. Cause undetermined, no injuries.
Dec. 27, 1953	17505	426	Shemya	Landing being attempted at night in adverse conditions with crosswind to 50 knots and blowing snow. Aircraft touched down but was blown off the runway and broke up. Aircraft had made the trip on a marginal flight plan, the weather being poor, and had been held in a holding pattern due to radio and navigation aid problems at Shemya until its fuel was critically low. No injuries to eight aboard.
Dec. 30, 1953	17503	426	Vancouver	Aircraft returning to Vancouver after shutting down No. 4 engine. Aircraft iced up en route and on approach to land stalled on final. Nose and starboard wingtip dug in. Starboard wing broke off and aircraft flipped over, but did not burn. Minor injuries to some of the 54 aboard.
May 8, 1954	17522	4 OTU	North Luffenham	While executing a precautionary landing, aircraft allowed to touch down 75' short of the runway. Aircraft struck a mound of clay and concrete. The landing gear was driven up into the wing and the wing badly damaged. Though the aircraft remained intact, it was a write-off. No injuries to the 17 aboard.
Mar. 19, 1956	17513	CEPE	Dorval	Destroyed in hangar fire along with C-119 22124.
June 21, 1959	17525	426	Athens	On approach to land, aircraft landed short, striking a sea wall and damaging the undercarriage. A fuel leak resulted and a fire broke out. The aircraft burned to destruction but no injuries.
Mar. 24, 1960	17509	426	Trenton	Damaged by fire on the ground at 6 Repair Depot. Aircraft jumped the chocks during run-up, struck ground equipment and caught fire. Later ferried to St. Jean, Quebec, and struck off strength.
Feb. 2, 1962	17520	412	Hall Beach	About two minutes after takeoff at 1000' No. 2 engine ran rough when oil pressure dropped to zero. Seconds later No. 3 also malfunctioned, then No. 4. A single-engine wheels-up landing was made on the tundra near the airstrip. Visibility very poor in ice fog. Accident investigation notes, "Lack of oil dilution caused oil starvation on start-up initiating bearing breakdown on all engines but severe on engines 2 and 3." No injuries among 23 aboard. Spar cracked during salvage attempt.

Leapin' North Stars! Pilots Ron Baker and Herb Seagrim haul back with everything they've got in a dramatic life-and-death moment at Calgary on July 21, 1948. They were carrying out a three-engine takeoff at gross weight during crossover exhaust trials when a truck suddenly drove in front of them just as they passed 105 mph IAS. After breezing past the truck, CF-TFG climbed away normally. J.T. Dyment notes of the amazing incident, "The response of the aeroplane throughout the whole procedure was, in fact, remarkable." Aboard the North Star, besides the pilots, were engineers Ian S. Macdonald and J.M. Grisdale. The photos originated from timed 16mm footage taken for certification purposes. (Air Canada Archives)

July 21, 1948.

CALGARY TRUCK INCIDENT

At Take-off: Aircraft TCA No. 207, with standard antennae, de-icer shoes on wings, propellers, etc., for line service.

Type . . . . . . North Star M2
Engines . . . . Rolls-Royce Merlin 624
Propellers . . . Hamilton Standard, 4-bladed, 12'8" dia., #6813A-6 blades equipped with Goodrich electric de-icer shoes.

Gross Weight at Take-off . . . . . 71,266 lb
Airport Pressure Altitude . . . . 3,555 ft A.S.L.
Airport Density Altitude . . . . . 5,600 ft
Outside Air Temperature . . . . 78°F
R.P.M. 3000
Manifold Pressure . . . . 67.5" Hg.
Charge Temperature . . . . . . 77°C
Chart Power for Above . . . . . 1590 BHP

Take-off towards South, on Runway 16.
Truck proceeding towards West, on Runway 25.

	Time-From Start (Seconds)	Distance-From Start (Feet)	Airspeed (mph)	Picture No.
#1 Engine Cut	32.4	3020	105 IAS 118 TAS	-
Commenced Evasive Action	33.1	3130		#1
Over Truck	34.7	3400		#5

Note: Clock in prints reads to 1/100 sec. (i.e., 100 represents 1 second, 1 rev. of sweep hand represents 3 seconds).

One engine and controlability speed 126 mph IAS at 80,200 pounds or approximately 115 mph IAS at 71,266 pounds.

Summary of C-4 and North Star Serious Accidents and Ground Incidents

Estimated hours flown	2 million
Total accidents	18
Total fatal accidents	5
Total writeoffs	17

Date	*Aircraft*	*Operator*	*Location*	*Fatalities*	*Damage*
12-8-48	CF-TEL	TCA	Sydney, N.S.	0	Destroyed
9-2-50	CF-CPR	CPA	Tokyo	0	Destroyed
15-3-51	17523	RCAF	Resolute	0	Destroyed
29-1-52	G-ALHH	BOAC	Tokyo	0	Substantial
22-7-53	17501	RCAF	Ottawa	0	Destroyed
27-12-53	17505	RCAF	Shemya	0	Destroyed
30-12-53	17503	RCAF	Vancouver	0	Destroyed
8-4-54	CF-TFW	TCA	Moose Jaw, Sask.	37	Destroyed
8-5-54	17522	RCAF	North Luffenham	0	Destroyed
21-9-55	G-ALHL	BOAC	Idris	15	Destroyed
19-3-56	17513	RCAF	Montreal	0	Destroyed
24-6-56	G-ALHE	BOAC	Kano	32	Destroyed
9-12-56	CF-TFD	TCA	Mt. Slesse, B.C.	62	Destroyed
21-6-59	17525	RCAF	Athens	0	Destroyed
24-3-60	17509	RCAF	Trenton	0	Destroyed
1-2-62	17520	RCAF	Hall Beach, N.W.T.	0	Destroyed
11-4-62	VP-KNY	EAAC	Nairobi	0	Destroyed
4-6-67	G-ALHG	BMA	Stockport	72	Destroyed

Oops! These notes from *Flight Comment* describe three RCAF North Star incidents.''

It was a miserable night, dark, wind from the Northwest at 17 mph and gusting. But, regardless of weather, the North Star's engines had to be tested so two qualified technicians were detailed for the job. The aircraft was positioned, chocked, and all necessary checks were made. When the four engines were running, the airmen on fire extinguisher duty returned to the hangar.

No. 2 throttle was advanced to check the engine at high rpm. After the check No. 2 was set to idle at 1800 rpm, the same as the other engines. No. 4 throttle was advanced and when the rpm reached 2600 to 2700 the aircraft lurched forward and No. 3 engine burst into flame. When the smoke had cleared away, the aircraft and an energizer were severely damaged.

It was a miserable night.

(Sept.-Oct. 1960)

The first officer of a North Star told the captain that smoke was coming from No. 4 engine. The captain continued to taxi and asked the first officer to watch for a definite indication of fire. Sparks were soon dropping out of the rad flaps and the pilot was convinced.

A North Star made a normal approach to a runway 14,000 ft. long, and into a 22 kt. wind. Touchdown was normal, about 1000 ft. from the approach end. The nosewheel contacted the runway almost immediately, and the pilot commenced braking, with ''regular gentle pumping action of the brake pedals.''

The tower at this time informed the pilot that the starboard wheels appeared to be on fire, and a loud report, indicating a blown tire, was heard in the aircraft.

Investigation revealed that the starboard outer tire had blown and marks on the runway indicated this wheel had ''locked'' on, at, or shortly after, the point where brake had first been applied. Apparently the pilot's ''gentle pumping action'' was not in fact releasing the brake on the starboard wheel.

Poor airmanship was displayed here. With 13,000 ft. of runway ahead of the aircraft, plus a 22 kt. wind, the pilot's hurry to begin braking action as soon as the nosewheel was on, does not seem warranted. It would seem more reasonable in this situation to allow the aircraft to slow down of its own accord, with only minimum braking required, at a very much reduced speed.

(Nov.-Dec. 1961)

During a primary inspection on a North Star the shear bolts in both main undercarriage uplocks were found sheared. The uplocks were operated manually from the wheel wells and a brief inspection carried out. Then, because the aircrew had not reported a malfunction, the groundcrew assumed that the bolts had sheared due to turbulent flight conditions. They installed steel bolts, informed the pilot of this temporary measure, and said the aircraft was ready to go.

When the North Star was brought in for the next landing both port and starboard undercarriage lights showed unsafe. A visual check by the F/E verified the unsafe condition. Several selections were made without result, and the F/E checked the hydraulic compartment and the cables through the dump valve hatches but could not discover the cause of the trouble. Finally he removed a panel from the pedestal and discovered that the uplock release actuating rod was broken at the lower end. The F/E could just reach the end of the rod with pliers and was able to release the uplock. A normal landing was carried out...

In this case, if the F/E had not known his aircraft thoroughly or if the failure had been in an inaccessible part of the system, the aircraft would have been committed to a wheels-up landing...

(Sept.-Oct. 1962)

"Cayó al Mar Avión de Carga Salvándose sus Tripulantes"

July 6, 1962 was a sad day for North Star YV-C-LBV. Formerly of TCA, the old workhorse had made its way to Venezuela via the UK to haul freight. At 2:15 that Friday morning it had departed Miami for Maiquetia, Venezuela, via Kingston. A reported 24,000 lb. of cargo was aboard, considerably more than TCA's North Stars were certified for.

Shortly before nine in the morning, a distress call was heard from the airplane and it was soon clear that it had gone down within 15 minutes of its destination. Aboard were three LEBCA pilots, Marcos Lopez Olivares, Manuel Guillermo Silva and Leopoldo Murillo.

Almost immediately, a DC-3 took off to search for the North Star and after only 12 minutes located it, ditched and broken in three. One man could be seen in a life vest while another was keeping afloat on a piece of wreckage. A Naval helicopter was dispatched and picked up Murillo but the other two men were unable to lift themselves aboard it. A fire department launch later arrived to pull them safely from the water.

According to *El Nacional* in its Saturday front page coverage of the incident, the ditching had been a rough one. As the North Star hit, it broke up and water rushed in so fast that it popped most of the windows. Captain Olivares was temporarily trapped in the cockpit with an injured head but the other two men managed to escape through windows. An Aeropostal DC-3 was first on the scene and dropped a life raft which landed too far from the North Star. Then a government DC-3 arrived and also dropped a raft. This time Olivares managed to swim to it, then tow it back to the others. In preliminary statements, Olivares indicated that en route head winds, aggravated by a fuel leak, had resulted in fuel starvation and failure of all four engines while cruising at 7500 feet. Lima Bravo Victor thereby managed to become the only North Star or Argonaut ever to have ditched at sea.

Canadair C-54GM/DC-4M/C-4/C-5 Specifications

Variants

C-54GM	RCAF non-pressurized aircraft serial nos. 101-124, RCAF nos. 17501-17523, 17525. Six aircraft delivered to TCA on loan and redesignated DC-4M1 (17518/CF-TEK, 17519/CF-TEL, 17520/CF-TEM, 17521/CF-TEO, 17522/CF-TEP, 17523/CF-TEQ).
DC-4M1	See above.
DC-4M2-3	TCA, Merlin 622 or 722, 4-blade propellers. Pressurized.
DC-4M2-4	TCA, Merlin 624 or 724, 3-blade propellers. Pressurized.
DC-4M2C	TCA, Merlin 724, 3-blade propellers. Cargo conversion. (Total DC-4M2s: 20 aircraft including CF-TFA through CF-TFT. Cargo aircraft: CF-TFB, CF-TFH, CF-TFJ serial nos. 125-144).
C-4	BOAC "Argonaut", 22 aircraft. Pressurized. Merlin 624 or 724-C1. 3-blade propellers. Serial nos. 145, 146, 151-170.
C-4-1	CPA, four aircraft. Similar to C-4 except to CPA specifications. Merlin 724, 3-blade propellers. Serial nos. 147-150.
C-4-1C	Cargo aircraft, two ex-CPA aircraft converted by TCA, CF-TFU and CF-TFV.
C-5	One only, Pratt and Whitney R-2800 engines. RCAF.

Basic Dimensions

Wing span	117′6″
Length	94′9.5″ (C-54GM, radar nose)
	93′6.2″ (DC-4M2)
	93′7.5″ (C-4)
Height	17′7″ (top of fuselage)
	27′6.3″ (tip of tail)
Fuselage height	11′6″
Fuselage width	10′6″
Stabilizer span	39′5″
Wing chord at root	228.4″
Wing chord at tip	70.49″
Stabilizer chord at root	133.3″
Stabilizer chord at tip	65″

Cabin Dimensions

Main cabin length	46′6″
Cabin floor width	8′8″
Cabin height	7′

Areas

Wings including ailerons	1457 sq. ft.
Ailerons including tab	92.5 " "
Wing flaps	214.6 " "
Horizontal tail surfaces	324.9 " "
Elevators	86.5 " "
Elevator tabs	6.84 " "
Vertical tail surface	179.3 " "
Rudder	47.6 " "
Rudder tab	4.75 " "

Weights in Pounds

	M2-3	*M2-4*	*C-4*	*C-4-1*	*C-54GM*	*C-5*
Operating weight empty	49,950	50,955	45,654	49,230	44,000	49,475
Max. gross takeoff weight	79,575	79,850	80,200	79,850	76,500*	86,000
Max. landing weight	72,000	72,000		72,000	67,000	
Max. zero fuel weight	68,000	68,000		68,000		

Max. weight, 3 engine takeoff: 60,000

North Star Freighter max. payload: 18,970 lb.

*78,000 for passenger version with modified nose gear.

Fuel Tank Capacities, DC-4M2

Max. fuel capacity,	8 tanks	2908 Imp. gal. (22,582 lb)
	6 tanks	2350 Imp. gal. (16,506 lb)
Oil (4 engines)		100 Imp. gal. (900 lb)

General Performance, DC-4M2

Stalling speeds (IAS)

TO weight, TO power, TO flaps	93 knots
TO weight, TO power, flaps & wheels up	108 knots
Landing weight, approach flap, power off	89 knots
Landing weight, full flap, power off	78 knots

Flap settings

Takeoff	15 degrees
Approach	15 degrees
Landing	40 degrees
Cruise, climb, descent	0 degrees

Cruising speeds @ 900 BHP/75,500 lb

5,000′	167 IAS knots	184 TAS knots
10,000′	158 " "	189 " "
15,000′	151 " "	196 " "
20,000′	141 " "	207 " "

Max. range (TO max. weight, full fuel, cruise @ 5000′)
3060 miles miles (2620 nautical miles)

Max. range (full payload, cruise @ 5000′)
1580 miles (1370 nautical miles)

Normal fuel flow @ 900BHP (lb/hr)

@ 5,000′	1588
@ 10,000′	1644
@ 15,000′	1680
@ 20,000′	1772

Speed Limitations, C-54GM

Max. level flight	223 knots IAS
Max. glide or dive	254
Flaps extended: 20 degrees	170
30 degrees	134
40 degrees	132
Max. gear down	152
Max. for fuel dumping	186
Stall (dirty)	74
Min. control speed (1 propeller idling, 3 engines at TO power, flaps TO position, gear retracted)	100
Min. TO speed	103

Speed Limitations, DC-4M2

	IAS Knots	*TAS Knots*
Normal operating speed limit (V_{no}) (Above 17,000′ reduce by 5 knots/1000′)	215	222
Never exceed speed (V_{ne}) (Above 16,000′ reduce by 5 knots/1000′)	245	255
Manoeuvring speed (V_a)	155	157
Lowering undercarriage (V_{lo})	170	175
Flaps extended (V_{fe})		
20 degrees	170	175
30 degrees	143	145
40 degrees	140	141
Min. control speed (V_{mc}) (1 engine out, propeller windmilling, flaps in takeoff position, gear up, TO power on 3 engines)	88	89
Speed in turbulence	150	153
Max. speed for dumping fuel	186	191
Min. takeoff climb speed		
73,000 lb	101	
79,850 lb	108	
Min. approach speed (same conditions as for min. control speed)	105	
Max. permissible mach. number	.54	(412 mph at sea level, 382 mph at 20,000′)
Takeoff distance (max. weight, zero wind, 4 engines)	4500′	
CAA takeoff distance	5800′	
Landing distance	3200′	
CAA landing distance	3200′	
Rate of climb:		
4 engines, max. weight, METO power	850 fpm	
3 engines, max. weight, TO power, flaps	400-480 fpm	
3 engines, max. weight, METO power, no flaps	370-380 fpm	

Pressurization, DC-4M2

Differential	4.16 psi	
Cabin altitude @	10,000′	1000′
	15,000′	4600′
	20,000′	8000′
	25,000′	11,300′

Notes from TCA Fleet Serviceability, April 1956

	Viscount	*North Star*	*Super Constellation*	*DC-3*
Delays per 100 revenue departures	2.89	3.9	8.8	.65
Featherings	0	7	8	0
Unscheduled engine changes per 1000 hours (over 12 months)	.08	.55	1.92	.36
Engine overhaul period (hours)	1400	1250	1100	1300

TCA Fleet Utilization 1959

North Star	8 hrs. 30 min. per day
Super Constellation	9 hrs. 30 min. per day
Viscount	8 hrs. 6 min. per day

TCA North Star Revenue Passenger Miles Flown 1947-1961.

1947	50,850,000*
1948	195,375,000*
1949	272,810,000*
1950	317,420,000*
1951	393,352,441
1952	492,500,147
1953	585,730,459
1954	600,339,344
1955	469,707,152
1956	424,697,365
1957	376,590,286
1958	376,797,322
1959	382,827,978
1960	325,917,793
1961	27,443,054
Total	5,292,358,341

*Approximated

Merlin Serviceability, 1947 - 1948, TCA

Excerpts from the Rolls-Royce report on the Merlin engine in the first year of North Star operations. The report does not indicate the amount of maintenance required to achieve these results.

Aircraft	*Commenced service*	*Aircraft hours*	*Engine hours*	*Single crossings*	*Utilization*
CF-TEK	28-5-47	2450	9800	158	7:45 hours
CF-TEL	24-6-47	2262	9048	139	7:47
CF-TEM	15-4-47	2758	11032	170	7:31
CF-TEO	18-4-47	2776	11104	176	7:35
CF-TEP	2-5-47	2127	8508	130	7:32
CF-TEQ	21-4-47	2815	11260	168	8:04
	Total	15188	60752	941	7:42
CF-TFE	2-4-48	105	420	6	
CF-TFG	6-4-48	30	120	2	
CF-TFH	6-4-48	30	120	2	
CF-TFI	5-4-48	60	240	4	
CF-TFJ	14-4-48	30	120	2	
	Total	255	1020	16	

TCA engine performance April 15, 1947 - April 15, 1948:

	Without failure	*Requiring minor repair*	*Requiring major repair*	*Total*
No. of engines over 2000 hours	2			2
1750	7	2	2	11
1500	3	3		6
1250	5	4	3	12
1000	2	3	1	6

No. time expired engines overhauled Winnipeg, 76 (Av. time 654:20 hours)
No. failed engines overhauled, 20 (Av. time 384:21 hours)
No. failed engines, non-major, 12 (Av. time 230:20 hours)

Loading

The following loading data dated November 15, 1960, is taken from the TCA North Star Operating Manual.

8 Passengers

.10 Seating Arrangements: The following seating arrangements are available in North Star airplanes. Ref. Illus. Page 25 of this chapter.

Service	No. of Seats
North American First Class	52
North American Tourist	62
Southern Tourist	50
Southern Tourist	60

.11 Average Passenger Weights:

1 NORTH AMERICAN SERVICES:

a WINTER OPERATIONS:

i Adult Male: 170 lbs.
ii Adult Female: 170 lbs.
iii Children 12 years of age and above - considered as adults.
iv Children 3 years and less than 12 years of age - actual weight.
v Infants under 3 years of age - no weight allowance.

b SUMMER OPERATIONS:

i Adult Male: 160 lbs.
ii Adult Female: 160 lbs.
iii Children 12 years of age and above - considered as adults.
iv Children 3 years and less than 12 years of age - actual weight.
v Infants under 3 years of age - no weight allowance.

2 SOUTHERN SERVICES:

a YEAR ROUND OPERATIONS:

i Adult Male: 165 lbs.
ii Adult Female: 145 lbs.
iii Children 12 years of age and above - considered as adults.
iv Children 3 years and less than 12 years of age - 80 lbs.
v Infants under 3 years of age - 20 lbs.

.12 Football Teams: The following description applies to all adult football teams. Players are considered at an average weight of 205 lbs. When ten or more players are carried, the maximum rear cargo compartment values of the Tabular Load Chart are to be reduced by 690 lbs., and the minimum values to be raised by 70 lbs., in order to provide balance compensation. In the case of school teams, special instructions will be issued on request.

.13 Passenger Baggage - Average Bag Weights: The total baggage boarded shall be accounted for as the product of the number of bags and the applicable average as shown below for each service.

1 SOUTHERN SERVICES: 28 LBS.

2 NORTH AMERICAN SERVICES: 25 LBS.

Except Goose Bay and Vancouver, Victoria-Seattle, which are to be accounted for as follows.

a GOOSE BAY: Actual Weight

b VANCOUVER, VICTORIA-SEATTLE: 22 LBS.

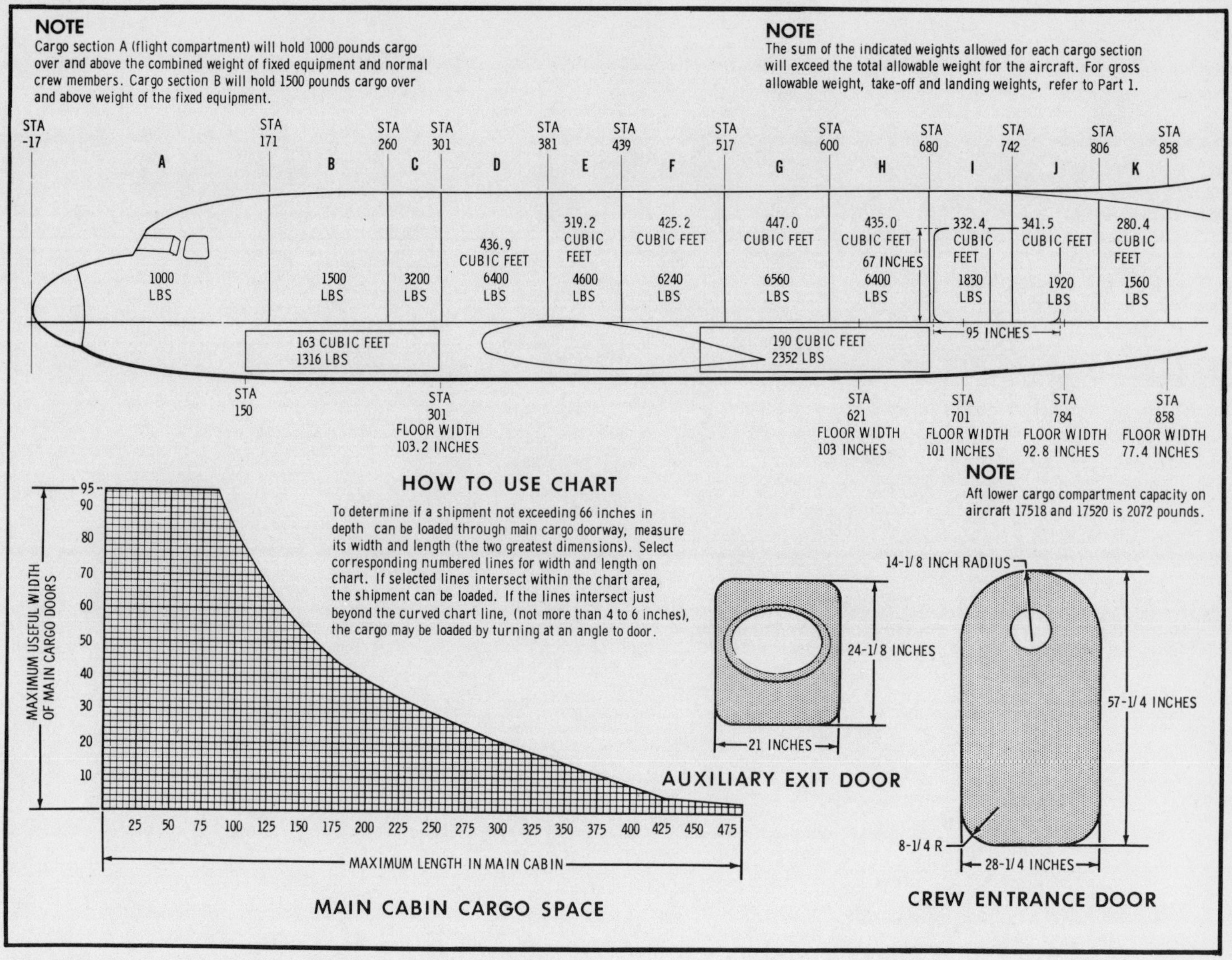

Cargo capacity and size limitations, RCAF North Star M1.

Flying the North Star

Excerpts from TCA's Pilot Operating Instructions for the North Star M2. These pages reflect a constant updating as shown by page layout variation. (via Harry Clarke and John Bosma)

1 Ramp Check (Cont'd)

.24 Control Pedestal: (Cont'd)

18 Hydraulic System Valve .... ON (fully down)

.25 Floor:

1 Control Lock .... ON

2 Hydraulic Hand Pump Selector .... BRAKE SYSTEM

2 Starting Engines

.10 Before Starting Engines:

1 Heater Fuel and Ignition Switch .... OFF during STARTING

2 Door Warning Lights .... CHECK OUT

3 Acknowledge Parking Brake .... ON
in response to hand signal from ground personnel.

4 Ignition Switches .... ON

5 Main Booster Pumps .... ALL ON
Check fuel Pressure 8-12 psi, indicating lines are primed.
Switch Booster Pumps .... OFF

.11 Starting Procedure:

1 Starting Order: 3,4, 2 and 1.

2 Primer Pump .... RIGHT
Check Primer Pump Warning Light ...... ON

3 Booster - Starter Switch .... ON - NO. 3

CAUTION: Do not exceed 40 seconds maximum engagement, then disengage for a minimum of 30 seconds before attempting further starting. This is important to prevent burning out starter.

4 Primer Switch .... NO. 3

NOTE: Priming and starting are to be simultaneous. Extreme care should be exercised to avoid overpriming. If engine is inadvertently overprimed, or difficult to start when hot, probably accompanied by exhaust torching, open throttle and hold idle cut-off switch to ENGINE OFF position until engine fires. When engine fires, release booster - starter switch and assist engine to run smoothly by use of primer, if necessary.

5 Oil Pressure.................................... CHECK NORMAL
within 30 seconds after starting. If not, stop engine immediately.

6 Fuel Pressure................................... CHECK 5 - 13 PSI

7 Coolant Loss Indicator Light..................... CHECK ON - below approx. 1900 RPM

8 Radiator Flaps:

a Main.. AUTO

b Auxiliary..................................... OPEN
for rapid warm up at low OAT.............. CLOSED

NOTE: Both should be OPEN at high summer temperatures.

9 Hydraulic Pressure.............................. 2650 - 3050 PSI

10 Wing Flaps..................................... LOWER TO 40°

11 Repeat starting procedure for 4, 2 and 1 in that order.

3 After Starting Engines

.10 Before Taxiing: These items should be completed while ground power is being removed, before signal to taxi is given.

1 Ground Power Warning Light....................... OUT

CAUTION: It is important to check the ground power light is OUT since the generators can indicate charge with the ground power unit connected, if it has a low output.

In the following response check, the F/O will call the items, the Captain will carry out the required action, and the Captain (C) and/or the F/O will call the response, as shown.

2	C	Primer Pump............................	OFF
3	C	Main Booster Pumps.....................	ON & PRESSURES NORMAL
4	C/F/O	Generators.............................	NORMAL CHARGE
5	F/O	Door Warning Lights....................	OUT
6	C	Suction................................	NORMAL
7	C	Directional Gyro.......................	UNCAGE, SET
8	C	Nose Lock Pin..........................	DISPLAYED

SIGNAL MAN WILL GIVE CLEARANCE TO DEPART FROM RAMP

PARKING BRAKE MUST NOT BE RELEASED UNTIL RECEIPT OF PROPER CLEARANCE FROM GROUND PERSONNEL.

.11 Starting:

1 THROTTLES: Open to approximately................ 1,000 RPM
to start rolling, then
Reduce Inners to..................... IDLING RPM or AS REQ'D

DO NOT OPERATE ENGINES IN RANGE 1,100 - 1,400 RPM.

.12 During Taxiing:

1	C/F/O	Gross Weight and V_1, V_2	CHECK
2	C/F/O	Turn and Heading Indicators............	CHECK WHILE TURNING
3	C/F/O	Wing Flaps.............................	RAISE TO UP POSITION WHEN CLEAR OF RAMP AREA

4 Ground Check:

.10 Positioning of Aircraft: Head as much into wind as possible, particularly in hot weather.

.11 After Stopping Aircraft:

1 Parking Brake ....ON

2 Idle Engines ....1100 RPM

.12 Pre Run-Up Response Check:

1	C	Anti-Collision Lights	.........AS REQ'D
2	C	Pitot Heat	.........ON
3	C	Master RPM Control	.........FULL INCREASE
4	C	Intercool Switch	.........FULL
5	C	Synchronizer	.........AUTOMATIC
6	F/O	Wing Flaps	.........15°

.13 Run-Up and Check Prior to Take-Off:

1 THROTTLES - Advance to 2000 RPM (approx. 30" M.P.)
Check: Coolant Loss Warning Light ....OUT
above approximately 1900 RPM.

2 All Engine Temperatures, Pressures & Fuel Flows ..... CHECK NORMAL

3 Propeller Controls - Check synchronizer and propeller governors at 2000 RPM by:

a Master Control Lever ....FULL DECREASE RPM
check RPM decreases to minimum (approx 1700 RPM)
and that Governor Warning Lights
come ON at ....FULL DECREASE

b Master Control Lever ....FULL INCREASE RPM
check Governor Warning Lights
come ON at ....FULL INCREASE

4 IGNITION .... CHECK

NOTE: The normal RPM drop on single ignition should not exceed 200 RPM. If this is exceeded the individual engine should be operated for a short period at 40" M.P. with the

a Induction Heat .... HOT

b Intercooler .... OFF

c RPM Synchronizer Switch .... MANUAL

d Master RPM Control .... FULL DECREASE RPM

Recheck Ignition (at 40" M.P.) with

e Induction Heat .... COLD

f Intercooler .... ON

If the drop does not exceed 150 RPM at 40" M.P. and the engine is smooth it can be considered satisfactory. This procedure is followed to endeavour to clear plugs which may have fouled, especially due to prolonged idling.

5 DELAY BEFORE TAKE-OFF: If there is a delay between the run-up check and take-off, place

a Master RPM Control .... FULL DECREASE RPM

b Intercooler Switch .... OFF

c RPM Synchronizer Switch .... MANUAL
NOTE: If delay extended, raise wing flaps for maximum cabin cooling.

immediately before take-off

d Master RPM Control .... FULL INCREASE RPM

e Intercooler Switch .... FULL

f RPM Synchronizer Switch .... AUTOMATIC

6 INDUCTION HEAT:

a At Ground Outside Temperatures below 0°C.. FULL HOT

During the cold weather season this prevents the engine developing power in excess of its rated value at maximum RPM and M.P. with possible resulting detonation.

b At Ground Outside Air Temperatures
above 0°C .... OFF

7 MAIN RADIATOR FLAPS:

a At Ground Outside Air Temperatures
Below +15°C .... AUTO

b At Ground Outside Air Temperatures
Above +15°C - Set Manually 15° .... AS DESIRED

8 CONTROL LOCK .... REMOVE

9 FLIGHT CONTROLS .... CHECK for full travel & freedom of movement

.14 Response Take-off Check:

1	C	Induction Heat	.... AS REQ'D
2	C	Main Radiator Flaps	.... 15° or AUTO
3	C	Trim Tabs	.... SET for TAKE-OFF
4	C	Wing Flaps	.... 15°
5	C	Master RPM Control	.... FULLY FORWARD
6	C	Intercooler Switch	.... FULL
7	C	Synchronizer	.... AUTO
8	C	Control Lock	.... OFF
9	C	Controls	.... FREE

5 Take-off to 1,000 Feet

.10 Take-off:

1 Line up with centre of runway - nosewheel straight.

2 Open throttles smoothly to full throttle. Depending on atmospheric conditions, RPM will vary slightly, but should be 2900 - 3000 at full throttle settings.

3 Use nosewheel steering. The rudder starts to become effective at approximately 45 knots IAS.

4 Maintain nosewheel on ground until 85 knots IAS. Raise nosewheel just clear of ground and fly airplane off at 100 - 110 knots IAS depending upon the gross weight.

5 During complete take-off, the First Officer will devote his attention to maintenance of required power settings and observe that all engine instrument readings are normal.

RPM	Maximum 3,000
Manifold Pressure	Maximum 71.5" Hg
Coolant Temperature	Maximum 135°C
Oil Temperature	Maximum 105°C
Oil Pressure	Maximum 120 psi
Charge Temperature	Maximum 105°C

6 When airborne, the Captain will order the First Officer to raise the landing gear by raising his thumb and calling GEAR UP.

NOTE: Although it has a centring mechanism, the nose gear can be deflected beyond the self-centring limits. This can occur whenever excessive nosewheel steering is required due to crosswind, engine failure or slippery runway. Pilots should ensure that the nosewheel is centred before retracting the landing gear to prevent damage to the nosewheel well doors.

7 When landing gear is fully retracted and landing gear red warning light goes OUT, the First Officer will indicate this to the Captain by saying GEAR UP. Leave landing gear selector UP.

8 On a signal from the Captain, the First Officer will retract flaps and leave flap selector UP. Flaps are normally to be retracted at or above 125 knots IAS.

9 Attain climb speed of 135 knots.

10 Reduce power settings to 2800 RPM and 54" M.P.

11 At a convenient time between 500 and 1,000 feet check nacelle appearance.

12 At an altitude of 1,000 feet, further reduce RPM to maintain normal climb power settings:

2650 RPM and 54" M.P. in MS gear.

NOTE: To avoid excessive cylinder pressure, reduce M.P. before RPM so that the M.P. does not exceed the value for the corresponding RPM set out below:

M.P.	RPM
71.5" or full throttle	3000
58.5"	2800
54.0"	2650

13 Climb at 135 - 150 knots IAS for the best rate of climb, consistent with air conditions and engine coolant temperatures.

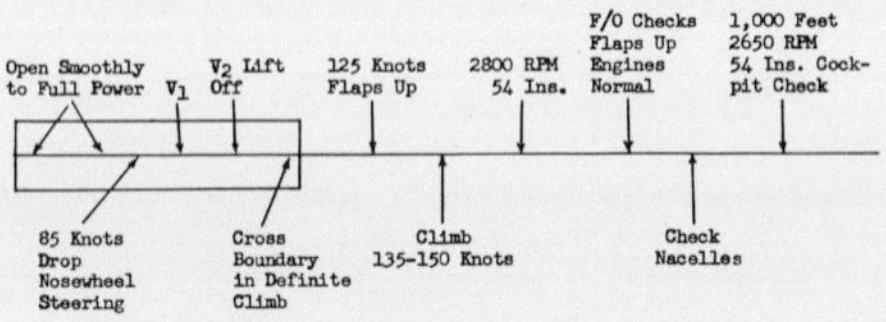

.11 **1,000 Foot Check: At a convenient time, but not below 1,000 feet, check the following:**

1 Fuel Booster Pumps .... OFF

2 "NO SMOKING " &"SEAT BELT" Sign Switches ...... OFF

3 Landing Lights .... OFF & RETRACTED

4 Auto-pilot Switch .... ON

**5 Captain's and First Officer's
Static Selector Valves .... ALTERNATE SOURCE
then return to Static Tube, checking normal change in airspeed indicator, altimeter and rate of climb.**

6 Hydraulic System Valve .... OFF (Fully up)

6 Climb

.10 **Power Settings:**

1 Normal climb power settings are 2650 RPM and 54" M.P. in MS and FS gear. At full throttle, the manifold pressure will drop off as altitude increases.

2 When cruising altitude is above 13,000 feet, change to FS gear when the manifold pressure drops to 50" Hg. at full throttle (approximately).

3 When cruising altitude is below 13,000 feet, climb to cruising altitude in MS gear at full throttle.

4 Gear change from MS to FS can be made at any RPM and M.P. However, to prevent the changeover being so noticeable to passengers, it is preferable to throttle back two corresponding engines to 30" Hg., switch from MS to FS then advance the throttles, repeating the procedure for the other two engines. When in FS gear, ensure that the supercharger warning lights are ON.

5 Climb power settings above the normal 2650 RPM and 54" M.P. in MS and FS gears are not prohibited and may be used in cases of operational necessity. Absolute maximum climb power settings are 2850 RPM and 58.5" M.P.

NOTE: When power settings above 2650 RPM and 54" M.P. are used for climb the actual settings and duration of their use are to be recorded in the log book.

.11 Airspeed: Climb at 135 - 150 knots IAS for the best rate of climb, consistent with air conditions and engine coolant temperatures.

.12 Coolant Temperatures: Check coolant temperatures during climb. Control radiator flaps to maintain coolant temperatures 100 - 120°C at normal power settings, preferably near the lower limit. Normally, main radiator flaps should be controlled in AUTOMATIC, endeavouring to keep them closed by inching the auxiliary flaps manually to maintain the desired 100°C. Under warm weather conditions, the auxiliary flaps will be wide open and the main flaps will then control automatically to maintain the coolant temperatures within limits. Alternatively, main radiator flaps can be controlled manually.

.13 Charge Temperature: Charge temperature should be maintained at not less than 40°C by selecting:

1 Intercooler Switch .... OFF, HALF or FULL

2 Induction Heat .... AS REQUIRED

.14 Pressurization: .... CHECK NORMAL FOR CLIMB

NOTE: Refer to Chapter 10, Section 15 for details on operation.

.15 External Check: Make a perodic visual inspection of the inboard sections of the nacelles from the cockpit especially after use of take-off and climb powers. This inspection is to check the cross-over exhaust and applies particularly to Cargo Aircraft.

NOTE: There are no fire detectors in the cross-over duct because of the high temperatures involved, and since the engine is protected by stainless steel sheet in all vulnerable areas.

7 Cruise

.10 Levelling Off:

1 Level off at cruise altitude.

2 Maintain climb power settings until desired cruising speed is attained.

3 When cabin rate of climb returns to ZERO set:

a Cabin pressure change limit hands, 50 ft/min ...... UP & DOWN respectively.

b Start Marker .... SET 500 FT ABOVE ALTITUDE OF DESTINATION

.11 Cruise Control: Consult Cruise Control Charts, Chapt. 8, to obtain correct power settings for the operating conditions.

1 For most effective overall operation during cruise, FS gear should not be used below 13,000 feet, density altitude (Ref. Power Control Chart, Chapt. 8).

2 When setting up engine powers the RPM should be adjusted to the value given in the Power Control Chart. However, because fuel flowmeter readings usually differ, the RPM should be further adjusted so that the average of the fuel flowmeter readings will be equal to the specified fuel flow value.

.12 Engines:

1 Nacelles and Wings .... CHECK - NORMAL APPEARANCE

2 Main Radiator Flaps .... AUTO

NOTE: Maintain coolant temperatures at 100 - 105°C by progressively inching up the auxiliary flaps.

3 Charge Temperature: Minimum 40°C - Maximum 120°C. Maintain charge temperature as close to but not less than 40°C during all operations by:

a Intercooler Switch .... OFF, HALF or FULL

b Induction Heat .... AS REQUIRED

c Increase Power Settings .... AS REQUIRED
to restore fuel flow up to value existing prior to the application of induction heat or the adjustment of the intercooler. (This fuel flow value should be the same as the cruise control chart value shown for the cruising density altitude.)

4 Coolant Temperature	- Minimum	40°C
	Normal	100 - 105°C
	Maximum	115°C
5 Oil Temperature	- Minimum	15°C
	Normal	50 - 80°C
	Maximum	90°C
6 Oil Pressure	- Minimum	55 psi
	Maximum	120 psi

NOTE: An engine may be kept running when oil pressure is between 45 and 55 psi. When oil pressure is below 45 psi, then engine must be feathered. When oil pressure is below 55 psi, the fact must be entered in the aircraft log book accompanied by a full set of existing engine reading.

.13 Fuel Management:

1 Fuel Quantities .... CHECK & RECORD

2 Fuel Tank Selectors: (Refer to Chapt 10, Sect. 8.)

3 Fuel Pressure (Booster Pumps OFF) .... 5 - 13 PSI

NOTE: Booster pumps ON if required for satisfactory engine operation at altitude.

.14 Pressurization: .... CHECK

NOTE: Refer to Chapter 10, section 15 for details on operation.

8 Descent

.10 Airspeed: Instructions for cruising conditions apply in general for a cruising descent. Descend at an indicated airspeed not greater than 220 knots.

.11 Power Settings: Descend at a normal RPM of 2,350. Maintain a manifold pressure of 35" or above in both FS and MS gears. Do not exceed 900 BHP/engine during descent - at this power the fuel flow will average approximately 425 lbs/engine/hr in FS and 400 lbs/engine/hr in MS gear. Change over from FS to MS at approximately 10,000 feet. If descent conditions require lower power, the RPM may be reduced to a minimum of 2,200 RPM in FS and 2,000 RPM in MS gear, maintaining a minimum of 35" M.P. If a high rate of descent is required at low airspeeds, it may be necessary to lower flaps (and wheels) to retain this minimum manifold pressure.

CAUTION: To prevent leading and plug fouling at low powers, maintain charge temperature at not less than 40°C, by means of induction heat and/or controlling and intercooler switch as required.

NOTE: At part throttle conditions during descent through precipitation in possible icing conditions, maintain induction air temperature at a minimum of +10°C.

.12 Pressurization:

1 Start Marker .... 500 FT ABOVE ALTITUDE OF DESTINATION

2 Cabin Pressure Change Limit Hands ...... Both UP & DOWN SET 300 FT/MIN DOWN

.13 Holding: Circuits for holding patterns should be flown with 15° of flap and at 120 knots IAS.

9 Approach and Landing

.10 In-range Check: Prior to initial approach the following check should be carried out:

1 Induction Heat .... AS REQUIRED

2 Main Fuel Booster Pumps .... ON

3 SEAT BELT & NO SMOKING Sign Switches .... ON

4 Lights .... AS REQUIRED

5 Auto Pilot Switch .... OFF

6 Superchargers .... MS
Supercharger Warning Lights .... OUT

7 Static Selector Valve .... ALTERNATE SOURCE
return to .... STATIC TUBE

8 Fuel Tank Selectors .... MAIN

9 Cross-feed Selectors .... OFF

10 Parking Brake .... OFF

11 RPM Synchronizer .... AUTOMATIC or AS REQUIRED

12 Wing De-icers .... OFF

13 Hydraulic System Valve .... ON (fully down)
check Hydraulic Pressure .... 2650 - 3050 PSI

14 Hydraulic Hand Pump Selector .... BRAKE SYSTEM

15 Wing Flaps (initial setting) 15° at not over 170 knots IAS.

16 RPM Control .... SET 2350 RPM

17 Recommended minimum airspeed for initial approach - 120 knots IAS.

.11 Before Landing Check:

1 Landing Gear .... DOWN at not over 170 knots IAS

2 Verbal Response .... PRESSURE UP - THREE GREEN LIGHTS

NOTE 1: If desired, in other than engine out or holding procedures, wing flaps and landing gear may be lowered at the same time. This will reduce the tendency for the aircraft to balloon.

NOTE 2: Maintain radiator temperatures by leaving auxiliary radiator flaps CLOSED until final approach, as necessary. Maintain charge temperatures above 40°C.

3 Landing Lights .... EXTEND and use as req'd at not over 175 knots IAS

.12 Final Approach and Landing:

1 Final Approach Airspeed .... 115 KNOTS IAS

2 Wing Flaps (final setting) .... 40°

a Leave Flap Selector Lever .... DOWN

b Do not exceed Flap Placard Speed ...... 140 KNOTS IAS

3 Master RPM Selector .... FULL INCREASE RPM (fully forward) just prior to landing.

4 Airspeed Over End of Runway .... 98 - 100 KNOTS (depending on conditions)

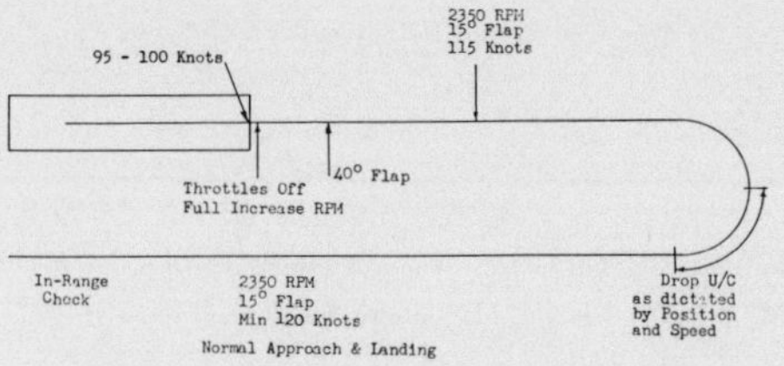

Normal Approach & Landing

10 After Landing

.10 After Landing Check: On order from Captain.

1 Scoop Heat and Pitot Heat .... OFF

2 Anti-collision Lights .... OFF

3 RPM Synchronizer Switch .... MANUAL

4 Master RPM Control .... DECREASE RPM (fully back)

5 Intercooler Switch .... OFF

6 Wing Flaps .... UP

7 Hyiraulic Pressure .... CHECK 2650-3050 PSI

8 Controls .... CENTERED & LOCKED

.11 At Ramp:

1 Parking Brake .... ON

2 Fuel Booster Pumps .... OFF

3 Idle Engines .... 1,000 RPM

CAUTION: Avoid 1,100 - 1,400 RPM range. Ref. Limitations Chapter 4, Section 5.16.

4 Idle Cut-off Switches .... ENGINE OFF

5 Ignition Switches .... OFF after propellers stop turning.

6 Signal to Service Personnel from open left Cockpit Window .... SWITCHES OFF & CABIN DEPRESSURIZED

NOTE: This signal confirms for Service Personnel that the airplane is depressurized and it is safe to open cabin and cargo doors.

7 All Switches and Selectors not required......OFF

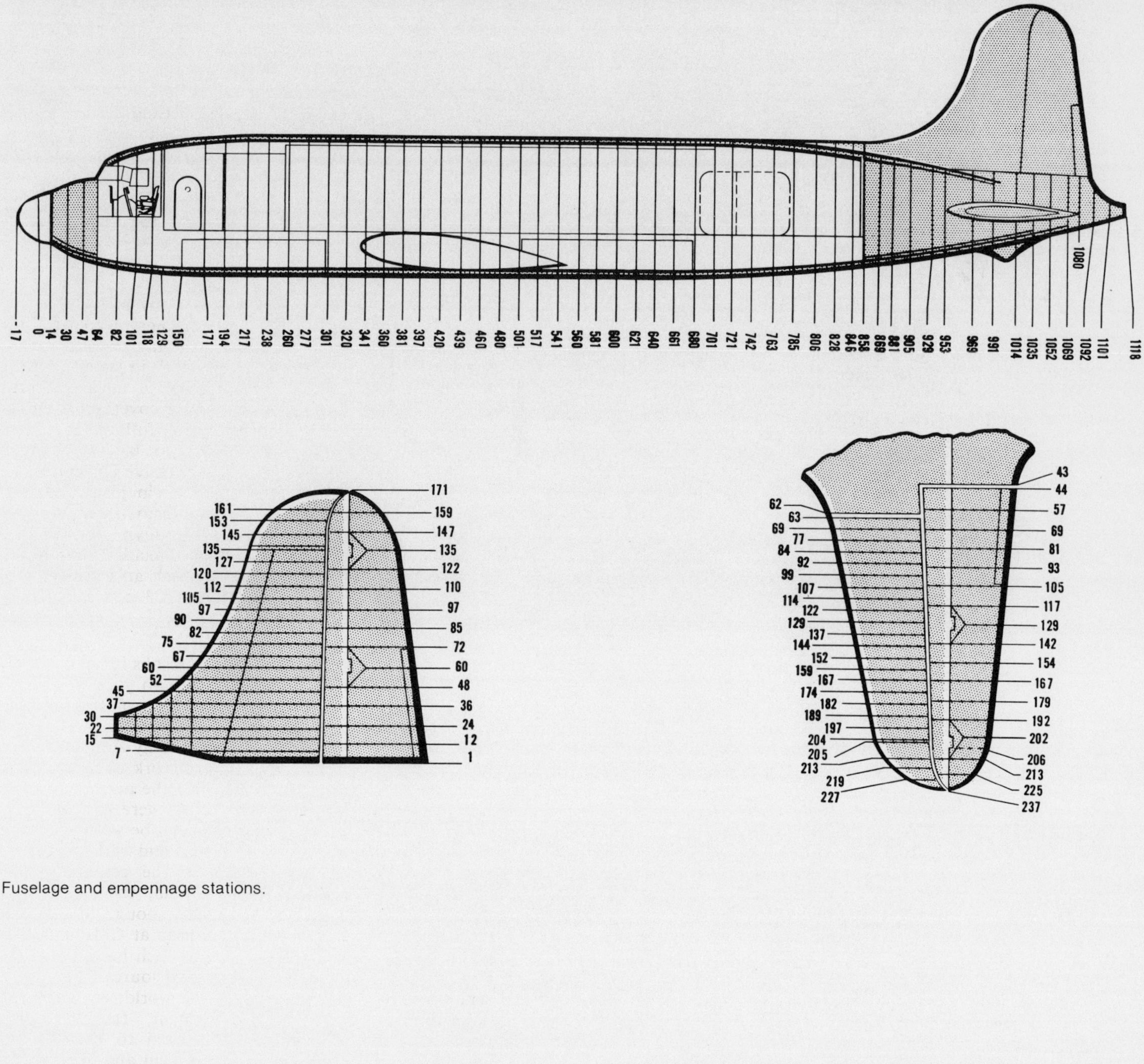

Fuselage and empennage stations.

North Star RCAF Squadron Commanders

426 Squadron	W/C C.A. Willis, DFC	August 1946-February 1947
	W/C C.G.W. Chapman, DSO	March 1947-February 1949
	W/C C.H. Mussells, OBE, DSO, DFC	February 1949-March 1951
	W/C J.K.F. MacDonald, DFC, CD	June 1951-August 1952
	W/C H.W. Lupton, AFC, CD	September 1952-July 1955
	W/C A.J. Mackie, DFC, CD	July 1955-November 1959
	W/C J.O. Maitland, CD	November 1959-October 1961
	W/C A.J. Mackie, DFC, CD	October 1961-September 1962
412(T) Squadron*	W/C W.H. Swetman, DSO, DFC	April 1947-August 1949
	W/C B.H. Moffitt, DFC, AFC	August 1949-August 1950
	W/C R.I. Trickett, DFC	September 1950-February 1952
	W/C H.A. Morrison, DSO, DFC, AFC, CD	February 1952-January 1955
	W/C W.G.S. Miller, CD	January 1955-March 1958
	W/C W.K. Carr, DFC, CD	March 1958-August 1960
	W/C J.W. Borden, DFC, CD	August 1960-August 1963
	W/C M.G. Bryan, CD	August 1963-June 1966
	W/C D.C. MacKenzie, CD	January 1966-January 1968

*Includes era of the C-5.

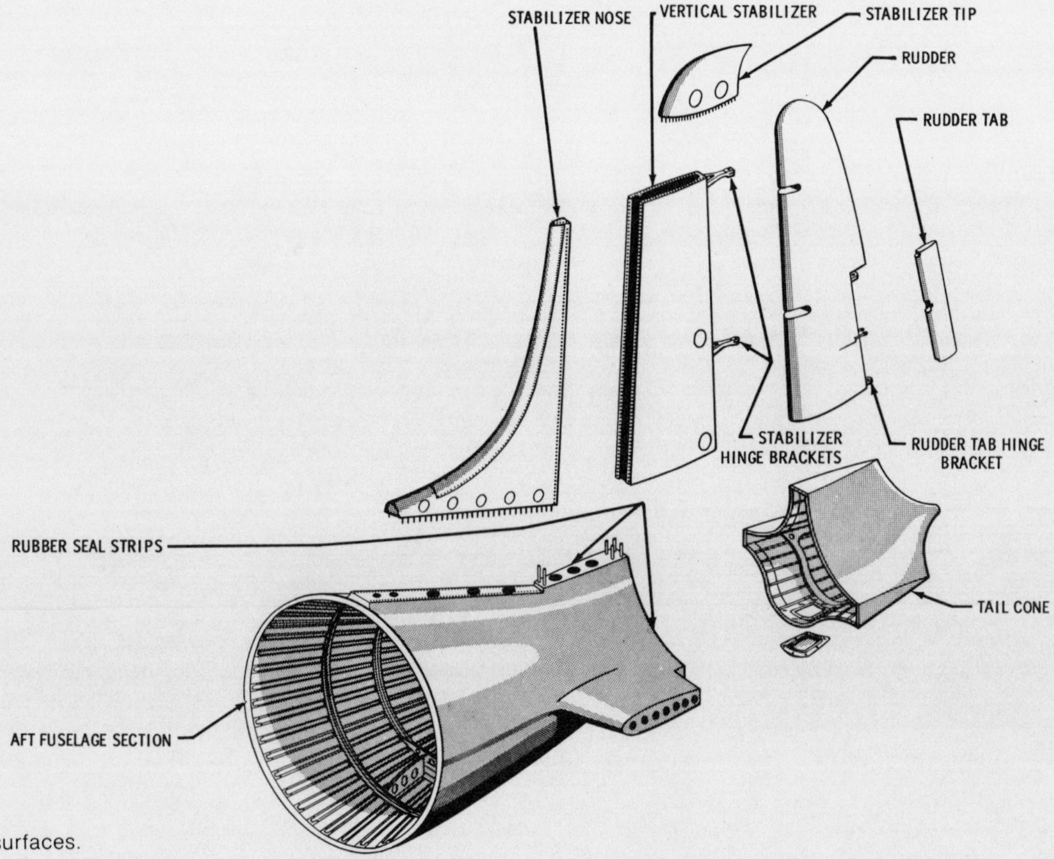

Vertical tail surfaces.

Bibliography

Books

Childerhose, Chick, *Wild Blue*, Hoot Publications, Victoria, B.C., 1978.

Griffin, John, *Canadian Military Aircraft, Serials and Photographs,* Canadian War Museum, Ottawa, 1969.

Keith, Ronald A., *Bush Pilot with a Brief Case,* Doubleday Canada Ltd., Toronto, 1972.

Kostenuk, Samuel and Griffin, John, *RCAF Squadron Histories and Aircraft, 1924-1968,* Samuel Stevens Hakkert and Co., Toronto, 1977.

Milberry, Larry, *Aviation in Canada,* McGraw-Hill Ryerson, Toronto, 1979.

McGregor, Gordon R., *The Adolescence of an Airline,* Air Canada, Montreal, 1970.

Vincent, Carl, *The Liberator and Fortress,* Canada's Wings, Stittsville, Ont., 1975.

Periodicals and other Publications

The Aeroplane, Air Pictorial, Between Ourselves, The Douglas DC-4 (Air-Britain), Flight, Flight Comment, Propliner, The Roundel.

Index

Figures in italics refer to photographs.

DANGER
CAUTION
PULL OFF